VAMPIRE'S GRASP

Rubbing of gravestone of Simon Whipple Aldrich, Union Cemetery Annex, North Smithfield, Rhode Island. *Courtesy of Cyril Place.*

VAMPIRE'S GRASP

The Hidden History of Consumption in New England

Michael E. Bell

To Carole

© 2024 Michael E. Bell

All Rights Reserved

ISBN: 979-8-89372-352-6

Published by:
Warren History Press
Warren County Historical Society
50 Gurney Lane
Queensbury, NY 12804
www.wcnyhs.org

Contents

Map and Chart: Vampire Incidents in New England, iii

Acknowledgements, vii

Part I: The Vampire's Grasp
    Introduction: Medical Mysteries, 3
    Chapter 1: From Consumption Rituals to Vampire Narratives, 9

Part II: Eyewitnesses
    Chapter 2: Signs of the Dead Preying on the Living, 25
    Chapter 3: The Disinterment Was Done under My Personal Supervision, 37
    Chapter 4: The Exhumation Will Take Place on Saturday at 10 AM, 49

Part III: Newsworkers
    Chapter 5: Wrath of the Devil's Dark Angel, 69
    Chapter 6: The Above Are Facts ... Sickening and Horrible, 81
    Chapter 7: Eliza Heard and Believed These Stories, 93
    Chapter 8: My Sisters Were Drawing My Life Away, 107

Part IV: Family and Media Workers Talk about Mercy Brown
    Chapter 9: She Had Turned Over in the Grave, 123
    Chapter 10: What Really Happened to Mercy Brown?, 135

Part V: Historians
    Chapter 11: Heathenish Mutilations of the Dead, 149
    Chapter 12: Her Heart Should Be Consumed for the Benefit of Her Sisters, 159

Part VI: Physicians
    Chapter 13: The Charm Worked No Good and the Patient Died, 179
    Chapter 14: The Village Witch Told Them ... Cut Her Heart Out and Bury It, 189

Part VII: Evolutionists

    Chapter 15: A Peculiar Kind of Vampirism, 199

    Chapter 16: In New England Consumption is a Spiritual Visitation, 211

    Chapter 17: Explain It? I Can't. It Is an Old, Old Belief, 219

Part VIII: Community-Based Storytellers

    Chapter 18: A Strange Superstition Was Whispered in Robinson Hollow, 233

    Chapter 19: Sarah Came Every Night, Causing Great Pain and Misery, 243

Part IX: Authors

    Chapter 20: The Dead Gnawing and Feeding upon the Living, 255

    Chapter 21: Thoughts of a Horrid Nature Arose in the Parents, 263

    Chapter 22: They Drove a Stake through Her Breast, 273

Conclusion: Disease and Death, Fear and Hope, 287

Notes, 295

Index, 326

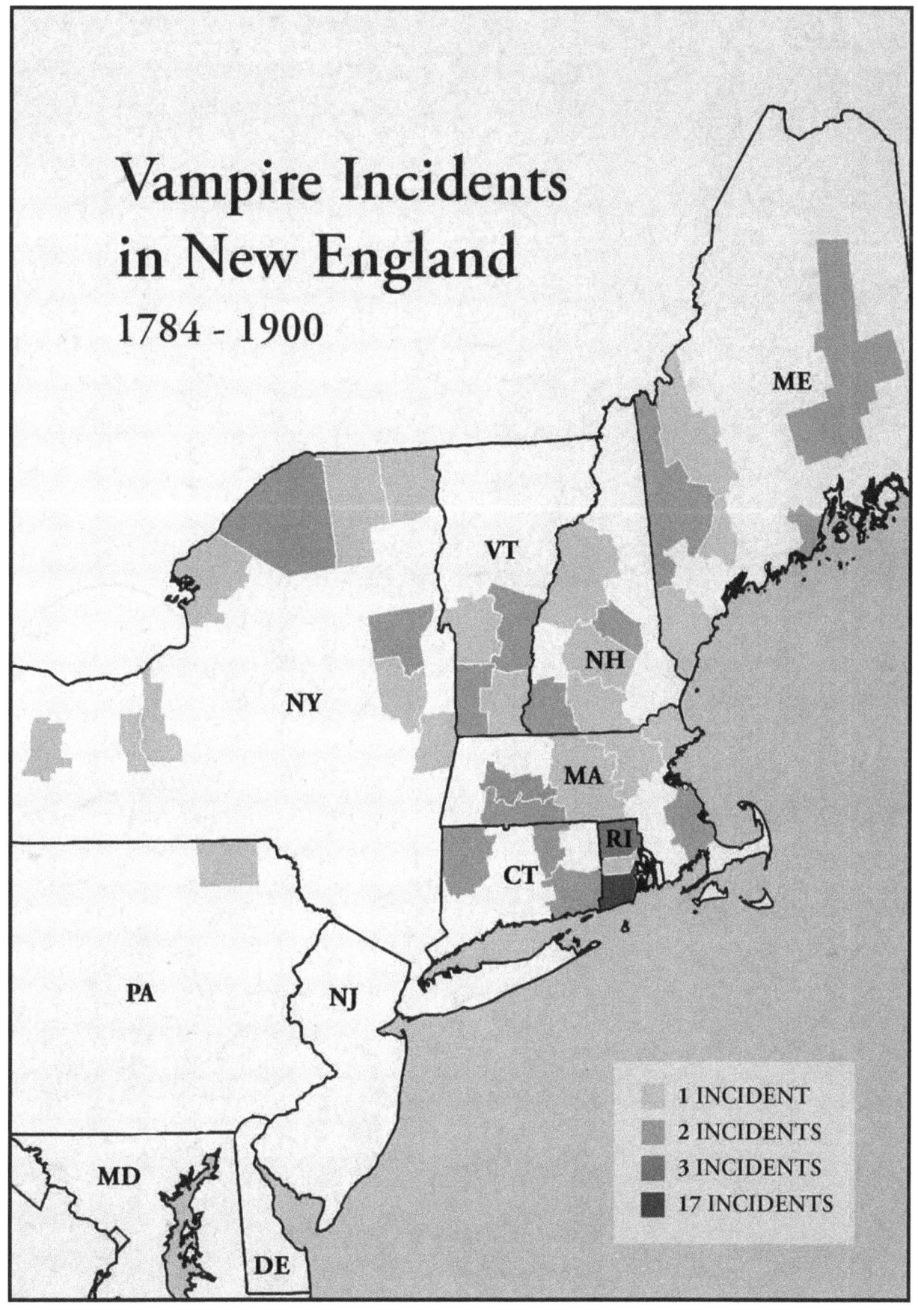

STATE	COUNTY	TOWN	DATE	CH/P
Connecticut (6)	Litchfield	?	<1851	6
		Cornwall Hollow	1869	7
	New London	Griswold	c1826-36	concl
			1854	6
	Tolland	Willington	1784	4
		West Stafford	<1871	20
Maine (11)	Androscoggin	Lewiston	1874	7
	Franklin	Jay Hill	1869	7
	Knox	Union	1832-3	11
		Vinal Haven	c1883	7
	Oxford	Norway	1832	5
		Oxford	?	3
	Penobscot	Bangor area	1827	12
			c1828	12
	York	Saco	c1852-62	13
	?	?	<1832	4
			<1854	3
Massachusetts (11)	Essex	?	c1812	4
	Hampden	Chicopee	1852-72	14
		Springfield	1814	5
	Hampshire	Belchertown	1788	2
		Hatfield	1788	2
		Chesterfield	c1847	18
	Middlesex	Waltham	<1853	12
	Plymouth	Kingston	1807	18
			<1822	18
	Worcester	Harvard	c1794	12
central	?	?	<1831	5
New Hampshire (9)	Belknap	Barnstead	1810	3

			New Hampton	c1860-70	14
		Cheshire	Keene	1822	3
			Jaffrey	c1838	4
		Grafton	Grafton	<1891	15
		Hillsborough	New Ipswich	c1810	13
		Merrick	Loudon	c<1810	3
		?	?	1869	7
?				c1880s	3
New York (13)		Cayuga	Kelloggsville	1851	6
		Clinton	Chazy	1818	12
		Franklin	Malone	c1807-35	15
		Jefferson	?	<1854	p. V
		Livingston	Mount Morris	c1853	6
		Rensselaer	Grafton	1867	6
		St. Lawrence	?	<1853	p. V
				<1853	p. V
				<1853	p. V
		Saratoga	Ballston	1790	5
		Seneca/Schuyler	Seneca Lake	?	19
		Warren	Glens Falls/Albany	c1859	13
			West Glens Falls	1858	6
Pennsylvania (3)		Susquehanna	Lenox	1871	4
		?	?	<1871	13
				<1871	13
Rhode Island (19)		Kent	Coventry	c1838	6
		Newport	Newport	1875	7
		Providence	Cumberland	1796	1
			Foster	1827	19
			Smithfield	?	3
		Washington	Exeter	c1799	19

			1892	9-10
		North Kingstown & South Kingstown	1872	8
		Westerly	c1858	12
		?	1877-1902	17
			1877-1902	17
			1877-1902	17
			1877-1902	17
			1877-1902	17
			1877-1902	17
			1877-1902	17
	?	? "country town"	1890	7
			<1900	21
			1900	21
Vermont (6)	Bennington	Manchester	1793	11
		Winhall	1859	6
	Rutland	Middletown	1839	5
	Windham	Dummerston	c1799	11
	Windsor	South Woodstock	1817	1
		Woodstock	c1829	15
New England (7-8)	?	?	<1788	2
			<1834	13
			<1834	13
			<1834	13
			<1834 (?)	13
			c1848	3
			c1830-50	7
Canada (1)	Quebec	Stanbridge East	c1809-11	5
86-87 INCIDENTS				

Acknowledgments

Throughout the years that I devoted to this project, I've had the good fortune to be helped by many organizations and individuals. I am especially indebted to the librarians and archivists who have compiled, catalogued, and made available the enormous amount of data that I accessed for this book.

The following scholars read and reviewed most or all of the manuscript and offered valuable insights: Simon Bacon, John Blair, Richard Hite, Anthony Hogg, and Richard Sugg. Their comments were a great benefit. Of course, I am responsible for any errors or omissions in this publication.

For decades I have been interacting with three extraordinary individuals—Nicholas Bellantoni, Katie Gagnon, and Faye Ringel—whose investigative skills and extensive knowledge have guided me along the vampire trail. I think (fondly) of the four of us as the "Vampire Research Partners."

Several other people have provided me with documentary sources for vampire incidents, including John Edgar Browning, Jeanne Grandchamp, Paul Grzybowski, Marla R. Miller, Bill Page, Marybeth Reilly-McGreen, and Chris Stonestreet.

My son, Brendan Hyatt Bell, designed the book's front and back covers, and the figures and charts. My foster son, Gavin Castleton, created the map of vampire incidents. I would admire their skills in graphic design even if they were simply acquaintances and not close kin. The photographer, Cyril Place, and I have been visiting cemeteries and other vampire-related locations throughout New England since 2002. His work is consistently both atmospheric and informative.

I was on the verge of self-publishing *Vampire's Grasp* when I encountered Don Rittner, who attended a presentation I gave at the College of St. Rose in Albany, New York. When he asked if I had a publisher for my new book, my frustration with the current state of book publishing must have become palpably evident. But my demeanor brightened considerably when he, the Executive Director of the Warren County (New York) Historical Society, said, "We want to publish this book." I cannot thank them enough for coming forward at just the right time.

Philip Turner, who edited the first edition of *Food for the Dead* for Carroll & Graf, agreed to act as my agent and editor for *Vampire's Grasp*. His editorial talents would jump out at anyone who compared my first draft to the published version. I will always value my professional ties to, and friendship with, Philip.

Carole, my wife—always my best friend—has contributed more than it's possible to enumerate. Although she recently retired as an environmental scientist, she has also been an editor, folklorist, and English teacher. So, when she read, edited, and commented on every iteration of this work, she brought to it an invaluable and unique perspective. And she loved me enough to break my heart every time she said something along the lines of, "Just cut this entire section. It's irrelevant. Okay, it might be interesting, but it doesn't advance the narrative." If you find there's still too many needless words, you may blame me for sometimes ignoring Carole's cogent advice.

Part I The Vampire's Grasp

The story of New England's vampires begins with a scourge whose tragic trail is visible in cemeteries throughout the region. Incredible as it may seem to contemporary Americans, vampires preyed upon their not-so-distant ancestors. Vampire attacks increased dramatically during the eighteenth century and remained the leading cause of death in New England throughout the nineteenth century. But this unseen killer did not resemble the clever Count Dracula of Bram Stoker's imagination. Indeed, it was so small that it was undetectable. New England's authentic vampires, you see, were pathogenic microbes ("bacteria with fangs," as a nurse once described them).

Prior to the twentieth century, a diagnosis of consumption (as pulmonary tuberculosis was called at that time) was a virtual death sentence. As the coronavirus crisis was firmly grabbing the world's attention in 2020, it struck me that if Americans in the early nineteenth century had somehow discerned the benefits of distancing, wearing a mask, and getting vaccinated, they might have stemmed the spread of pulmonary tuberculosis and I could not have written this book. Unfortunately, as I described that era of consumption in my first book on this topic, *Food for the Dead: On the Trail of New England's Vampires* (2001), the science that might have generated an effective strategy for flattening the curve of the consumption epidemic did not exist until after 1882, the year that Robert Koch proved that tuberculosis was a bacterial infection. By that time, most of the tragic events examined in this book had already occurred.

Gravestone of Simon Whipple Aldrich, Union Cemetery Annex, North Smithfield, Rhode Island. *Courtesy of Cyril Place.*

Introduction: Medical Mysteries

Our species (*Homo sapiens*) has always been stressed by pathogens. The archaeological record documents what may be the first plague among *Homo sapiens*, in Africa during the Paleolithic Era, about 100,000 years ago.[1] A prehistoric epidemic occurred approximately 5,000 years ago in northeastern China, where archaeological evidence indicates that an entire village was extinguished in a very short time. Historically documented plagues and epidemics from the ancient world range from the Plague of Athens (430 B.C.E.) to those in Rome (A.D. 165-180) and the Byzantine Empire (bubonic plague, A.D. 541-542). A bacterium caused the Black Death, which devastated Europe from 1346 to 1353; it returned to kill fifteen percent of London's population three centuries later (1665-1666). Both Marseille (1720-1723) and Moscow (1770-1772) lost about 100,000 citizens to plague epidemics in the eighteenth century.[2]

Consumption was just one in a long list of infectious diseases assaulting early Americans. As John Duffy wrote in *Epidemics in Colonial America* (1953), "Epidemic disorders visited death and destruction upon the American colonies with relentless regularity."[3] Duffy's chapter headings vividly express what these early Americans had to endure: smallpox; diphtheria and scarlet fever; yellow fever; measles, whooping cough, mumps; respiratory disease; agues, fluxes, and poxes (including malaria, dysentery, typhoid fever, typhus, and venereal disease). Fast-forwarding to the twentieth century, the persistence of plagues in American history is exemplified by the influenza pandemic of 1918 and the polio epidemic of the 1950s, when public swimming pools were avoided, and children were put into iron lungs.

In the Covid crisis, rapidly mutating strains outpaced health systems, reminding us of a pervasive vulnerability that earlier human civilizations demonstrated over and over. Effective therapies and vaccines are now widely available, but the absence of lifesaving treatments reminds us of what our forebears had to contend with—and over a much longer period. During the tuberculosis epidemic of the nineteenth century in New England, people asked the same questions we now ask. What is going to happen to us? Will we be safe? What should we do? What *can* we do? These questions, generated by uncertainty and fear, accompany all epidemics, regardless of time or place. If these concerns are not adequately addressed by medical and government or other officially sanctioned, authoritative cultural organizations—if the answers provided do not lead to acceptable resolutions (primarily, not dying!)—then people look elsewhere. Folk medicine offers answers.

In his online essay, "The Values—and Dangers—of Folklore during a Global Pandemic," folklorist James Deutsch made just this point in the context of the Covid pandemic: "One of the folkloric genres that is especially relevant at this time is folk medicine, which includes folk remedies and cures to combat illnesses, especially when more conventional medicine has been ineffective." Only days into this pandemic, seemingly plausible treatments began circulating, as Deutsch noted, "transmitted more informally from person to person or group to group—nowadays often via text messages or social media."[4] The fact-checking website Snopes.com began collecting and evaluating these newly relevant examples of folk medicine, including:

- Sipping water every 15 minutes will prevent you from becoming infected.
- Using a hair dryer to breathe in hot air can cure COVID-19 and stop its spread.
- Gargling with salt water or vinegar will eliminate the COVID-19 coronavirus.
- Eat bananas to prevent coronavirus infection.[5]

Folk medicine is not the only alternate strand of medicine to step into the void when official systems cannot offer effective treatments. Untrained practitioners, who deliberately misrepresent their medical qualifications for profit (often called "quacks" or "snake oil salesmen"), peddle remedies and treatments that have not been tested or approved by the official health-care system. Instead of snake oil, for example, one herbalist company offered "immunity oil" as a cure for Covid; and Covid treatment with Ivermectin, an antiparasitic drug, has been widely touted and debunked.

Those who were diagnosed with consumption prior to the twentieth century had every reason to be uncertain and fearful, for they had little hope of being cured. Throughout most of the nineteenth century, the medical community, unaware that the disease resulted from a pathogen, could provide no scientifically based efficacious treatment. In his history of illness and death in New England, Alan C. Swedlund wrote: "In city and country it was a constant, daily reminder of chronic illness and of the mortal self. And although it was well known that death from consumption could take one at any age, people were beginning to understand that a particular segment of the population in New England, as elsewhere, was unusually vulnerable: adolescents and young adults, especially young women—people normally considered to be in the 'prime of life.'"[6] The inscription on the gravestone of Simon Whipple Aldrich, who died on May 6, 1841, in North Smithfield, Rhode Island, condenses this tragic aspect of consumption:

Altho' consumption's vampire grasp

Had seized thy mortal frame,

Thy ardent and inspiring mind

Untouched, remained the same

Simon's gravestone is flanked by those of his two sisters (each of the three siblings died at age twenty-seven). True horror is knowledge without power. As youth dissolves into death, the intact mind, having witnessed this trajectory in others, knows there is no negotiating with consumption's grasp.

Yet there was a folk medical practice circulating in New England during the late eighteenth century and throughout the nineteenth century that promised to banish the scourge that was destroying entire families. While symptomatic of consumption, the folk belief hinted, the deaths were caused by a mysterious evil (sometimes labeled a "vampire") residing in the corpse of a family member. Although there were several major variants of this remedy, the typical therapeutic ritual prescribed exhuming the bodies of deceased relatives and looking for a corpse with fresh blood in its heart or other vital organs. The target organ was cut from the body, burned to ashes, and often, to complete the cure, administered as medicine to the afflicted.

In *Food for the Dead*, I related my twenty years of scholarly research into this topic. The reader accompanied me as I uncovered clue after clue in old newspapers and long-forgotten manuscripts, on crumbling gravestones, from the mouths of interviewees, and even in the rearranged bones in a newly opened grave. At that time, I was certain that the twenty incidents I described were the tip of a larger iceberg. Now, aided by online databases, this book documents more than eighty vampire events (see Vampire Incidents in New England). The evidence shows that in New England during the late 1700s and throughout the 1800s, the consumption ritual was known, accepted, endorsed, and sometimes actually performed, by the community-at-large, by town authorities, by doctors, and even by clergymen.

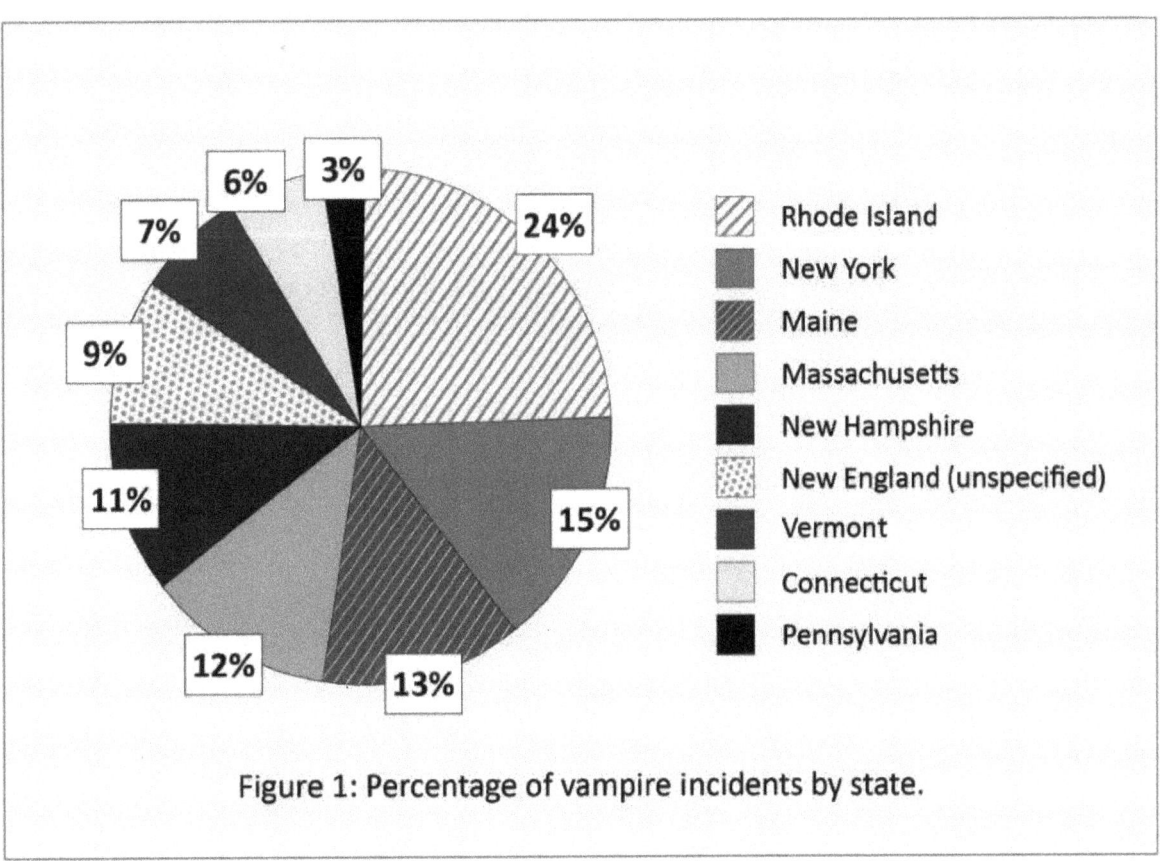

Figure 1: Percentage of vampire incidents by state.

Dire predicaments with fatal outcomes drive people to seize on any measure that offers a glimmer of hope. This was true in the eighteenth century, and, as we have witnessed with the coronavirus pandemic, it still holds. Viewed in this light, we might grant that despondent people who take desperate measures may be acting reasonably, bearing in mind that one does not have to believe to act. The therapeutic consumption ritual was a collaborative action that mustered a community in the fight against a common enemy. Facing almost certain doom, doing *something* was more empowering than doing *nothing*. Moreover, as John Duffy noted in his history of epidemics in Colonial America, "contemporary medical practices were unable to alleviate the high mortality from epidemic infections." He pointed out that the survival rate for those infected remained basically the same no matter what choices they made regarding intervention. Indeed, as Duffy also observed, people who resorted to medical practitioners sometimes actually increased their chances of dying. The treatment of these afflictions, he wrote, "justified the aphorism of Dr. William Douglass that 'more die of the practitioners than of the natural course of the disease.'"[7] An afflicted person, understandably, had little reason to reject any possible cure.

The finger-pointers who cried "superstition" when confronted with the consumption ritual had an earnest argument, which will be evident in this book—as will my opposition to it. Branding a belief or practice a "superstition" or "survival from a lower level of culture" essentially removes it from serious consideration, saying, in effect, "I do not believe in what you are doing" or "your practice is an irrational relic from the past." Such judgments shed more light on the labelers than on the targets of their disapproval.

While it did not happen overnight, and there were missteps and failures along the way, empirical science finally demonstrated the best way forward in confronting infectious disease—eventually even tuberculosis. Americans today place their trust in the biomedical paradigm (now simply called *medicine*) to successfully address their routine health care needs, as well as their more serious afflictions. During most of the nineteenth century, however, such trust was not warranted, especially regarding tuberculosis. The symptoms of fast-burning diseases, such as smallpox and yellow fever, manifested early. It was obvious that infected people could pass it to others, so isolating them was effective. John Duffy writes, "The seriousness of smallpox attacks and the obviously contagious nature of the disease led to a whole series of legislative measures, all of which aimed at isolating the sick. Practically all the early colonies enacted laws requiring ships from infected ports to await ten to twenty days' quarantine. In general, these measures were administered adequately and served to reduce outbreaks of smallpox, yellow fever, and other epidemic contagions."[8] Another series of laws was directed toward isolating infected individuals within their homes. During the smallpox epidemic of 1730-31 in Massachusetts, for example, the general court passed a law whose purpose was "to prevent persons concealing the small pox and requiring a red cloth to be hung out in all infected places." A third measure of control used by authorities was the construction of quarantine hospitals, termed "pest houses" at that time. Duffy concludes that, "In general, the quarantine measures and the recourse to pesthouses worked fairly effectively in New England."[9]

As a point of contrast, Covid, unlike smallpox, apparently takes longer to manifest, even to one who is infected. A significant number of those infected manifest no symptoms, but still may infect countless others and never realize it. Now, imagine a communicable disease a great deal slower to manifest than Covid, with symptoms even more ambiguous. One that did not blast through a population—leaving in its wake the dead and those who survived through good fortune or natural immunity—and then disappear or become latent. A disease that, instead, once it grasped a person, could go in and out of remission over a period of months, or years, or even decades. This was consumption in New England throughout the nineteenth century: more *endemic*—a constant presence that ebbs and flows—than *epidemic*—a rapid, but transitory spread of disease within a confined area. Adding to its mystery, consumption seemed capricious in choosing victims. Some families escaped intact while others were decimated. It seemed apparent that consumption "ran in families." Some people wondered: *Is there a "hereditary predisposition" to contract the disease?*[10] Yet it also took hold in particular locales, so others asked, *Is there something in the soil, water, or air that leads to consumption?* Another possible explanation arose from the observation that people in close contact with those who had the disease were likely to contract it, prompting the question, *Is consumption a contagious infection, like smallpox?*

These questions troubled the medical community as well as the general population. Two major factions regarding consumption's cause began to emerge early in the nineteenth century: heredity versus contagion. But versions of the contaminated-place theory also lingered. One posited that damp soil, especially if cold, caused or promoted consumption. Another, shared by physicians and laymen alike, blamed bad air or *miasma* (noxious atmospheres). Folk ideas of contagion were broader than those of medical science: death, itself, was contagious. The evil spirit that contaminated a corpse brought death (consumption) to each member of the family, one by one, until it was ritually destroyed. In this sense, death (consumption) was also hereditary.

Some performances of the consumption folk ritual were deemed sufficiently story-worthy to be cast into narratives that appeared in several contexts. In this book, narratives (of rituals)—their creators, tellers, and audiences—intersect with a disease (consumption)—mysterious, pervasive, and seemingly invincible—and a label (*vampire*)—now universally known, continually evolving, and liberally applied, yet passionately contested. The basis of this book's narrative could be termed *forensics*. Evidence from historical records, genealogy, and biography blend to create portraits of "vampires," families, communities, narrators, and audiences. Insights offered by folklorists, anthropologists, and historians (among others) provide an interpretive framework. Consumption's destructive impact on families and communities in New England was the foundation for the performance of the rituals and, in turn, their narratives.

I often think of my research into these events as an assigned "impossible task."[11] I have been given a large jigsaw puzzle, but there is no box to hold the pieces, and there is no picture that shows what the completed puzzle should look like. The pieces are scattered; many will never be found. I am allowed to fill in some of the blank spaces, but only if I do so judiciously, using the surrounding pieces as clues.

One area of this puzzle is quite distinct, for consumption is no longer a mystery (even though our ability to contain the global endemic of tuberculosis remains a challenge). The consumption/vampire incidents of New England demonstrate that disease and death transcend time and place and that, despite the great accumulation of scientific knowledge since the nineteenth century, we still have our own "vampires"—our own mystifying, fatal diseases—to confront and defeat. The coronavirus pandemic is the most recent reminder of this unfortunate reality, joining a list of viruses that began devastating the world just as the puzzle of consumption had been solved.

Folklorists have noted that folk beliefs are recruited when people perceive they are at risk yet have limited control of the outcome. Official cultural systems tend to dismiss such beliefs as irrational superstitions. But the consumption ritual was judged more harshly because it included corpse mutilation and, sometimes, cannibalism. For families facing almost certain death, however, yielding to the possibility that the dead could prey on the living empowered them to confront disease and death armed with hope.

1 From Consumption Rituals to Vampire Narratives: Folklore, Context, and Agency

While people today cannot observe the actual consumption rituals of New England, they can experience them through the *stories* of others. The texts of these stories are the foundation of this book. A ritual is like a *staged performance*: It has a script, a cast of characters, a theme, and a plot.[12] In the New England rituals to halt the spread of consumption, the most obvious key cast members were the troublesome dead (the "vampires"). Other dramatis personae included family members whose lives were dwindling away and who, thus, required healing, as well as those who, though not presently ailing, were at risk and needed protection. A consumption ritual was initiated when an afflicted family questioned if a deceased relative could be responsible for their illness and death, a question frequently resolved through a consultation that included the head of the family and other male kin. Leaders in the community, outside of the family, also may have participated.

Folklorist Lauri Honko described this situation: "A person who has experienced a supernatural event by no means always makes the interpretation himself; the social group that surrounds him may also participate . . . . In their midst may be spirit belief specialists . . . whose opinion, by virtue of their social prestige, becomes decisive."[13] Those who tell stories of a consumption crisis and its successful resolution play an essential role in this drama, for it is through their narratives that communities "grasp the living essence of folk belief." As Honko observed. "Belief in the existence of spirits is founded not upon loose speculation but upon concrete, personal experiences."[14]

A decision to perform the consumption ritual initiated four basic acts, with the potential to be combined in various ways:

1. *Inaugural* acts. Ritual practitioners confront the corpse(s). Exhumation was almost invariably the inaugural act.

2. *Diagnostic* acts. Practitioners determine that a corpse is either a threat to the living or lifeless. Variations in diagnostic acts are grounded in phenomena that can be interpreted as revealing signs of continuing life, usually liquid ("fresh") blood in the heart or other vital organs ("vitals").

3. *Transformational* acts. Practitioners eliminate, or at least neutralize, the threat. The most common transformational act is burning one or more vital organs. In some cases, the entire corpse may be burned. In other instances, the corpse might be turned face down and reburied.

4. *Healing* acts. Practitioners restore the well-being of the afflicted. Ill family members might be instructed to ingest the ashes from the incinerated vital organs, or to inhale, or be fumigated by, the smoke from the burning corpse.

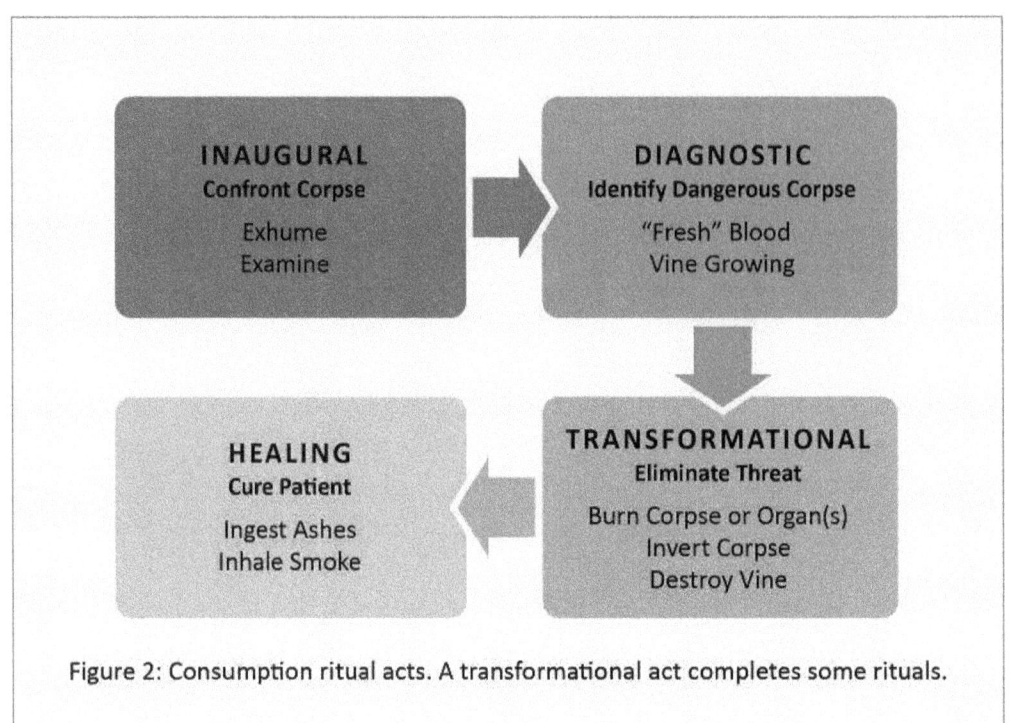

Figure 2: Consumption ritual acts. A transformational act completes some rituals.

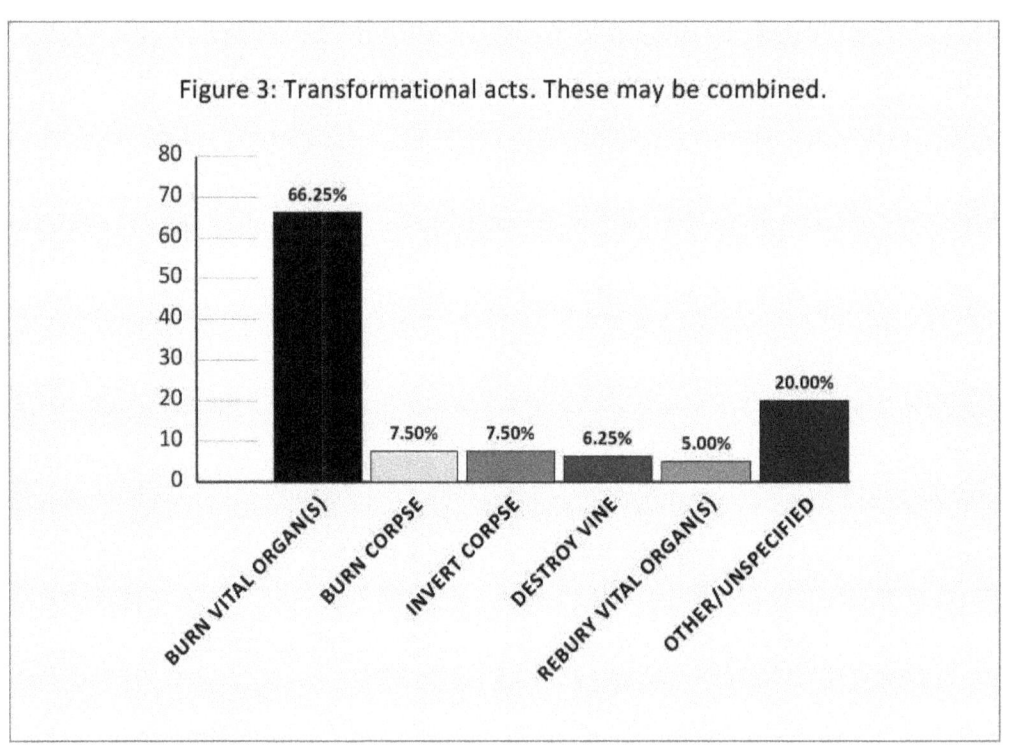

Figure 3: Transformational acts. These may be combined.

What is the meaning of these dramatic performances that were set apart from everyday life? Many anthropologists have asserted that ritual, in general, makes manifest that which is abstract, transforms the unknowable into the known, controls the uncontrollable, changes the unchangeable, and renders natural that which is supernatural. Some have argued that funereal rituals separate dangerous corpses, marginalize them, and then reaggregate them into society.[15] For better and worse, all human cultures have maintained connections to the dead.[16] Vampires are a devastating link to the living that must be destroyed so that the family and community can return to their proper order. To end the perceived harmful relationship, a secondary mortuary ritual must be performed (burial, itself, was usually the primary mortuary ritual in New England). The primary and secondary disposal of the remains of the dead—termed "double obsequies" by anthropologists—affirms that "the corpse has fully deteriorated or has been destroyed by human means," so mourning comes to an end and the spirit of the deceased is firmly established in its new existence."[17]

The inaugural and diagnostic mortuary rituals carried out in New England were intended to identify which corpse had not "fully deteriorated" and therefore was yet to be "firmly established in its new existence." Transformational and healing acts, such as burning the corpse's vital organs and ingesting the ashes, were supposed to end the unnatural relationship between dead and living kin and restore health. When successful, the ritual completed the mourning, so that the living could leave behind "otherworldly concerns" and return to their "more mundane concerns."[18]

Keith Thomas called attention to the psychological function of ritual to provide comfort, solace, and deliverance. He observed that the common folk of seventeenth-century England noticed that people die while under the care of the orthodox medical system. "If this is true today, when medical technique has made such striking advances," Thomas wrote in 1971, "we can hardly wonder at the attitude of seventeenth-century villagers, when medical therapy still proceeded along its traditional paths of purging and blood-letting." People are inclined to "stick to their traditional remedies, some of which afford a degree of psychological release and reassurance not to be found in Western medicine. They cherish the dramatic side of magical healing, the ritual acting-out of sickness, and the symbolic treatment of disease in its social context."[19] Finally, simply *experiencing* ritual may be efficacious.[20] Even if the deaths do not end, the family may be symbolically healed through a performance that ritualizes the way things ought to be.[21] In a similar vein, anthropologist Michael Lambek argued that, "As an anthropologist you shouldn't try to explain the symbol or ceremony, you interpret it and try to locate its power."[22]

Granting that the consumption ritual had various social uses and cultural functions, was it effective in curing the disease? Here are some statistics. In almost 46% of more than eighty documented cases, the outcome of the ritual is either unknown or cannot be determined, based on the information provided. Nearly 35% of the time, the ritual was deemed unsuccessful, as the disease continued its fatal course. In about 7% of the cases, the ritual was either not performed or remained incomplete; in most of the latter, the signs that would indicate a troublesome corpse or suspicious grave were absent. Just slightly more than 8% of the rituals were deemed successful, in that the ailing person recovered. A little more than 4% of the time, the results were considered mixed, as one more family member died, but others recovered or continued in good health. In Litchfield County, Connecticut, for example, as reported in the *Boston Herald* (September 1, 1851),

"The consumption which had commenced in the living subject, however, was not arrested, but went on to its termination. Some of the younger members of the family escaped, are still living."

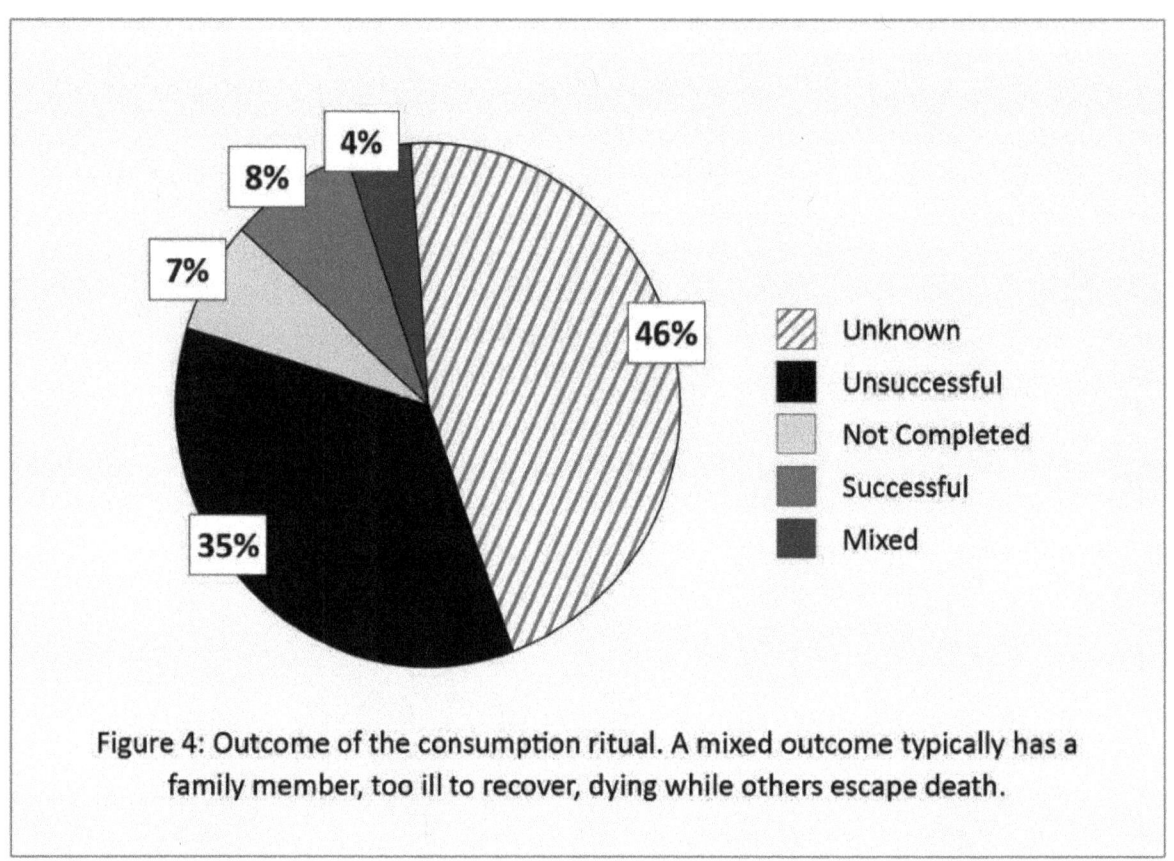

Figure 4: Outcome of the consumption ritual. A mixed outcome typically has a family member, too ill to recover, dying while others escape death.

***

What folk ideas inform the scripts of the exhumation rituals? When Reverend Justus Forward of Belchertown, Massachusetts (1788), wrote in a letter to a friend that he had "consulted many about opening the graves of some of the deceased, to see whether there were any signs of the *dead preying on the living* [italics in the original]," he was presupposing that it was *possible* for the dead to prey on the living. But, nowhere in his letter does he hint at *how* it was possible.[23] The author of the Robinson Hollow, Massachusetts (ca. 1847) text, Adaline M. Tirrell, provided some detail that hints at an underlying principle. She wrote that a daughter of Dr. Robinson, who was ill with consumption, "had heard the story of the sinister work of 'vitals.' The story said that, when one member of a family died with consumption, his or her 'vitals,' meaning by that term the lungs, heart, and liver, became animated after burial and came back to earth in invisible form to prey upon the 'vitals' of others in the family until they, in turn, wasted away with the same disease."[24] The best explanation that Francis Gerry Fairfield, from West Stafford, Connecticut, could offer was that "the vital organs of the dead still retain a certain flicker of vitality, and by some strange process absorb the vital forces of the living."[25] The anonymous author of a text from New Marlborough, Massachusetts, wrote that "the reason assigned" for consumption killing off families "was that there was a sort of vital current existing between the living and dead—that those organs in the dead body that contained fresh blood and appeared to be alive, would continue to live until the vitality of the living subject was exhausted, unless said organs were taken out and consumed by fire."[26]

"Vital current" seems analogous to the theory of electricity, which—at least in the popular imagination of the mid-nineteenth century—was viewed as a powerful but invisible force of nature. Yet the talk about "vitals," "flicker of vitality," and "vital current" has deeper cultural roots, embedded in a concept aptly termed *vitalism*. Simply stated, vitalism posited that blood, the "paramount humor," contained the essence, or *vital spirit*, of the creature in which it flowed, which explains why exhumers were looking for "fresh" blood in the vital organs of suspicious corpses.[27]

There was a medical tradition, practiced in both Europe and New England, that explicitly postulated a connection between a universal vital spirit and human blood (as well as other body parts, including flesh, heart, skull, and bone marrow). These Paracelsians—as they were called, after Philippus von Hohenheim (1493-1541), the Swiss-German chemist and physician known as Paracelsus— believed that medicines healed by the occult "influences" or "spirits" they contained. Anthropologist Karen Gordon-Grube noted that the ancient ideas of "like cures like" and "sympathetic action" (or "action at a distance") were fundamentally important to Paracelsians, whose "spiritual" interpretation of healing was "fundamentally opposed to that of the official, Galenist school of medicine."[28] For the Paracelsians, "the body as a symbol" was "*not* holy" but "merely a worthless piece of matter in which the life-spirit dwells."[29]

While those who participated in consumption rituals, or spread knowledge of them through stories, made no explicit connections to the spiritual presence in human corpses postulated by Paracelsians, there are similar underlying folk ideas. The spiritual connection between the living and the dead, central to the vampire tradition, is analogous to the "sympathetic action" that is fundamental to Paracelsian theory. The nature of this spiritual connection is quite opposite in these two traditions: the "quasi-divine" vital spirit has healing power for the Paracelsians, whereas in the vampire tradition the in-dwelling spirit is lethal. In the consumption rituals, the connecting spirit must be burned (purified or rendered inert) before healing can take place.

***

When examination revealed "fresh" blood in the vitals of a corpse, can we justifiably label the corpse a *vampire*? A word has no meaning apart from its contexts of use.[30] People using words create meaning. An ongoing debate over the appropriate use of the word *vampire* exemplifies the rich variety of contexts that can accrue to a single word. It also points to the privilege and power associated with the "ownership" of words.

The word *vampire* (in any of its inflected forms) is conspicuously absent in the narratives of participants in New England's consumption rituals. Even among outsiders, the word was used sparingly, occurring in just twelve of the eighty-some collected texts. It first appeared fifty-eight years after New England's first (thus far) documented consumption ritual (1784), debuting in a narrative that was likely a fictionalized account of the authentic tradition. In her 1842 essay, "Popular Prejudices," Mrs. S. E. Farley capped what she claimed was her next-door neighbor's exhumation story with the following admonishment: "Surely when such things are believed and practised in our own country, we need not wonder at the Vampyre tales of other lands."[31] Eight of the twelve narratives in which the word appeared were recorded in 1885 or later. As it was used by outsiders, the term *vampire* was a virtual synonym for "horrible superstition."[32] Bruce A. McClelland made this point in a discussion of the vampire's paradoxical relationship to life and death in the Balkans, writing that "the ambiguous vampire may even serve as a social metaphor for residual pagan customs themselves, which both 'departed' as they were supplanted or taken up by Christian belief, and did not depart, since the converted pagans did not fully let go of their own traditional rites."[33]

Franklin Benjamin Hough, in his *History of St. Lawrence and Franklin Counties, New York* (1853), included "searching for vampires" in a short list of superstitious activities he encountered. While he summoned the courage to condemn vampire hunters, he shrank from specifying what they did. Hough wrote, "It had been our design to enumerate some of the evidence of superstition, as evinced in various enterprises," noting that "the annals of St. Lawrence County afford at least three instances" of "searching for vampires." But, he concluded, "our space forbids the details, revolting to humanity, and regard for the living, leads us to pass unnoticed these heathenish mutilations of the dead."[34]

The absence of a simple "emic" (or insider) generic name, comparable to the precise "etic" shorthand of *vampire* that was used by some outsiders, suggests that insiders did not conceptualize the life-threatening phenomenon as an entity that needed a label. If they had blamed consumption on a corporeal being with an independent existence that left its grave to attack the living, for instance, surely they would have labeled this devastating foe.[35] When insider Reverend Forward wrote about opening graves and looking for "signs of the dead preying on the living," he seemed to approach this vampire concept. But the harmful connection between the living and dead was formless and obscure: its nature consisted of "some strange process" that could "animate" the "vitals" of deceased relatives and "absorb the vital force of the living."

Considering these etic vs. emic distinctions, are there grounds for equating a phrase such as *the dead preying on the living* with the word *vampire*? That question returns us to contexts of use. In his study of vampires, Paul Barber began by taking observers' descriptions of suspected vampire attacks at face value, and then looking for scientific evidence that verified the reported phenomena. Barber is clear that on-site observations are often supported by today's forensic science, particularly knowledge of what happens to human corpses undergoing the process of decomposition. He arrived at a definition of *vampire* that suited all of the cases that he had considered, writing: "It is probably not going too far to suggest that a vampire might be defined as a corpse that comes to the attention of the populace at a time of crisis and is taken for the cause of that crisis."[36] Some scholars are convinced that Barber *has* gone too far, leaving the vampire door open for any corpse that becomes a scapegoat. The arguments between the vampire "splitters" and "lumpers" (to borrow a dichotomy used by taxonomists) will be taken up in more detail in chapter 16. For now, I have adopted Barber's definition as a working shorthand.[37]

Why did New Englanders who performed consumption rituals *not* label troublesome corpses *vampires*? These rituals almost certainly are rooted in Europe (in particular, Eastern and Central Europe), as a comparison of the two traditions establishes. How did a tradition from continental Europe become established, by the late eighteenth century, in a region populated mainly by people of English descent? One clue is in the incident from Willington, Connecticut (1784), in which a town official complained about "a quack doctor, a foreigner" who was promoting the consumption ritual and had, indeed, induced a townsman to exhume the bodies of his children. This, with other evidence, leads me to conclude that the ritual was introduced to New Englanders by traveling healers whose roots were in areas of Europe where the vampire tradition was practiced. These healers were selling quick-fix remedies, not an entire tradition and its context. As New Englanders learned and employed these remedies, they told others about them. These local stories focused on a medical practice, so the *vampire*—as a word, concept, or context—was absent.[38]

***

Was the New England consumption ritual a folk tradition? Although folklorists define and describe folklore in many ways, and the American Folklore Society (the professional organization that serves the field of folklore studies) has no official definition of *folklore*, there is a common core of agreement:

> Folklore is the traditional art, literature, knowledge, and practice that is disseminated largely through oral communication and behavioral example. Every group with a sense of its own identity shares, as a central part of that identity, folk traditions—the things that people traditionally believe (planting practices, family traditions, and other elements of worldview), do (dance, make music, sew clothing), know (how to build an irrigation dam, how to nurse an ailment, how to prepare barbecue), make (architecture, art, craft), and say (personal experience stories, riddles, song lyrics). As these examples indicate, in most instances there is no hard-and-fast separation of these categories, whether in everyday life or in folklorists' work.[39]

Folklorist Jan Brunvand emphasizes that folklore is "the traditional, unofficial, non-institutional part of culture."[40] Folklore persists through time in stable, recognizable forms. "Its center holds," as Henry Glassie put it. This is the conservative aspect of folklore, where "changes are slow and steady." But folklore is dynamic, too, as Glassie also observed: "The tradition remains wholly within the control of its practitioners. It is theirs to remember, change, or forget. Answering the needs of the collective for continuity and of the individual for active participation, folklore . . . is that which is at once traditional and variable."[41] Barre Toelken, in *The Dynamics of Folklore*, summarized this interplay between conservative and dynamic aspects of folklore: "Folklore is made up of informal expressions passed around long enough to have become recurrent in form and content, but changeable in performance."[42] Attempting to synthesize the central core of folklorists' concerns, Simon Bronner recently stated, "Simply put, folklore is '*traditional knowledge put into, and drawing from, practice.*"[43] Glassie gets the last word: "Customarily folklorists refer to the host of published definitions, add their own, and then get on with their work."[44]

Forty years of investigating this topic leads me to conclude that, by the late eighteenth century, consumption exhumation rituals had, indeed, become part of the folklore of New England: the rituals were traditional cultural expressions, performed mostly apart from institutional nurturing and support. Knowledge of them was transmitted by word of mouth and practice; and, although their "form and context" was "recurrent" (traditional and conservative), they were "changeable in performance" (dynamic). The *narration* of the ritual events, however, has been performed in contexts beyond that of the folk process.

***

If one attribute sets humans apart from the rest of the animal kingdom, it is the ability to tell stories. Before we can formulate our own stories, however, we hear, process, and store for future use, not only unique or familiar tales, but also models. As one writer of supernatural fiction remarked, "I have always thought that the oldest profession was that of storyteller—in particular, the teller of supernatural tales."[45] It is not surprising, then, that newspaper publishers, medical practitioners, and authors of local histories, among others, offered their own stories of the vampire rituals.

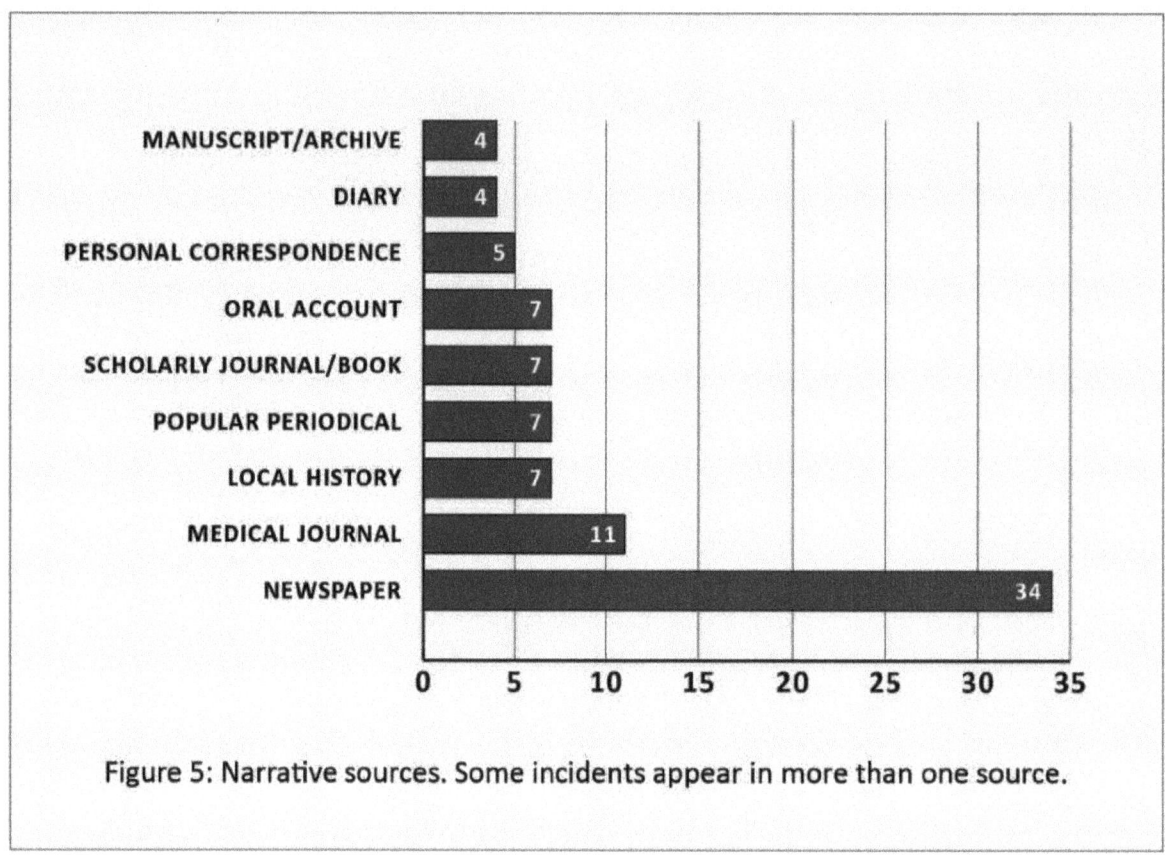

Figure 5: Narrative sources. Some incidents appear in more than one source.

Eyewitnesses: Participants and Observers. First-person narratives are the most compelling texts, especially those contributed by participants. In an extraordinary letter to a young family member, an elderly man described his direction of a consumption ritual many years earlier in an unspecified New England town. He wrote, "The disinterment was done under my personal supervision, as carefully and respectfully as such things should be." He described, in minute detail, the exhumation of the young relative's grandmother, grandfather, mother, and brother. Participants, such as this anonymous letter writer, often expressed uncertainty as well as hope. Their accounts testify that the consumption rituals were actually performed by real people.

Family Storytellers. Family vampire narratives generally have functioned to connect kin across generations, humanize predecessors, and explain family customs. Family stories that were initiated soon after the event and have continued into the present are rare. The case of Mercy

DATE	NARRATOR	ROLE	LOCATION	CH
1784	Moses Holmes	Observer	Willington, CT	4
1788	Rev. Justus Forward	Participant	Belchertown & Hatfield, MA	2
c 1807	Joseph R. Chandler	Observer	Kingston, MA	18
1810	Rev. Enoch Hayes Place	Observer	Barnstead, NH	3
1812	Dr. Samuel Preston Hildreth	Observer	Essex County, MA	4
1818	Maria Carver Gregory	Observer	Chazy, NY	12
1822	William Sheldon Briggs	Observer	Keene, NH	3
< 1832	Seba Smith	Observer	ME	4
c 1838	A.O. Warren	Observer	Jaffrey, NH	4
c 1848	Anonymous Family Head	Participant	New England	3
1851	Dr. William H. Cook	Observer	Kelloggsville, NY	6
1871	Spectator	Observer	Lenox, PA	4
nd	Anonymous	Participant	NH (?)	3
nd	Anonymous Old Man	Observer	North Kingstown, RI	17

Figure 6: Eyewitness narratives.

Brown, which we encounter in Part IV, is an extraordinary exception. Daniel Ransom's account is moving testimony to consumption's power to decimate a family. In 1894, eighty-year-old Ransom wrote about the exhumation of his older brother, Frederick, some seventy-seven years earlier.[46] While Daniel surely did not attend the ritual at the cemetery, he must have heard family conversations about the event. How much can a three-year-old understand and then remember? Daniel addressed those questions in his unpublished memoirs: "My remembrance of him [Frederick] is quite limited as I was only three years at the time of his death and I date my remembrance of anything at a visit of Dr. Frost to Frederick in his sickness. Keeping shy of the Doctor, fearing he would freeze me." Daniel's memory of Frederick's consumptive condition, possible treatments, exhumation and, finally, the outcome probably was augmented by family-experience stories which began during the crisis and continued until well after the results of the ritual were realized.

The Ransoms were an early and distinguished family in the town of South Woodstock, Vermont. Richard Ransom and Elizabeth (Betsy) Mather Ransom, Daniel's parents, were born in Lyme, Connecticut—situated on the Connecticut River in southeastern Connecticut—where both of their families had established roots. Their son, Frederick, was a twenty-year-old student at Dartmouth College when he died of consumption on February 14, 1817. In light of his subsequent exhumation, the final phrase of his obituary is imbued with unintended irony: "At Woodstock, on Friday evening last, Mr. Frederick Ransom, aged 20 years; a youth of most accomplished manners; which endeared him to a large circle of relatives and friends; and of real piety, which rendered him highly esteemed by the church of which he was a member, and enabled him to endure the pains of a lingering disease with christian [sic] fortitude, in a well-grounded faith of a happy resurrection from the grave."[47] Two of Frederick's younger siblings had preceded him to the grave: Stillman, who died December 4, 1811, at the age of eleven, and Louisa, who died January 5, 1812, at the age of four.

Daniel Ransom does not identify the family member needing the remedy. The family's oral tradition noted a susceptibility to consumption, and a possible remedy, for Ransom wrote:

> It has been related to me that there was a tendency in our family to consumption, and that I, who now in 1894 am over eighty years old, would die with it before I was thirty. It seems that Father shared some what in the idea of hereditary diseases and withal had some superstition, for it was said that if the heart of one of the family who died of consumption was taken out and burned, others would be free from it. And Father, having some faith in the remedy, had the heart of Frederick taken out after he had been buried, and it was burned in Captain Pearson's blacksmith forge. However, it did not prove a remedy, for mother, sister, and two brothers died with that disease afterward.[48]

For the record, Daniel's mother, Betsy, died July 10, 1821, at sixty-five years of age; sister, Eliza Ely, died May 19, 1828, at twenty-four years of age; brother, Richard Mather, died April 10, 1830, at thirty-six years of age; and, brother, Royal Makepeace, died June 17, 1832, at thirty-one years of age. His father, Richard, lived to the relatively ripe age (for this family, at least) of sixty-nine. A statement that followed Frederick's obituary summarized the community impact of losing four of the five Ransom sons: "It has been said, that the death of a young man is a public calamity, for the fair prospect of a long course of usefulness is at once blighted forever."[49] Daniel, the sole surviving sibling, died at the age of eighty-one, on October 24, 1894.

Newsworkers. Newspaper stories of New England's vampires exist in a commercial environment. While newspaper writers, editors, and publishers may have their individual reasons for telling these stories, they share an interest in selling newspapers. Sensational vampire stories attract readers and require little in the way of context development. Because vampire practices deviate from acceptable mainstream behavior, journalists feel free to make light of them. In the recent past, vampire stories allowed both writer and reader to congratulate themselves for having traveled so far along the road to civilization.

Historians. Unlike newsworkers, authors of New England's local histories mostly ignored consumption rituals. Many of these town histories were written during the latter half of the nineteenth century, when the "old New England" appeared to be slipping away; it survived in the memories of the very old and in whatever tangible artifacts from the past had been preserved. But old New England, itself, was an idealized artifact. Local chroniclers routinely skirted topics, such as ritual exhumations, that detracted from the cherished image of idyllic pastoral towns. Franklin Benjamin Hough's history of St. Lawrence County, New York (1853) exemplifies the deliberate exclusion of "searching for vampires." Chroniclers who *did* address these events used them to illustrate the "progress" their communities made in becoming "free from superstition," or to counterbalance their normative "good old days" narratives.

Medical And Scientific Establishment. Beneath the surface of the romanticized folk lurked the savage. The theory of cultural evolution, embraced by social scientists in the latter part of the nineteenth century, provided a "scientific" rationale for censuring behavior of those deemed incompletely civilized. As human culture advanced from savagery through barbarism to civilization, the theory went, beliefs and practices carried over from lower stages hindered civilization's consummation. These *survivals*, including vampire rituals, needed to be identified so that they could be eradicated. America's nineteenth-century medical establishment, pushing to consolidate its authority, used such practices to marginalize competing medical systems. In the emerging sovereignty of the biomedical paradigm (now simply *medicine*) physicians joined historians, newspaper publishers, and other establishment authorities in condemning the consumption ritual.

Community-Based and Fiction Authors. Some authors who wrote about New England's vampire tradition straddled a line between condemnation and romanticism. Vampire rituals provided plot and character, as well as local color, for community-based authors, whose narratives wavered between fiction and nonfiction. Where sticking to the facts might narrow a story's appeal, some writers chose aesthetics over authenticity.

Storytellers make choices as individuals while performing within a framework of social consensus and cultural obligation. Narratives are cultural resources that people create to achieve goals with both individual *and* collective significance. Feedback loops, incorporating social relationships and interactions between storytellers and audiences, shape emergent stories. Sometimes a narrator's voice may be explicitly collaborative, such as when a newspaper reporter's story is revised by editorial colleagues. Yet, ultimately, narrators direct the storytelling process; they choose, which gives them agency.

***

Access to the channels of storytelling provides *agency*, a privileged vantage that gives narrators some control over *how* events are interpreted. Since storytelling contexts are neither isolated nor self-contained, I have attempted to sort out the conceptual backgrounds that give stories meaning. Most who narrated New England's consumption rituals were not interacting closely with the ritual's performers. Distanced by culture, locale, or time, outside narrators marginalized the ritual and othered its participants.[50]

Newsworkers, historians, medical practitioners, and scholars were communicating to different audiences within distinct contexts, but they were routinely unitary in characterizing the rituals as delusional superstitions. What is *not* in a narrative may be as telling as what is. Some local history chroniclers (we cannot know how many) chose to exclude events, such as vampire exhumations, that did not advance (or detracted from) their larger narrative. Recovering the perspectives of marginalized ritual performers is a salvage operation, as muted voices must be rescued from the bits and pieces of scattered sources. With few exceptions, reading between the lines is the only way forward.

Consumption rituals and their participants were not *invariably* marginalized. At times, the ritual was sanctioned (sometimes implicitly) by official cultural systems. The best documented example of official endorsement comes from the Cumberland, Rhode Island, Town Council that, on February 8, 1796, gave Stephen Staples permission to exhume the body of one of his daughters, Abigail, to "try an experiment" in order to save the life of another, married, daughter Lavina Chace.[51] In the town of Manchester, Vermont, an eyewitness related that, in 1793, five hundred to one thousand people were present when the heart, lungs, and liver of Isaac Burton's first wife were burned on Jacob Mead's blacksmith forge in hopes of saving Hulda Powel Burton, Isaac Burton's second wife.[52] If forty to eighty percent of the people in a town (Manchester's population was 1,276 in 1790) turned out to witness a ritual, it would strain credibility to deem it beyond the bounds of acceptable behavior. In a similar display of public access (if not acceptance), the following notice appeared in the *Montrose* (Pennsylvania) *Democrat* on April 25, 1871, under the heading, "Bodies to be Exhumed":

> The subscriber being the only survivor of a large family of brothers and sisters, all of whom have died of that dread disease, consumption, and whose children are following them by the same disease, had consented that the bodies of some of the first who died may be exhumed to satisfy the belief entertained that it will arrest the further progress of the disease, which is destroying the remaining survivors. The exhumation will take place on Saturday, April 29th, at 10 o'clock A. M., at the burying ground on David Whitney's land, in Lenox township, this county. This notice is given that all interested may be present.

The man who wrote this invitation, it should be noted, had relocated to Pennsylvania from Rhode Island. A long-time acquaintance of the family attended the ritual and then gave his account to the same newspaper. He estimated that 100 people, including two physicians, also attended the exhumations.

These and other examples demonstrate that consumption rituals engaged a broad cross section of community members, including town officials, medical doctors, and clergymen, sometimes involving large numbers of citizens, both formally and informally, in making decisions and carrying out the ritual. Writing about what he terms "the magic world view in early America," D. Michael Quinn argued that the supernatural was more socially integrated than is generally acknowledged by those who interpret history: "Occult text, occult folklore, and occult practices appealed to Harvard graduates as well as the barely literate. . . to community leaders as well as the nondescript, to devout Christian believers as well as non-believers, to members of churches as well as unchurched believers in privatized, folk religion."[53]

Scholars are also part of the narrative calculus. Fulfilling my role as a public sector folklorist in Rhode Island, which included fieldwork, public presentations, and publications, has had an impact on the narration of New England's vampire tradition. In my discussion of Everett Peck's narrative of Mercy Brown, I have much more to say about such issues, including the burden of storytelling, the relinquishing of agency, and the appropriation of narrative.

Consumption rituals and their stories raise resonant issues, including concepts of disease and healing; ritual patterns and processes; relations between the dead and the living; folk processes of transmission and oral tradition; storytelling sources, contexts, and meanings; agency, social status, and marginalization; and, of course, vampires. These issues are interwoven into the two major threads of this book: the consumptive families (and other participants) who performed the vampire rituals, and those who formulated, communicated, and heard or read narratives about these rituals. The next chapter begins where these two threads merge, as the head of a family recounts directing the exhumations of his own kin.

Gravestone of Palmyra Frost, Old Burial Ground, Jaffrey Center, New Hampshire. *Courtesy of Cyril Place.*

Part II   Eyewitnesses

The next three chapters examine ten firsthand encounters with vampire exhumations recorded by eyewitnesses. Three of them were active participants; the others were at the scene observing for various reasons. Most of these eyewitnesses occupied, or later assumed, respected roles in their communities, including minister, physician, lawyer, judge, editor, and town official. So, their accounts probably would have been accepted at face value. Some of the narratives were set down soon after the events: Reverend Enoch Hayes Place recorded an exhumation ritual from Barnstead, New Hampshire, in his daily journal, on the day of its performance, in 1810; In 1788, Reverend Justus Forward described his family's exhumations in a letter to a friend, written the next day; the accounts of both Moses Holmes (1784) and a "spectator" (1871) were published in newspapers just a few days later. William Briggs' account, however, was published many decades after his boyhood encounter in a cemetery. Since stories of exhumations by eyewitness are scarce, each is a valuable contribution to New England's vampire chronicles.

2   Signs of the Dead Preying on the Living

On an unusually mild, sunny Saturday three days before Halloween, in 2017, three parties converged on Belchertown's South Cemetery. Had the residents of this small Massachusetts town on the eastern fringe of the Connecticut River Valley known the plot of this rendezvous, they surely would not have driven so heedlessly past their "Old Burying Ground." Of its nearly 1500 graves, dating from 1736 to the present, only one drew the attention of the film production crew from Bristol, U.K., the "Bone Finder" with his ground penetrating radar (GPR), and me, enacting my role as a vampire-hunting folklorist. If we did, indeed, find an "anomaly" about a foot above the coffin of Martha Dwight, interred in 1782, what would that mean? A letter penned by Martha's father holds the answer [italics original].

Monday, July 21, 1788

Respected Sir: While I was on my journey to Stratford, my daughter was taken with bleeding inwardly, at Hatfield, and has raised blood several time since. You must think that these things excited great concern in a parent, whose family was so wasted with consumption—three dead with it, and two more in imminent danger of death.

I had consulted many about opening the graves of some of the deceased, to see whether there were any signs of the *dead preying on the living*; and though many advised to it, and most thought it awful, yet Dr. Williams of Hatfield and some others spoke in such a manner about it, that some of the family were not soon reconciled to it. However, they consented, and last Friday mother Dickinson's grave was opened. She had been buried almost three years. Nothing appeared like what was represented in Mr. Smith's son. She was wasted away to a mere skeleton when she died. The coffin had moisture in it towards the foot, face fallen in to the bones, the lungs, consuming as fast as any part, did not properly adhere together but seemed like meal a little wettish. Dr. Scott of this town opened the body. We did not try to separate the lungs from the body, but buried it again. It was suggested that perhaps she was not the right person. Since I had begun to search, I concluded to search further, and this morning opened the grave of my daughter Dwight, who had died the last of my three daughters, almost six years ago. She was considerable fleshy when she died; quite so, six or seven weeks before death. On opening the body, the lungs were not dissolved, but had blood in them, though not fresh, but clotted. The lungs did not appear as we would suppose they would in a body just dead, but far nearer a state of soundness than could be expected. The liver, I am told, was as sound as the lungs. We put the lungs and liver in a separate box, and buried it in the same grave, ten inches or a foot, above the coffin. As I never saw any grave opened, save to receive the dead, before, I am unable to judge how long after burial it is before bodies usually are reduced to dust, and these instances do not determine it—one being, as to the lungs, more reduced in three years than the other in six. I shall leave you and others to make what speculations you think proper upon this matter, only observing that the soil in which the persons were buried was very different; Hatfield, between a sand and loam, the other, sand and gravel, with many roundish stones.

Your obe't. servant,

Justus Forward.

P. S. Since writing the above, I have conversed with Dr. Scott, who opened the body. He said the lungs and liver appeared to him much in such a state as he should suppose they would in a creature which was opened and hung up till it began to taint; there was blood in the lungs, perhaps several spoonfuls together, which appeared to him much like the blood drawn from a person's arm that has stood 26 hours.

J. F.[54]

GPR was just one of several avenues of investigation stimulated by Justus Forward's detailed first-person account, published in the *Greenfield Gazette*, in 1877, under the head, "A Curious Old Letter. Among the papers of the late Dr. Harwood of Whately, was found the following letter, written by Rev. Justus Forward of Belchertown to Col. Elijah Williams of Stockbridge, in 1788." To understand how this Congregational minister might have arrived at the decision to open the graves of his family, we need to examine several questions, particularly:

- Who were the people involved in this event?
- What were Justus Forward's behavioral and emotional characteristics?
- How did he enact his roles as family head, minister, healer, and scientist?
- What interrelations prevailed among religion, magic, healing, and science?
- What beliefs and attitudes were associated with disease, death, and the dead?

***

The families named in the letter—Forward, Williams, Dwight, and Dickinson—were pillars of their communities and descendants of the founders of their towns in the Connecticut River Valley. Since these were anything but ordinary folks whose existence fades into anonymity over time, adequate information about them was available, beginning with the nine people mentioned in the letter. Justus Forward was born May 11, 1730, in the portion of Simsbury, Connecticut, that later became the town of Granby, in or near a small territory known as "the Notch," that swung from Massachusetts to Connecticut, then back again, during border disputes between 1774 and 1804. Forward graduated from Yale in 1754, wining a faculty prize for his mastery of the classics and knowledge of languages. He taught school for a short time in Hatfield, Massachusetts, where he also studied theology with the town's minister, Reverend Timothy Woodbridge. In 1756, Forward settled in the nearby town of Cold Spring (renamed Belchertown in 1761); the town's population was just over three hundred at that time. He served as minister of the Congregational church in Belchertown for more than fifty-eight years, until his death at the age of eighty-four on March 8, 1814.

Reverend Woodbridge officiated at the wedding of Forward and Violet Dickinson (1738-1832), on December 8, 1756. They had eleven children. They were, in order of birth, as recorded in the Belchertown vital statistics:

Joshua Dickinson (1757-1765)

daughter, stillborn (1759)

Martha (1760-1782)

Violet (1762-1781)

Charity (1764-1782)

Mercy (1766-1789)

Lucinda (1768-1856)

Joshua Dickinson (1771-1776)

Justus, Jr. (1774-1855)

Pamela (1776-1806)

Eunice (1780-1809).

Forward's twenty-one-year-old daughter, Mercy, almost surely was the daughter who began bleeding internally while traveling with her father to Stratford, for she was the next to die, following her sisters Violet, Charity, and Martha to the grave. Two of Forward's remaining three daughters eventually succumbed to consumption, as well. Pamela, who married William Graves in 1797, died at age thirty, in 1806. Eunice was twenty-nine years of age when she died three years later, leaving her husband, Abner Phelps, with two young children. Her obituary in the *Old Colony Gazette* (April 7, 1809) noted the family's consumption plague: "In Belchertown, 21st ult. of consumption, Mrs. Eunice, wife of Mr. Abner Phelps, and youngest daughter of Rev. Justus Forward; the dispensations of Providence have been somewhat remarkable toward Mr. Forward's family; of 11 children, 9 have been removed by death, 5 of them died of consumption." The two who escaped consumption lived long. Justus, Jr. married, had five children, and died at age eighty-one in 1855. Lucinda died the next year at eighty-eight years. She had married Dr. Francis Harwood of Whately, a small town about five miles north of Hatfield. Among their several children was Dr. Myron S. Harwood. The letter to Col. Elijah Williams from Reverend Forward (or, perhaps, a copy or draft of it) ended up in the papers of Dr. Harwood (Forward's grandson) and was published in the *Greenfield Gazette* (September 10, 1877) nine months after his death. The town of Greenfield is the Franklin County seat and a trade center on the Connecticut River some ten miles north of Whately. Reverend Forward's letter had been addressed to Col. Elijah Williams, son of Ephraim Williams, Sr., the founder of West Stockbridge, Massachusetts.[55]

The identities of the two corpses Reverend Forward had exhumed are apparent: "Mother Dickinson," exhumed on Friday, July 18, 1788, in Hill Cemetery, Hatfield, was Justus Forward's mother-in-law, Martha Morton Dickinson. "Daughter Dwight," exhumed three days later, was his daughter, Martha Dwight, who was interred next to her husband, Pliny Dwight, in Belchertown's South Cemetery.[56] Dr. Scott, "who opened the bodies," probably was Dr. Amasa Scott, of Belchertown. According to Everts' *History of the Connecticut River Valley* (1879), Scott "is mentioned in the records of 1786. He lived in the east part of the town, and occupied the 'old Fenton house,' which stood near the old burying-ground."[57]

Another participant in the exhumation ritual, although unwittingly as he undoubtedly was deceased at the time, was "Mr. Smith's son." The outward appearance of his corpse apparently was the gauge for assessing the state of Martha Dickinson's remains. Commenting on the latter, Forward wrote to Williams, without further elaboration, "Nothing appeared like what was represented in Mr. Smith's son"—strong evidence that Williams shared in Forward's knowledge of the Smith family's (implied) exhumation. A good hypothesis is that the person Smith performed the ritual to save did not die soon afterward. Otherwise, one would have to question why Forward would choose to enact a remedy that seemed to fail. I have found no convincing evidence from any of the three towns relevant to Forward's narrative (Belchertown, Hatfield, and Stockbridge) that would establish the identity of this Smith family or their son.

A single entry in the Wells and Wells' history of Hatfield (1910) comprises everything I have found regarding the identity of "Dr. Williams of Hatfield." The authors included excerpts from a diary kept by a seamstress, Miss Rebecca Dickinson, from 1787 to 1802. In her diary entry for November 15, 1787, she described taking a sudden cold during a visit to a friend. The ailment, she wrote [rendered as transcribed], "has confined me for a week with a most distressing Collick I thought my life to be a going—the day of my illness sent for doctor wiliams who opened a vein which has given me ease."[58] The authors described "Aunt Beck" as an "interesting person" who "traveled from house to house about her work" and "acquired a fund of information concerning her neighbors that was unequaled by any other person. A gift of making pithy, epigrammatic remarks caused her to be regarded as something of an oracle."[59] In context, the last observation suggests that Aunt Beck was viewed by her neighbors as a person who, though somewhat odd, had an uncanny ability to see into the future and give advice, most likely delivered through proverbs and other concise folk sayings that express "wisdom, morals, and the common sense of life's experiences and observations."[60] Aunt Beck was a solitary woman—unmarried with no children—who earned her own keep. In that time and place, an independent woman like Aunt Beck might have evoked ambivalent feelings of envy, awe, fear, and suspicion.

Wells and Wells, the Hatfield historians, obviously had access to Aunt Beck's diary, or at least a portion of it. If they knew about her entry of August 3, 1788, I could imagine why they chose not to include it in their history. Fortunately for our narrative, Marla R. Miller vividly portrayed the life of Rebecca Dickinson (1738-1815), gleaned largely from her journal. Miller (2014) included this second-hand exhumation narrative by Aunt Beck, providing corroborating evidence that supports Forward's first-person narrative:

> "The corpse of the wife of Joshua Dickinson" (no relation), she wrote in summer 1788, "was taken up in order for the cure of her granddaughter after she had lain in her grave three years" (August 3, 1788). Her friend, tavernkeeper Lucy Hubbard, had passed on to Rebecca the "sad relation" of these tragic events, and given that Lucy's house stood next to the town burial ground, she was in a position to know. Perhaps she even watched as the grisly business unfolded. They exhumed the body of Martha Morton Dickinson. Aunt Beck reported, "She was not what they had sought for her inwards was consumed or putrified." The body did not hold the hoped-for cure.
>
> "The week before last," Aunt Beck added, the heart "was taken out of the wife of Pliny Dwight who had been dead near six years." The "heart was whole after she was dead

six years with fresh blood, but what use to be made of it I am not able to write." The phrase, "but what use to be made of it I am not able to write," might at first suggest that she did not know what use would be made of it, but she probably meant that she lacked the stomach to commit to paper what she heard had occurred. According to the folk belief at the time, the heart, if found to be bloody, must be burned and the ashes fed to the patient to ensure recovery. It seems that the Dwights, in their sorrow and desperation, pursued this horrific course to the end, though "it appears folly to me," Aunt Beck observed.[61]

I can understand why, without knowing of Forward's letter, Miller would assume that Martha Dwight's heart was burned, and the ashes fed to Mercy, as those are common transformational and healing acts of this ritual in New England. (Miller cited my book, *Food for the Dead*, for her statement regarding the prevailing folk belief at that time.) In his letter, Forward wrote that the lungs and liver were removed and buried separately; he does not mention the heart, burning vital organs, or ingesting ashes.

***

What appeared to Aunt Beck as "folly" was, to Justus Forward, nothing short of a life-or-death predicament. The underlying tension in this Congregational minister's first-person narrative is his questioning if a dead relative was preying on others in the family. (Figure 7 suggests that Reverend Forward's question was certainly no outlier in the context of the consumption ritual.) Key to understanding how he addressed this question is his character: What frame of mind guided him when he faced difficulties, had to make choices, and arrive at decisions? Mark Doolittle, in his *Historical Sketch of the Congregational Church in Belchertown* (1852), devoted several pages to describing Forward and his ministry. Doolittle's assessment suggests that Forward was a man of measured judgement who made decisions and formed opinions objectively, authoritatively, and wisely. He wrote that "Mr. Forward's character was strongly marked by the stern, faithful, unassuming, considerate traits, showing his puritanic lineage."[62] Yet, he applied the prescribed tenets of this heritage in ways that were personal and concrete: "In his public discourses, he dwelt much on the practical influences of these doctrines on the heart and life. He did not fail to preach the doctrines, but did not make them the exclusive theme of his discourses to so great an extent as some did."[63] Doolittle observed that Forward "possessed a well balanced mind; if it did not take so wide a range, or so high a flight as that of some other men, it moved in sure, safe and well directed courses. He possessed much of what is sometimes called forecast of thought in drawing just conclusions from a given course of measures. He was preeminently a matter-of-fact man."[64]

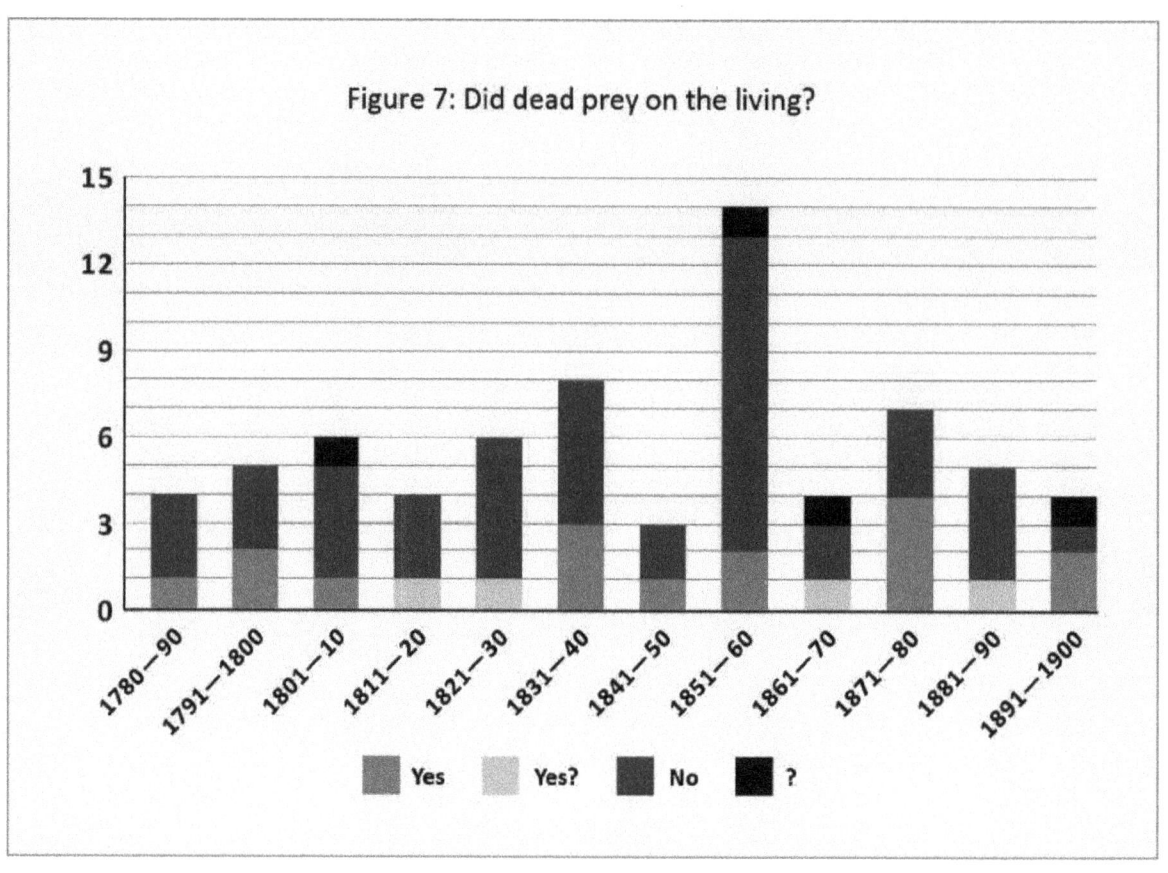

Figure 7: Did dead prey on the living?

Forward's pragmatic approach to religion was not idiosyncratic. Well before his tenure as a Congregational minister, communities had been established increasingly far from Boston. The resulting need to be self-reliant created an attitude of independence that was incorporated into various cultural systems. Congregationalism, having no centralized hierarchy, was based on local authority. Harry S. Stout made this point in his book on religious culture in colonial New England (1986): "For God's Word to function freely, and for each member to feel an integral part of the church's operations, each congregation must be self-sufficient, containing within itself all the offices and powers necessary for self-regulation. New England's official apologist, John Cotton, termed this form of church government 'Congregational,' meaning that all authority would be located within particular congregations."[65] To maintain his congregants' support, a Congregational minister had to accommodate the realities of their daily lives. Although the religious and civil systems cooperated closely, they were not—in the terminology of early Puritan leaders, such as John Cotton—to be "confounded." What we now think of as "the separation of church and state" was established early in many parts of New England, including Massachusetts.

Forward's wide-ranging interests and open-mindedness are rooted in the ascetic Protestantism of seventeenth-century Europe and New England that helped lay the groundwork for an "enhanced cultivation of science and technology."[66] In his elaboration of Max Weber's ideas, sociologist Robert Merton argued, "The deep rooted religious interests of the day demanded … the systematic, rational, and empirical study of Nature for the glorification of God in His works and the control of the corrupt world." The "hardly disguised utilitarianism" of Protestantism embraced "a thoroughgoing empiricism" that sanctioned free thinking and the questioning of authority. Such values, Merton asserted, "were congenial to the very same values found in modern science." Protestantism and scientific-technological interests thus "were well integrated and, in essentials, mutually supporting."[67]

A capacity to entertain competing ideas and achieve consensus was also established in the political system. When the terrible reality of consumption cast its shadow over Forward's family in 1788, the new country was hard off the Revolutionary War, which in large measure was a repudiation of central authority. A spirit of independence and democracy was in the air, reinforcing the local autonomy already established through Congregationalism. As Oscar and Lilian Handlin observed in their book on the American Revolution (1982), "People who defined justice by communal consensus rather than by code kept its administration in their own hands rather than delegating it to officials."[68]

**

In the early days of Forward's ministry, Mark Doolittle wrote, "there was no practising physician in the place and none nearby," so Forward "gained considerable knowledge of medicine" through "reading and practicing" and "became very useful to his people in sickness."[69] Richard D. Brown, in his article on medicine in colonial Massachusetts (1980), addressed this syncretism, writing that "the common practitioners of the healing arts—clergymen, self-taught 'empirics,' midwives, and ordinary people in general—were eager to learn from their betters, and if they could not afford sustained medical instruction, they did buy, borrow, and read popular medical books, and clip from the newspapers reports of new treatments."[70] Forward's dual role, born of necessity, was typical, as medical historian James H. Cassedy (1980) observed: "The medical practice in which many an early New England preacher engaged is a well-known phenomenon. It was an activity that filled an essential need in society during a time when trained physicians were few and far between."[71]

The clergyman's dual role had practical and social value. Brown wrote, "Visiting the sick was one of a minister's primary responsibilities—to lead prayer in the afflicted household, to pray with and over the sick person, and to recommend appropriate medical treatment."[72] Forward's tending to the sick, Doolittle noted, "gave him a two-fold advantage for doing good; while administering to their physical wants, he gained a knowledge of their spiritual wants, and was always ready to meet them."[73] "Often the merits of treatment existed largely in the eyes of the beholder," Brown argued. "At the patient's bedside the chief distinction might well be social rather than medical."[74] Given the state of established medicine at that time—when physicians, such as Dr. Williams who treated Aunt Beck, were purging the digestive tract with cathartics and emetics or bleeding patients to "balance the humors" and restore good health—Forward's palliative treatment was at least as efficacious, certainly less damaging, and probably more congenial.

Magical medicine was in New England's cultural DNA, flowing from an ancient wellspring of folk beliefs and practices. The supernatural ideas that early New Englanders wholeheartedly participated in had a firm footing in Tudor and Stuart England (1485-1714), where, perhaps surprisingly, the flow of learning frequently went from folk to official or orthodox systems.[75] Even scientists and intellectuals opined that herbalists, wise women, and cunning folk were often more efficacious in their cures than learned physicians.[76] Medicine at that time, wrote Keith Thomas (1971), "was essentially a mixture of common-sensical remedies, based on the accumulated experience of nursing and midwifery, combined with inherited lore about the healing properties of plants and minerals. But it also included certain types of ritual healing, in which prayers, charms or spells accompanied the medicine, or even formed the sole means of treatment. Magical healing of this kind might sometimes be attempted by the patient himself or a member of his family."[77]

In his exploration of British folkways in America (1989), David Fischer linked "magic ways" in New England to those in the counties of eastern England from which early New Englanders had migrated. Taking witchcraft as an example, Fischer wrote, "In England, every quantitative study has found that recorded cases of witchcraft were most frequent in the eastern counties from which New England was settled." He cited a conclusion of John Demos that "Essex was beyond doubt a center of witch-hunting within the mother country; and Essex supplied a disproportionately large complement of settlers for the new colonies across the sea."[78] New Englanders were thus bequeathed a realm inhabited by a host of apparitions and specters—harmful, helpful, providential, or simply benign—from demons and witches to ghosts and revenants. Many New Englanders experimented with occult practices, such as alchemy, astrology, divination, and dowsing.[79]

Richard Godbeer, in his study of magic and religion in early New England (1992), wrote, "More often than not people combine the two strategies: either they adhere to a system of beliefs that includes both magical and religious elements—usually on the grounds that divine authority does not exclude human power as a subordinate agency—or they ascribe simultaneously to two separate traditions, one magical and the other religious, switching back and forth between them as convenient."[80] It was within this magico-religious atmosphere that Reverend Justus Forward, predisposed to counterbalance religious doctrine with worldly concerns, questioned if his deceased kin were preying on those still living.

Forward's ongoing duties as a clergyman included delivering sermons as well as eulogies for the dead. Preparing for death, the meaning of human existence for Congregationalists, was a focus of both his sermons and eulogies. In this Calvinist faith, a person's fate after death was predetermined, so salvation was beyond one's control. Most were preordained to suffer eternal damnation. This dark fixation on death, David Fischer wrote, "led New Englanders to adopt … lugubrious deathlore."[81] Children were forced to read gruesome Bible verses and literally stare death in the face at open graves. Even in a community where death and dying were constant companions, Forward's own personal encounters with death, from the time of his youth and through his entire adult life, were unusually grim.[82]

Yet there was nothing in Forward's orthodox Congregationalism that granted the dead an ability to prey on the living. His community's folklore established that possibility, resting on a vague notion of demonic corpse possession. This was no indictment of the deceased, as the soul already had vacated the earthly vessel, leaving behind only physical remains that might be seized by some evil spirit. The corpses of those who had died of consumption, while inert and blameless, were open to possession. To determine if that had indeed occurred, it was necessary to open graves and inspect remains. Those that showed signs of incomplete putrefaction were not normal and required supranormal (supernatural) treatment.

***

A curious note included in a brief biography of Forward provides additional evidence of his wide-ranging interests, especially as connected to death: "In the course of his ministry there were seven hundred and ninety-eight deaths, of whom fifty died of the consumption."[83] This is probably an allusion to Forward's article, "A topographical description of, and bill of mortality for Belchertown, for thirty years; with remarks on pulmonick [sic] consumptions." (Forward's mortality figures show that more than 6% of deaths in that thirty-year span were attributed to consumption.) The *Cumberland Gazette* (June 16, 1789) noted that Forward donated his article to the American Academy of Arts and Sciences in 1789, less than one year following his family's exhumations. Bills of mortality were an English tradition in which parish clerks kept a record of burials. These death records became especially useful for monitoring virulent epidemics, notably those of the plague that began in London in the late sixteenth century. As bills of mortality became more detailed (adding, for example, cause of death, sex of the deceased, and age at death) and more inclusive (also recording births, baptisms, and marriages) they naturally became a valuable statistical resource. The tradition continued in New England, and it properly fell to the minister to maintain the vital statistics that were linked to church-based rites of passage.

As trained physicians became more numerous throughout New England, the need for clergymen to practice medicine waned. However, as James Cassedy wrote, "their function as record-keepers ... of certain of their medical events" continued, and even expanded. Ministers became "unofficial but nonetheless de facto health monitors of their communities. During the long generations before civil provisions were made to gather information about the causes of death, these men were providing such services to their communities."[84] Cassedy suggests that these bills of mortality "represented a continuing tradition of concern for the collective disease and death in the community." As these records became more detailed during the latter half of the eighteenth century, they took on "demonstrable usefulness in community health considerations."[85] Cassedy points to Pastor Ezra Stiles, who kept mortality tables between 1760 and 1764, in Newport, Rhode Island, as

> one of a new generation of New England clergymen whose Enlightenment interests in science were affecting their theology. Specifically, the concept of a capricious and willful God intervening at random to influence events or punish mankind was for some giving way to that of a rational God governing the universe according to well-established laws. As part of this, mortality was increasingly coming to be seen not so much as a matter of special providences as something which occurred in more or less regular patterns according to God's deliberate plan.[86]

One year before Justus Forward commissioned the exhumations of his mother-in-law and daughter, the thirty-year mortality records of the First Church of New Haven, Connecticut, were published. They showed that "a third of the congregation's deaths during that time had been caused by four diseases: consumption, 'canker,' dysentery, and fits or convulsions."[87]

Given Forward's familiarity with medicine and death, and interest in geography (which is implicit in his article on topography and mortality), his conversation with Dr. Scott ("who opened the graves") regarding the physical condition of the corpses of his mother-in-law and daughter—almost clinical in its objectivity—seems entirely in character. His suspicion that differences in their rates of decomposition might be explained by differences in the nature of the soil in the two cemeteries reinforces a conclusion that his interest in science was purposeful. If Forward was not familiar with the writing of the German theologian and academic, Michael Ranft, their thinking converged. In Ranft's 1728 report on vampirism in Europe, he "sought answers to the vampire phenomenon in nature, which was in compliance with the philosophy and scientific methodology of the Enlightenment." He postulated that "the assumed lack of decay of corpses attributable to vampires is a result of the condition of certain types of soil and chemical processes in the soil in which the bodies of the alleged vampires had been buried."[88] Scientific evidence has confirmed the hypothesis of both Forward and Ranft concerning the relationship between decomposition and soil type.[89] And, if Forward's concern with Belchertown's mortality, especially as caused by consumption, seems to go beyond the mere routine duty of a minister to record the deaths in his parish, a potent stimulus might have been his own family's decimation from consumption during the years included in his bill of mortality.

***

When Justus Forward wrote that he wished "to see if there were any signs of the dead preying on the living," he was presupposing that such a thing was possible. At that time, he was both the town's Congregational minister and the head of a household in imminent danger of destruction. When he convened a process to decide if he should, indeed, open graves, he could have done so enacting either role, or a combination of both. As the author of a history of Harvard, Massachusetts observed in 1894, "The minister and his spouse" were the "indisputable arbiters in all social matters, the referees for all unanswerable questions."[90] Forward apparently deemed his appropriate role in this matter should be secular rather than ecclesiastical. The family cultural system in Forward's community, as in others in the region at that time, was the primary unit for local authority and social control— "an instrument of larger purposes," in the words of David Fischer.[91] Since the first corpse being considered for exhumation was Forward's mother-in-law, who had lived in the nearby town of Hatfield, we can surmise that at least one of her kinfolk was invited to partake in the deliberation. One wonders if Forward was referring to that portion of the family when he wrote that "some of the family were not soon reconciled" to the exhumations. It appears that "Dr. Williams of Hatfield" was among those opposed to exhumation.

Forward's inclusion of physicians in the debate and at the exhumations suggests that he conceived the possibility of a medical/scientific solution to the question of the dead preying on the living; or, at least, that a doctor might be able to provide a diagnosis of the corpse's condition vis-a-vis putrefaction. While today's dominant biomedical paradigm draws a clear line between scientific and folk practices, the medical field was not so "unadulterated" in late eighteenth-century New England. When Aunt Beck of Hatfield caught a cold, Dr. Williams "opened a vein"—an approach based on the centuries-old Galenist doctrine of humoralism. From both the scientific and folk perspective at that time, blood was regarded as the primary humor because it contained the essence, or vital spirit, of the creature in which it flowed. To both Dr. Williams and Reverend Forward—a self-taught medical practitioner, himself—looking for "fresh" blood in the vital organs, the typical diagnostic procedure in therapeutic exhumation cases throughout New England, was not the "bad science" that it became a few generations later.

An alternate medical theory rested on a spiritual rather than physical interpretation of healing. For these Paracelsians, anthropologist Karen Gordon-Grube asserted (italics original), "the body as a symbol is *not* holy, the body is merely a worthless piece of matter in which the life-spirit dwells."[92] A medicinally prepared mixture of human blood and flesh, termed *mummy* or *mummia*, contains "spiritus," the magical and medicinal "World Soul." The presence of a spiritual understanding among some New England Puritans may explain Forward's seemingly dispassionate description of his family's corpses. Richard Sugg, in his thorough treatment of corpse medicine (2011), noted that "Paracelsian mummy presents a distinctive mixture of science and religion. ... Paracelsians display an unflinchingly hard-headed, neutral attitude toward human bodies, depersonalised into so much organic raw matter." Yet, the core of such seemingly detached scientific observations, Sugg argued, "is emphatically religious," evoking a powerful "general reverence for God's natural resources."[93] Some of Forward's words and actions regarding the consumption ritual suggest a blending of science and religion. While cannibalism was not part of the ritual described by Forward (although it does play a central role in others that we will consider), the Paracelsian tradition provides an additional context for viewing Forward's conduct as reasonable.

Plainly, Forward's decision-making was not guided by a single overarching worldview or unitary belief system. "No society ever has just one organization or structure," anthropologist Murray Leaf asserted, and "no community ever has a single unified system of ideas and values or 'symbols and meanings,' at any level. There are always multiple, independent and often mutually opposed cultural idea-systems." In short, "All societies, regardless of scale, are pluralistic."[94] Forward's letter discloses that he was presented with several contrasting courses of action from which to choose. Each of the cultural idea-systems Forward apparently consulted in his community — particularly those of religion, folklore, kinship, science, and medicine—had its own premises and operational rules. Forward implies that the issues were hotly debated, as those consulted argued from various perspectives: "many advised to it"—despite thinking that it was "awful"—others were not "soon reconciled to it." Other voices evident in Forward's letter are not available. Ultimately, the family "consented," thus sanctioning Forward's decision to try the folk ritual. The decision was not taken lightly, nor was it unilateral or unanimous. With the reluctant acquiescence of his family (and others), he authorized the exhumations.

When the body of Forward's mother-in-law, Martha Dickinson, was unearthed, revealing its advanced state of decomposition, "it was suggested that perhaps she was not the right person." Forward decided to press on, and, three days later his daughter, Martha Dwight, was disinterred. Forward made it clear in his letter that he was unable to speculate on whether the state of her corpse was natural, as he had no experience with viewing corpses after interment. His postscript implies that Dr. Scott— "who opened the body"—also had no experience with human cadavers. Dr. Scott's supposition about the appearance of the lungs and liver— "in such a state as he should suppose they would [appear] in a creature which was opened and hung up till it began to taint"—reads like an analogy to dressing wild game or slaughtering farm stock. His only direct reference to human physiology is comparing the blood found in Martha Dwight's lungs to "blood drawn from a person's arm that has stood 26 hours." Dr. Scott undoubtedly was intimately familiar with drawing human blood, which, as we have seen, was a core medical procedure at that time. Perhaps a motive for Dr. Scott's agreeing to perform the autopsies on the two women was the rare opportunity to dissect a human cadaver. Needed for hands-on instruction and dissection in medical schools, cadavers were in such short supply that an entire industry, a shadowy cadre of professional body stealers known as resurrectionists (whom we shall consider in more detail later) arose to meet this need."[95]

The ambiguous organs were removed from Martha Dwight's corpse, but, contrary to the usual folk pattern in New England—burned to ashes—they were buried in a separate container, about a foot above the re-interred corpse. As rare as this procedure is, its inherent folk logic is plain: physically separating the possessed organs—the ones containing unclotted ("fresh") blood and therefore retaining some flicker of life or vitality—from the corpse, breaks the implicitly harmful spiritual connection between the two; the corpse is allowed to complete the natural process of decomposition, freeing it from whatever evil had possessed it. Did the ritual work? As we have seen, Forward's daughter, Mercy, died less than six months later; two other daughters survived an additional eighteen and twenty-one years; and a son and daughter lived into their eighties. Perhaps order and balance were restored in both the afflicted family and the disrupted community with the palpable evidence showing that the dead were, indeed, now completely dead.

Did the results of the ground penetrating radar add a tangible layer of evidence to that of eye-witness accounts and corroborative documents? Multiple scans over Martha Dwight's grave consistently revealed an outline of Martha's burial and an anomaly showing a ground void (where the box may have been), about twelve to eighteen inches above the grave.

3 The Disinterment Was Done under My Personal Supervision

Justus Forward's first-person exhumation account, while extraordinary, is not unique in the chronicles of New England vampire narratives. Another letter, also written by a person who oversaw the exhumation of family members, appeared—again, like Forward's—in a newspaper many years after the event. The significant differences between these two letters, however, are immediately obvious. The following correspondence was published in the *Chicago Daily Tribune* on October 24, 1885:

> Bloomington, Ill., Oct. 20—[Editor of The Tribune]—The following letter, written to a near relative in 1848, recently fell into my hands. The writer, an octogenarian, is still living, and I am induced to make it public, together with some other matters pertaining to this subject, hoping it may stimulate inquiry into so curious a topic. The letter related to the removal of a number of relatives from an old burying-ground in one of the cities of New England, and substantially read as follows:
>
> The disinterment was done under my personal supervision, as carefully and respectfully as such things should be. The first of the graves opened contained the remains of your grandmother, but, being too decayed for removal, she was placed in a new case. Your grandfather's coffin, however, was entire, and was taken up without in the least disturbing the contents; only the fastenings of the lid were gone, and the admirable preservation of the coffin induced me to contemplate for the first and last time the remembrance of my respected ancestor, who had departed this life nearly sixty years before. The skeleton was entire, and growing over it, inside the coffin, six feet under ground, in a dry gravel soil, was a little vine, which had year after year spread its meshes like a web over the entire corpse, and seemed to be still alive and growing! It was snowy white and cold as ice, a fit inhabitant of the dark and silent grave.
>
> In the coffins of your mother and brother the same little vine was found growing. If you remember, consumption has for many years made sad havoc in our family, and these three are said to have died of this disease.
>
> Now, there is a tradition among the Germans that if a descendant pluck from the corpse of the last victim of consumption this little vine, it will eradicate the disease from the family. If there be anything in it, which I hardly believe, consumption has been destroyed from ours, for I had the temerity to tear it away from the remains of your brother, the thought of which makes me shudder even now, disclosing, as it did, the ghastly skeleton which this growth had kindly endeavored to conceal.
>
> In another grave I found sprouts growing from the head, the hair of which was well preserved, though buried well nigh three-score years. This vine, or sprout, was also white, but sparkled an like [sic] ice-plant, and on examination it proved to belong to the common sorrel family, whose tenacity of life enabled it to grow four or five feet under ground.

Many years ago, in Smithfield, R. I., a family were fast becoming extinct from consumption. Through the advice of friends, the last victim of this disease was disinterred, and, on opening the coffin, this vampire or consumption-vine was found growing from the breast of the dead. The body was afterward burned, the friends hoping in this way to exterminate this dreadful malady. In an old abbey in Dunfermline, Fifeshire, Scotland, workmen, while excavating for laying the foundation of a new building, came upon well-preserved stone sarcophagi, tightly covered with slabs of stone. On removing the cover, seemingly this same curious vine or fungus was found growing over the skeleton. In another of these stone receptacles for the dead, which were evidently many hundred years old, the body of some great dignitary was found in perfect preservation. Crowds came from near and far to view these remains, and there was some controversy at the time as to the unusual growth in the coffin, and the identity of the unknown celebrity. The removal of my informant soon after the event prevented further particulars.

In conversation with an old cemetery superintendent, he admitted sometimes finding a net-work of rootlets in the coffins of those long dead and nothing more; suggesting the discovery of what was supposed to be the grave of Roger Williams, "in which, when opened, no remains of the skeleton even could be found, for the roots of an apple tree planted above had embraced the skull, trunk, and limbs, and preserved their shape." Nature, the wonderful alchemist, having completely transformed and utilized the poisonous elements, thus adding another proof of the countless resources of the Creator, who can turn to purpose even the refuse and decay of the grave. Many are the tales told and read of wonderful growths of hair found in the coffins of the dead, years after burial, which cover the unsightly remains as with a garment, clearly showing that Nature in its most repulsive forms is endowed with an artistic sense which would fain hide what is hideous and change ugliness into beauty and use. - Mrs. B. M. Prince[96]

This complicated text consists of four interrelated narratives: (1) an introduction of the letter, written by its submitter, Mrs. Prince; (2) a first-person account of family exhumations in an unnamed New England town, written by the unnamed person who supervised them; (3) a brief description of a similar exhumation case from Smithfield, Rhode Island, apparently narrated by Mrs. Prince; and (4) a discussion of vines growing in graves, with analogues, also written by Mrs. Prince.

Was this letter edited prior to being submitted for publication? The first sentence, alone, supports this hypothesis: "The disinterment was done under my personal supervision." It plops the reader down in the middle of a narrative, begging for context, a fine literary device that seems out of place in personal correspondence, unless this letter was part of a chain of correspondence in which the ritual had been previously introduced. Despite its empty spaces, there are elements in this complex narrative that can be deciphered. We go where evidence takes us, beginning with the following features:

- The original letter's author was born between 1796 and 1805, as he was termed an "octogenarian" in 1885.
- The exhumations occurred no later than 1848, the year the original letter was written.

- The bodies of the grandmother, grandfather, mother, and brother of the original letter's recipient were exhumed; there was at least one additional exhumation, for after describing the first four, the letter writer mentioned sprouts growing from a skull in "another grave."
- The letter recipient's brother was the last of the four family members to die;
- The exhumed grandfather died "nearly sixty years before" the exhumation, which would be about 1788, or earlier.
- Consumption "made sad havoc" in the family of the exhumed.
- The writer seemed willing to entertain the notion of the ritual's efficacy but was unconvinced ("If there be anything in it, which I hardly believe . . ..").
- The vine-analogue narrative was written after 1860, the year that Roger Williams' remains were exhumed and the tree root discovered.
- Mrs. Prince's expressed purpose in sharing the narrative was "hoping it may stimulate inquiry into so curious a topic."

My research showed conclusively that Mrs. B. M. Prince was Barbara Miller Prince (1835-1908). Born Barbara Maria Miller, in June 1835, in Bavarian Rhenish Palatinate, Germany, she immigrated with her parents to the United States at the age of thirteen, the year (1848) the original letter was written. Although little is known about her, the man she married left a substantial trail in the historical record. Barbara Miller and Ezra Morton Prince were married in Pittsfield, Illinois, on July 2, 1866. Ezra Prince was born in Turner, Maine on May 27, 1831, to Job Prince (1795-1875) and Zilpha Spaulding Prince (1799-1844). He had "a common education," attended Bowdoin College, studied law in Bath, Maine, and then entered Harvard to complete his law studies. One year later he was admitted to the bar. In 1856, Prince was persuaded by his cousin, Leonard Swett, to join him in Bloomington, Illinois, where he had practiced law since 1849. Swett introduced Prince to Abraham Lincoln, after which they became friends, often traveling the Eighth Circuit together to hear cases. Like Lincoln, Prince was a Republican and strongly anti-slavery. He was a founder and lifelong member of the Unitarian Church in Bloomington. Although Prince was a prominent lawyer in the city, he was never elected to the several public offices for which he stood. Still, he was an active member of the community, being among the founders of Bloomington's first public library and of the McLean County (Illinois) Historical Society. He wrote and edited several works on the history of McLean County, about his travels out West, and of his reminiscences of Lincoln. He also taught in the Law School at Illinois Wesleyan University. Barbara Prince died May 2, 1908, and Ezra Prince followed her on August 27, 1908.

The author of the letter probably was a kinsman of Ezra Prince, since Barbara immigrated to the United States from Germany, where, presumably, her ancestors lived. Mrs. Prince's introductory sentence is ambiguous concerning to whom the "near relative" was related: "The following letter, written to a near relative in 1848, recently fell into my hands." My guess is that the letter's recipient was a "near relative" of the letter's author. Mrs. Prince wrote that she had received the letter "recently," so it seems likely that its owner had died not long before or, being near death, was in the process of distributing material that might be of interest to others in the family. That person also was related closely enough to Ezra and Barbara Prince for them to have received the letter. Why Barbara, instead of Ezra, submitted it to the newspaper is open to speculation. Perhaps her attraction to the topic was greater than her husband's, for she pointed out that she submitted the letter to the newspaper "hoping it may stimulate inquiry into so curious a topic." I checked subsequent editions of the *Chicago Tribune*, hoping (along with Mrs. Prince) that her account stimulated a response. I found none.

The letter's author wrote that "consumption has for many years made sad havoc in our family." Consumption did, indeed, plague the family of Ezra Prince's mother, Zilpha Spaulding. Zilpha's sister, Sally, married their first cousin, Increase Spaulding. Sally died at the age of forty-two in 1836. Increase followed three years later. Cole and Whitman's *History of Buckfield* (Maine) included the following excerpt from a journal entry for April 22, 1839: "Increase Spaulding died this morning at his father's of consumption."[97] On January 30, 1844, Zilpha, herself, died. Ezra, who was just thirteen years of age at the time, later wrote of his mother's death:

> I remember very little of my mother. . . .. She was naturally delicate and died of consumption in the 45th. year of her age. She was sick three years and bore her suffering with uncomplaining fortitude. The picture of her that is most vivid in my mind is her last attendance at the Universalist Church at Turner, shortly before her death. It was communion Sunday and knowing the end was near she desired to attend to partake once more of the mystic elements, the pledge of her love to God and faith in immortal life. She was then very feeble and father carried her in his arms to the sleigh and from that into the church. I remember that she wore a green silk hood.[98]

Zilpha's brother, Captain James Spaulding, was born June 10, 1802. After his wife's death on December 11, 1864, he moved from Buckfield, Maine, to Earleville, Illinois. Did Captain James Spaulding write the letter? The author was described as an octogenarian and still living in 1885. James was eighty-three in 1885; he died the following year.

Captain Leonard Spaulding, Zilpha's father, was one of the first settlers in Buckfield, Maine, as was Job Prince, Ezra's grandfather, who settled there in 1788.[99] Since the name Leonard was used commonly in the Spaulding family, we have every reason to wonder if there was a genealogical connection to the Leonard Spaulding (1728-1788) of Dummerston, Vermont, whose family also was decimated by consumption. In 1798, the corpse of one of his sons was exhumed and its "vitals" were removed and burned. The exhumers apparently found a vine growing from the coffin and broke it, hoping "to destroy its influence or effect" on the living family members—an act very similar to that described in the letter. (The exhumation of this Spaulding family, from Dummerston, will be discussed in chapter 11.) Some genealogical forensics showed that there was indeed a link: Edward Spaulding (1596-1670) immigrated to Massachusetts between 1630 and 1633 from Spalding, Lincolnshire, England. From his first wife, Margaret Elliot (1613-1640), descended the Leonard Spaulding of Dummerston. From his second wife, Rachel Needham (1622-1670), descended Zilpha Spaulding of Buckfield. Not a close connection, but a family link, nonetheless.

Other possible vampire connections to this family are based on locality. Ezra Morton Prince's great-grandfather, Kimball Prince (1726-1814), was born and died in the small village of Kingston, Plymouth County, Massachusetts. Ezra's grandfather, Job Prince (1765-1831), was born in 1765, also in Kingston (relocating to Buckfield in 1788). Kingston was the hometown of Joseph Ripley Chandler (1792-1880), whose serialized article, "Superstitions in New England," appeared in a popular periodical. Chandler intimated that most of his published stories were circulating in the community as he was growing up in Kingston. One of his installments, republished in several newspapers in 1822, contained an account of a vampire exhumation that occurred about 1807. (We will examine Chandler's narrative in chapter 17.) The population of Kingston in 1765, the year Job Prince was born, was 759. Almost certainly, if there were exhumation stories circulating in the village, the Princes, as well as the Chandlers, would have been aware of them.

Two additional vampire links are associated with the town of Buckfield, Maine. Seba Smith was born in Buckfield in 1792. He became a newspaper editor, publisher, and well-known author, often writing under the pseudonym, Major Jack Downing. In 1832, two of his periodicals, the *Family Reader* and the *Portland Courier*, published an account of several exhumations that the editor (Smith, at that time) claimed to have witnessed. It is not certain if he attended the exhumations while he was living and writing in Portland, Maine, or when living in his home in Buckfield. A second exhumation narrative in the same article was located near Norway, Maine, less than twenty miles from Buckfield. Sylvanus Cobb (1798-1866), who was born in Norway, seems to have been a conduit for exhumation narratives, connecting Norway, Maine, to Waltham, Massachusetts. As a Universalist preacher in Maine, Cobb "was among the earliest occasional preachers of the new faith in Buckfield."[100] Ezra Prince wrote, "My father though reared in the Baptist Church, in his early manhood became a Universalist."[101] Exhumation narratives had ample opportunities to circulate in different communities as people moved about. (The narrative of Seba Smith is discussed in chapter 4; that of Sylvanus Cobb is treated in chapter 11.)

Community-based social networks and family connections were essential in communicating exhumation narratives. Genealogy establishes that the Spauldings were related to the Princes, but on a grander scale, it affirms that kinship networks in early New England were extensive and had a variety of uses. They were channels for information of all sorts, especially a narrative odd enough to resist decorum's call for silence. Folklorists have affirmed the family's crucial role in creating and sustaining supernatural narratives. "Regional American collections indicate that the family has been especially important in passing along legends," Ronald Baker and Simon Bronner wrote. "Many legends in the United States told in family settings are supernatural legends (memorates) dealing with premonitions, death and burial, ghosts, haunted houses, good and evil, special powers, witches, and monsters . . .. Often beginning as first-person accounts of encounters with the supernatural, as these personal experience tales are repeated by family members, they develop into true legends."[102]

Baker and Bronner stressed the important role of storytelling events in passing down the lore that is critical to the preservation of a community's shared sense of identity: "Older adults at social gatherings are often called upon to relate stories of 'the way things were' and to comment on healing and medical legends for the restoration of health and preservation of youth. Appealing to various groups as a narrative vehicle to convey (or explain) beliefs, disturbing moments, and social changes, legends are often the basis of 'folk histories' in which communities explain the events and people that are critical to the collective memory and imagination."[103]

The anonymous author of the exhumation narrative that was published in the *Chicago Tribune* many years after the events perhaps lacked the opportunity to tell his story to the letter's recipient in person. With no spreading elm tree or other face-to-face venue available, he put his family narrative into writing. The portion of the correspondence that introduced the grave vine as a German tradition apparently was written by him. But Mrs. Prince continued the discussion, inserting examples from Scotland and Rhode Island, the latter consisting of an indefinite exhumation narrative from Smithfield and the discovery of Roger Williams's remains entangled in the roots of an apple tree. It seems a bit odd that, given her German heritage, Mrs. Prince did not comment on the German connection raised by the letter's author. Some of her text appears to have been derived from conversations with informants but, like the exhumation story from Smithfield, lack of detail creates a dead end. The vine motif occurs in other New England exhumation narratives, but the meaning that Mrs. Prince finds in the grave vine is unexpected and unusual (and, I think, therapeutic). She concludes that "Nature in its most repulsive forms is endowed with an artistic sense which would fain hide what is hideous and change ugliness into beauty and use." The Creator uses the corpse as a canvas, transforming the ugliness of death into an aesthetic experience, proving that within even the most hideous aspect of Nature there is utility and beauty. I would have liked to know much more about Barbara Maria Miller Prince.

***

Medical doctor, Harold D. Levine, included the following narrative in his article, "Folk Medicine in New Hampshire" (*New England Journal of Medicine* 1941):

> With regard to tuberculosis, the following story, which appeared in the July, 1939, issue of the magazine, *Yankee*, is of interest:
>
> "You must be Ned's boy," an old man said to me once. "I knew your father. Yes, I knew your great-grandfather and all his family, and I'm going to tell you something that will surprise you.
>
> "You probably don't know it, but if a family is dying off of consumption, the disease goes no farther after one of the members of the family has been buried face down.
>
> "I was at the funeral of your great-uncle John. He died of old-fashioned consumption just as other members of the family before him. I was one of several who made up our minds to stop the run of consumption in that family, so we stayed in the cemetery until the relatives had gone, then we lifted the casket from the open grave and turned it over. So your great-uncle John was buried face down, and it ended consumption in that family."[104]

This story presumably was told by a participant in the ritual, but he omitted verifiable details, including place, date, and surname. We do not know why Levine situated the incident in New Hampshire. In the 1939 *Yankee* magazine article cited by Levine as his source, the place of the event is not given. The UCLA Archive of American Folk Medicine also places this event in New Hampshire, no doubt because Levine's article is its only citation.[105] Ernest Poole, in *The Great White Hills of New Hampshire* (1946), provides an extraordinary two-part analog from that state: "To stop 'consumption' in a family, the first victim was buried face down; and to kill the devils causing that dread disease, she brewed a concoction of ashes, cobwebs, cow and sheep turds, turkey dung and ground worms simmered well in lard."[106]

The *Yankee* story and Poole's prescription are rare examples of taking preemptive action, rendering exhumation unnecessary. Burial face down is the vampire ritual's most elegant, certainly least disturbing, remedy. The UCLA Archive of American Folk Medicine includes two additional entries for prone burial in New England, both from Maine. "A family in Oxford, Me., was afflicted with tuberculosis. After one member of the family died and others were sick his body was exhumed and turned over and the course of the disease was stayed."[107] The other example combines an initial, diagnostic examination of the heart with prone burial: "There was a belief up until 1854 that if tuberculosis was raging in the family to cure this disease, dig up one of the deceased family and cut the heart open, and if they found fresh blood, turn it on their face and bury them again. This disease would stop."[108] Both of these entries are from uncited sources published between 1930 and 1939.

Levine included another exhumation narrative from New Hampshire in his article. It contains enough information to allow some intriguing interpretation, which we pursue in chapter 14.

***

Enoch Hayes Place was born in Rochester, New Hampshire, on July 13, 1786. At the age of twenty-four, this Freewill Baptist Minister set out for Vermont from his home in southeastern New Hampshire to introduce his God to as many people as possible. On that same day, September 3, 1810, he began the daily journal that he would keep for fifty-five years. Place had begun preaching three years earlier, a mere month after being caught up in the sweeping religious revival known as the Second Great Awakening. A farmer, teacher, and strong advocate for education and the anti-slavery and temperance movements, Place remained, for most of his life, a minister in and around Strafford, New Hampshire.

While visiting the afflicted, he encountered many injuries and illnesses that included burns, drownings, being kicked by a horse, dropsy, cancer, rheumatism, cholick, fever, typhus, spotted fever, and, of course, consumption. Place reveals in his diary that often he did not know the cause of a death that he encountered (quotes of Place are as written). A person would become "quite ill" and then die. The constant and continuing battle to reach backsliders filled his diary. That people had visions and "strange feelings" that seemed to have no rational explanation was an ordinary occurrence. He wrote about his own encounters, describing "many strange things"[109] that he could not fathom, such as a mysterious light that appeared after an unexpected death.[110] Faith and prayer were the Reverend's fortress: he recounted a case where prayer healed an ailing woman before the doctor could arrive.[111] He believed that God protected him, offering the following as proof: "while returning home my horse fell and hurt me Some but I believe not materially, O how many thousand unseen and Seen accidents I have escaped through the tender mercy of the Lord our God."[112]

One of Place's first stops after embarking on his mission was Barnstead, New Hampshire. He was asked to visit Moses Denitt (Dennett, in vital records), who was dying of consumption. Later the same day, September 4, 1810, he was requested to "attend the taking up the remains" of Moses's daughter:

> I was then requested by Esqr Hodgdon, & others to attend the taking up the remains of Janey D. Denitt, who had been dead, over two years, (she died with the Consumption AE 21. She was the daughter of the beforementioned Sick brother— The people had a desire to see if any thing had grown upon her Stomach— Accordingly I attended. this morning wednesday Sep 5th a little after the breake of day with Br George, and a number of the neighbors. They opened the grave and it was a Solemn Sight indeed. A young Brother by the name of Adams examined the mouldy Specticle, but found nothing as they Supposed they

Should— Suffis to Say it was a melincolly sight to many I can Say of a truth I Saw Such a Sight, this morning, as I never Saw before. There was but a little left except bones and part of the Vitals. Which Served to Show to all what we are tending to. After the grave was filed up again, I went with Sister Wilson to visit Br Denitt and prayed with him where we had a good season to our souls Bless God for it.[113]

Moses and Elizabeth "Betsey" Nutter Dennett had ten children:

1. Polly (1782—1862)

2. Hannah (1785—1858)

3. Annie (1786—March 27, 1807)

4. Frances "Fannie" D. (1787—March 28, 1808)

5. Charles (September 28 or November 8, 1788—1867)

6. Oliver (November 6, 1790—July 11, 1865)

7. Elizabeth E. (November 28, 1792—January 7, 1873)

8. Olive (February 6, 1793—August 25, 1878)

9. Mark (November 5, 1795—May 10, 1843)

10. Eliza (1799—1872)[114]

My conclusion is that the "Janey D" mentioned by Place was actually Fannie D., who died at age twenty-one, two years and six months ("over two years") before the exhumation. The names "Fannie" and "Janey" are close enough in both sound and orthography that Place may have misheard her name or that the diary's transcriber misread Place's handwriting.[115] In any event, Frances D. Dennett seems to be the best match for the daughter who was exhumed. Unfortunately, Moses Dennett died on December 28, 1810, at the age of fifty-two, less than three months after the exhumation. Even though the attendants failed to uncover the sought-after suspicious growth, the ritual might have been judged at least partially successful, as the next death in family was in 1843—thirty-three years after Moses's death.

Later that day, Place traveled a few miles west to Loudon, New Hampshire, where he delivered a sermon. When he mentioned the Dennett exhumation in Barnstead, he received the following narrative:

The people there told me of an instant, in their neighbourhood simielar to this one. I mentioned in barnstead of a Woman that had ben dead eleven years. Who was taken up by the Shakers, they found the Bones and eleven Sprouts that had grown out of the bones, principally from the Stomach bone, one from thence grew out through the top of the Scull a number of others Stood up on the bone, all of them resembling potatoe tops when grown in the Seller, the persons that broke off those things Soon died. it was all to no purpose to the Sick relation.[116]

The death of the people who broke the sprouts suggests contagion, with the sprouts acting as a disease-spreading vector.[117]

Place does not judge the exhumation. He does not question why she was exhumed, interpret it as a horrible superstition, or exhort the exhumers to repent. Like the general tone of his diary,

Place is down-to-earth, accepting life's mysteries. Yes, the sight of Fannie's corpse was melancholy. Still, it manifested God's plan for everyone: the dissolution of the earthly vessel that houses the immortal soul. Reverend Place died at the age of seventy-eight on March 23, 1865, in Strafford, New Hampshire.

***

William Sheldon Briggs was a life-long resident of Keene, New Hampshire. He was a wood-working craftsman, as were his father, grandfather, and several uncles. His grandfather, Eliphalet, was a carpenter, joiner and cabinetmaker who built many of the first framed houses in Keene. His father, also Eliphalet, was a chairmaker, as well as a carpenter, joiner, and cabinetmaker. William learned those skills during his early years by working with his grandfather and father. William, like his father and grandfather, was known and liked in Keene. He served, volunteered, and donated. "From his long residence, observing turn of mind, and good memory," as one town historian noted, Briggs was "well versed in the history of his native place."[118] He shared this knowledge, for "he recorded many details of the history of the town in articles published in the local papers."[119]

During the first several months of 1873, Briggs wrote a series of eleven letters detailing his memories of the town—each entitled "Letter from New York" and signed "B."—to Keene's local newspaper, the *New Hampshire Sentinel*. Not knowing, at first, who "B" might be, I read the letters looking for clues to his identity. The textual evidence, reinforced by data from local histories and genealogies, lead me to conclude that the letter writer was, indeed, William Sheldon Briggs. Yet, federal census records and Keene city directories showed that Briggs lived in Keene his entire life, except at the very end, when he lived in Montpelier, Vermont, with his son and daughter-in-law. Not surprisingly, following his death on May 27, 1901, his body was returned to Keene for interment in the old burying ground, located on Washington Street. Briggs, himself, and other commentators, also wrote of a lifetime spent in Keene. We may never know why Briggs entitled his submissions as "Letter from New York."

In his first letter (February 3, 1873), Briggs was responding to a letter regarding the history of manufacturing in Keene, published the previous week. He begins, "Although some two hundred miles from Keene, I read the Sentinel every week, as I have done for more than forty years past. . . . Now I propose, with your leave, to take you through the streets of our beautiful town and show you the people and the houses they lived in as they looked to me."[120] In his wonderfully detailed narratives, Briggs led readers through the town as he remembered it, street by street and building by building—including homes, businesses, schools, and churches—filling in his broad sketch with specific anecdotes, often with named characters. In his eleventh and final installment, submitted on June 5, 1873, Briggs wrote. "I have fulfilled my promise to tell you how some of the people and places in Keene looked to me nearly fifty years ago. . . . I have had to trust *entirely* to my memory. I may have made many blunders; and if I have done injustice to any one, I am willing to be forgiven!"[121]

Briggs's second letter was published on May 22, 1873. He is walking up Washington Street, where he grew up: "I feel at home on this street, having . . . lived here over fifty years." Turning into the "old centre burying-ground" prompts Briggs to relate that, "with my own hands," he had "assisted in putting the remains of more than twelve hundred persons of Keene into their coffins, and most of them were buried in this yard. The first grave opened that I distinctly remember was for my grand-father, 1827." Briggs recalled that the "first grave made in the yard" was that of John Johnson, who died 1795. Briggs then related his eyewitness account of a consumption exhumation: "When a boy I once saw a sight in this old burying yard that would hardly be allowed in this enlightened day. A young man in a certain family in town was sick, supposed to be in consumption; his friends were told by some fortune-teller, that if they would disinter the remains of a relative that had been dead some time, take out the heart and burn it to ashes, and give those ashes to the sick man, he would get well. This was done, but the young man was buried in the old burying yard a short time after."[122] Who was this "fortune-teller"? Our contemporary image of a fortune-teller likely is a woman with a scarf wrapped around her head, sitting at a table, gazing at a crystal ball or tarot cards, predicting the future. What were fortune tellers at that time? About the same since that image has been around since at least the Renaissance. But, in New England, the term seems more inclusive: certainly, it could refer to a traveling "Gypsy" who claimed the gift of divination; but it might also designate the local wise woman or granny who had remedies, both practical and magical. We cannot be certain what kind of "fortune-teller" Briggs had in mind— a traveling, all-purpose healer and diviner, or a local person who had access to occult knowledge.[123]

A *New Hampshire Sentinel* article, in 1937, verified my conclusion that the letter writer was Briggs. But Clifford C. Wilber, in his weekly column, "The Good Old Days," did much more than that: he connected the Keene exhumation to a narrative that had appeared in more than a dozen newspapers in 1822. Wilber was an avid Keene historian, one of the three founders of the Historical Society of Cheshire County, in 1927. His column for May 28, 1937, the title of which— "Superstition Worthy of the Dark Ages"—was taken from the newspaper articles of 1822, began with a question (in the form of a statement) that, presumably, was submitted by a reader. The quotation marks in the following are as rendered:

Q. When passing the Washington street cemetery, I am sometimes reminded of a fantastic tale told me when I was young, of a body having been exhumed for the purpose of obtaining some part of it as a fetish against disease.

A. Writing in 1873, the late William S. Briggs said: "When a boy, I once saw a sight in this old burying ground that would hardly be allowed in this enlightened day.

"A young man in a certain family in town was sick, supposed to be "in consumption." His friends were told by some fortune-teller, that if they would disinter the remains of some relative that has been dead for some time, take out the heart, and burn it to ashes, and give those ashes to the sick man, that he would get well. This was done, but the young man was buried in the old burying yard a short time after."

"In the "Sentinel" of Nov. 2, 1822, is an item reprinted from the "New York Statesman," which says: "It is almost incredible that in this age of light and knowledge, an instance of such deplorable superstition as we are about to relate should have occurred.

"Lamentable and deplorable as it is, there is no doubt of the fact. The case occurred in a town not sixty miles from Boston, and the circumstances, as related in a letter from a highly respected individual to one of our friends, are as follows:

"A singular and disgraceful occurrence happened a short time since. A person, about 21 years of age, died of a consumption and was decently interred.

"About a fortnight after a person called on his father, and said that his other son, then in Boston, was sick with the consumption, and that he was sent to procure the heart of the deceased son to be pulverized and given to the living brother to cure him.

"The credulity of the father caused him to assent, and the dead child was disinterred, and his heart actually taken out and sent for the purpose above mentioned.

"It is to be hoped that the powdered heart will not afford a semblance of relief. If credulity should affect to believe in its efficacy, the grave would cease to be a resting place."

If the above item was intended to apply to the affair at Keene it fixes the date as 1822, when Mr. Briggs was only five years of age, and hardly old enough to have given such a matter any thought, but he, of course, could have received the details of what took place later from his elders.

Another story along the same lines, appears in the "Sentinel" of March 12, 1819: "On the 23rd ultimo, at Chazy, the body of Shepherd Woodward, who was buried about eighteen months ago, was exhumed, and burned on a pile previously erected for that purpose.

"The person whose body was thus consumed, died of a consumption and this measure was executed in pursuance of a superstition that burning the body of the deceased would prevent the further progress of consumption in that family."[124]

Keene is about sixty miles from Boston. Given the close correlation between these two narratives, Wilber's linking of the two exhumations is plausible.

Wilber may have appended the Chazy narrative to show that the practice was not singular (we consider this incident in chapter 12). As Wilber suggested, Briggs's memory may have been aided by later retellings. Wilber's column began with a reader's reminiscence, establishing that the exhumation narrative was still remembered more than a hundred years after the event. This recollection also illustrates how a community narrative is reinforced through visual cues: passing the cemetery stimulates the retelling of the exhumation tale, even if only in the mind of the beholder. Briggs narrated, for a published record, events that he had witnessed more than fifty years earlier. Surely, in the interim, he had talked about these experiences many times.

4 The Exhumation Will Take Place on Saturday at 10 A.M.

The 1837 volume of *Western Journal of the Medical and Physical Sciences* included a fascinating retrospective penned by an aging physician. His eyewitness account of a therapeutic exhumation was the centerpiece of "Extracts from the Diary of an Old Physician":

> The most singular case of deception and superstitious propensities of mankind, even in this enlightened land, took place in Essex County, Massachusetts, about twenty-five years ago. I was then studying medicine; a patient of my worthy teacher was lying in the last stages of consumption. She was a young woman, and of a family predisposed to this disease, of which several had already died—at this juncture a strolling Indian doctor came into the neighborhood, and hearing of her sickness called at the house. With great solemnity he told her father that the course of her disease, and the continual wasting of her flesh was occasioned by a brother who had last died of the disease, three years before, and whose heart he affirmed was still fresh and plump, and drew in some mysterious manner its support from her blood. The remedy proposed, he said was infallible; which was to take up the dead body of her brother, separate the heart from its attachment, burn it to ashes; and let the sick woman drink it in a decoction which he would prepare from "roots and herbs." On the strength of this declaration and the most implicit faith in its fulfilment, the body was exhumed and search was made for the sound and bleeding heart; but in vain, it had become a mass of corruption with the rest of the carcase [sic], and the poor girl saved from the trial of such a disgusting remedy. Led by curiosity, I was present with several others, in the burying ground at the time, and recollect the look of disappointment exhibited by the father, when no heart could be found. With such examples of credulity in our own land, we need not wonder at the disgusting and unnatural remedies of the days of ignorance and superstition.[125]

The meager clues regarding the author's identity—the article was signed, "H. - Marietta, April 27, 1836"—led me to Samuel Preston Hildreth.

Hildreth was born on September 30, 1783, in Methuen, Massachusetts, just north of the city of Lawrence, in Essex County. He attended school in nearby Andover. At the age of eighteen, he taught school for several months, then began studying medicine under the tutelage of his father, a physician and farmer. Hildreth acquired much of his medical education by reading. In 1803, Hildreth continued his medical training with an elderly physician, Thomas Kittridge, in Andover.[126] In the fall of 1804, Hildreth attended lectures at Cambridge College, a course of study that lasted only eight weeks, following which he was examined by "censors of the Massachusetts Medical Society and received a Diploma to practice Physic and Surgery." Hildreth spent the winter with Dr. Kittridge's family, then sought a suitable place to establish his own practice, settling finally in Hampstead, New Hampshire. But, in August, as Hildreth wrote, "I had fully made up my mind to leave the land of my father's and seek a new home in the western wilderness." He kept a detailed daily journal, including meteorological data, describing the journey to his new home in Marietta, Ohio.

Since Hildreth was the only physician listed among the journal's contributors from Marietta whose surname began with the letter H—not to mention one who was trained by an older physician in Essex County, Massachusetts—his identification as "the old physician" is solid. Hildreth's devotion to keeping detailed journals reinforces the connection. We only have a relative date of "about twenty-five years ago." Since all the evidence points directly to Hildreth as the "old physician," and he was studying medicine in Essex County with Dr. Kittridge in 1803, we might conclude that the entry was penned around 1828. A good hypothesis is that the exhumation occurred in or near Andover, Massachusetts, in 1803. After his death in Marietta, Ohio, on July 24, 1863, Hildreth was memorialized for his contributions to medicine and other scientific and historical pursuits.

The source of the remedy, a "strolling Indian doctor," suggests that traveling "quack doctors" and kindred "walk-abouts" were significant in disseminating the ritual. Hildreth attended the exhumation, out of curiosity—a choice that places him among many physicians who did likewise. (We will investigate the medical establishment's view of exhumation rituals in some detail in chapters 12 and 13.) The last to die of consumption, as usual, was deemed the troublesome corpse. The strolling doctor's theory is also familiar: the dead brother's heart would be found "still fresh and plump" because it "drew in some mysterious manner its support from her blood." Hildreth, as we might expect, offered his unequivocally negative view of such practices.

***

Between 1830 and 1857, the "Letters of Major Jack Downing" were read widely in dozens of American newspapers. Historian J. H. Schroeder wrote that the fictitious Downing, "combining Yankee common sense with a basic conservatism, . . . cleverly exposed what he considered to be the excesses of popular American Democracy during the late 1840's and 1850's."[127] Major Jack Downing was Seba Smith, Jr., born September 14, 1792, in Buckfield, Maine. (Buckfield was the early home of consumptive families we encountered in the last chapter, notably the Spauldings and Princes; clearly, the town folk knew about therapeutic exhumations.) Schroeder noted that, "In the three decades prior to the Civil War, Seba Smith of Maine was one of the most successful and popular satirists in the United States."[128] Through the character of Major Downing, Schroeder wrote, "Smith adroitly achieves the purpose of satire at its best: to instruct and entertain."[129]

After graduating from Bowdoin College in 1818, Smith briefly taught school before embarking on a career as a journalist in Portland, Maine. In 1828, he founded the *Family Reader*, a weekly paper promoted as "a useful, interesting, family miscellany with no political prejudice or vilification, with a condensed summary of world news and current literature, and hints on agriculture." A year later, he began publishing the *Portland Courier*, "a paper noteworthy as the first daily north or east of Boston," as Smith's biographer, Mary Alice Wyman, wrote. "The political aim of the *Portland Courier* was, like that of the *Family Reader*, to give a candid and independent representation of all parties."[130] It was in the pages of the latter periodical that the writings of Jack Downing first appeared, in 1830. Schroeder summarized the character of Major Downing that provided an early and true (if exaggerated) portrait of the homespun, but artful Yankee:

The Downing letters are rich in New England characters and Down-East dialogue. Again and again rustic anecdotes and parables drawn from Smith's Maine background are honed to lampoon a political event. Wielding this weapon good-naturedly is Major Downing himself, the classic Yankee. Downing is homely, provincial, and uneducated but, at the same time, innately sensible, shrewd, ambitious. As a Down-Easter, his rustic wisdom and common sense were at once at odds with the excesses of Jackson's Administration.[131]

One example shows how Smith was able to cut through the difficult issues of the era and reduce the tangled web of competing interests and intrigues to a simple proverb uttered by Major Downing. Smith understood "the serious difficulties involved in implementing a policy of aggressive expansionism" that faced President Pierce in 1853, so he has Downing remind the President that "in nine cases out of ten it costs more to rob an orchard than it would to buy the apples."[132] Seba Smith's wife, Elizabeth Oakes Smith (1806–1893), was a fiction writer in her own right, as well as a poet, editor, lecturer, and women's rights activist. Seba Smith died on July 29, 1868.

In the fall of 1832, the following narrative appeared under the heading, "Superstition," in the *Family Reader* and the *Portland Courier*, both of which Smith was publishing and editing at that time:

> One of the most popular superstitions of New England, and how many other countries we know not, is, that when a person dies of consumption some part of the body does not decay, but lives, and preys upon the relatives of the deceased till the whole family one after another sink under the same disease and drop into the tomb. The superstition goes further, and says, if the decaying body is taken up and the living part discovered and consumed by fire, the sick member of the family will recover. The editor of this paper was once present himself when two bodies were disinterred for this purpose, one after having been buried about a year and the other three years. They were a mother and daughter, both of whom died of consumption. Another daughter was fast sinking under the same fatal disorder, for whose benefit the exhumation was undertaken. The examination of the bodies were [sic] made by two regularly educated physicians; not because they believed in the superstition themselves, but for the satisfaction of the family. Nothing was discovered in the bodies however, more than the ordinary appearance of decay, and the sick daughter was soon laid by the side of her mother, and in a few days most of the members of the family followed.[133]

One cannot be sure why the editor, Seba Smith, was present at the exhumation. Was he present as a curious boy or young man who lived nearby, perhaps in the town of Buckfield? Or was he on the scene in his capacity as a newsman in the city of Portland or some nearby town? Without knowing when or where the event occurred, these are open questions.

Smith treated the exhumation narrative as Major Jack Downing might have, not shuddering at the horrible superstition, railing at the family's ignorance, nor openly poking fun at them. If he was using the text to expose human frailties, as he often did through Downing, he did it as the "thoughtful political critic" establishing "their folly without sermonizing."[134]

***

On June 1, 1784, Isaac Johnson of Willington, Connecticut, proceeded to the cemetery, under the advice of a foreign quack doctor, to exhume the bodies of two of his children to save the life of another child, who was dying of consumption. Also in attendance was Moses Holmes, whose letter describing the event was published three weeks later in the *Connecticut Courant and Weekly Intelligencer* (June 22, 1784):

> Whereas of late years there has been advanced for a certainty, by a certain Quack Doctor, a foreigner, that a certain cure may be had for a consumption, where any of the same family had before that time died of the same disease: directing to have the bodies of such as had died to be dug up, and further said that out of the breast or vitals might be found a sprout or vine fresh and growing, which, together with the remains of the vitals, being consumed in the fire, would be an effectual cure to the same family: — and such direction so far gained credit, that in one instance, the experiment was thoroughly made in Willington, on the first day of June instant: two bodies were dug up which belonged to the family of Mr. Isaac Johnson of that place, they both died with the consumption, one had been buried one year and eleven months, the other one year, a third of the same family then sick—on full examination of the then small remains, by two Doctors then present, viz. Doctors Grant and West, not the least discovery could be made; and to prevent misrepresentation of the facts, I being an eye witness, that under the coffin was sundry small sprouts about one inch in length, then fresh, but most likely was the produce of sorrel seeds which fell under the coffin when put in the earth. And that the bodies of the dead may rest quiet in their graves without such interruption, I think the public ought to be aware of being led away by such an imposture — Moses Holmes.

Isaac Johnson (1735-1808) and Elizabeth Beal Johnson (1734-1803) married in Willington on July 15, 1756. They produced eight children:

Caleb (March 6, 1757—January 17, 1823)

Isaac (April 15, 1758—March 31, 1777)

Amos (October 3, 1760—July 15, 1782)

Stephen (August 5, 1762—February 1, 1845)

Elizabeth (May 26, 1764—May 18, 1783)

William (May 25, 1766—September 4, 1785)

Elijah (December 24, 1771—October 10, 1798)

Lydia (June 22, 1774—November 1824)

Isaac was the first to die, on March 31, 1777, just two weeks before his nineteenth birthday. Five years elapsed, then Amos died at age twenty-one, followed ten months later by Elizabeth, who was almost nineteen. On the day of the exhumations, June 1, 1784, Amos had been dead for one year, eleven months and two weeks, and Elizabeth had been dead for one year and two weeks. The narrative stated that "one had been buried one year and eleven months, the other one year." Rounding down the two weeks, Amos and Elizabeth are a perfect match for the two who were exhumed.

Holmes is not clear whether the vines and vitals were burned after the doctors' examination revealed "small remains," which I interpret as the corpses in an advanced stage of decomposition. However, sprouts *were* found growing in the grave—which Holmes rationalized as simply seeds that had naturally fallen into the open grave. Still, his observation that "the experiment was thoroughly made" suggests that the ritual was completed. At the time of the exhumations, a third member of the family was said to be sick—undoubtedly William, who died fifteen months later at the age of nineteen. The "experiment" may have been judged at least partially successful, for it would be fourteen years before another family member died: Elijah, a few weeks before his twenty-seventh birthday, in 1798. Caleb, who had married and relocated to Williamsburg, Hampshire County, Massachusetts, died in his sixty-sixth year. Lydia married a widower, moved to Ohio, and lived to the age of fifty years. Stephen and his parents, Isaac and Elizabeth, apparently avoided the family's consumption plague, relocating to Hadley, Hampshire County, Massachusetts and surviving to old age.

The word "experiment" also appeared in the minutes of the Cumberland, Rhode Island, Town Council meeting of February 8, 1796, which granted Stephen Staples permission to exhume the body of his daughter, Abigail, to "try an experiment" that he hoped would save the life of another daughter, Lavina (chapter 1). Even though the official record does not specify what was to be done to Abigail's corpse after exhumation, surely some variant of New England's consumption ritual was to be performed. The etymology of the word *experiment* strengthens this interpretation. From the Latin *experimentum* ("a trial, test") through Old French, the word entered English as a magician's term, often used interchangeably with *magic*. Lynn Thorndike, in "Mediaeval Magic and Science in the Seventeenth Century" (1953), observed that "magic was still intermingled with science in the seventeenth century" and "medical cases and prescriptions were still spoken of as experiments."[135] As late as 1892, when the *Providence Journal* (March 20, 1892) reported on the Brown family exhumations, the consumption ritual was characterized as an experiment: ". . . if the young man for whose sake the experiment was made recovers from his illness, they will . . . consider the recovery due to the destruction of the spectral agent of his apparently fatal illness." We will encounter the term again (chapter 20) when we take up Mary A. Denison's apparently fictional vampire story (1853): "Various reasons are assigned as causes for belief in the efficacy of this curious experiment."[136]

A pertinent coupling of the word *experiment* with quack doctors appeared in the Second Annual Report of Boston's Board of Health (1872), which argued for requiring proper credentialing of "regular physicians" for the purpose of "preventing fraud" by "irregular practitioners" and "quacks":

> A sick man tries the quack's methods as an experiment in his search for restored health; he is disappointed in his hopes, grows worse, and, when in desperate strait, returns to the care of some regular physician, and under his care the case ends in death. It is thus that it happens that many quacks boast of their success in never losing cases, while the mortality lists of the best regular physicians attest the hold which they have on the confidence of the mortally sick.[137]

Another reading might be, if you are going to die anyway, it is best to have a "regular" doctor standing next to your bed.

Holmes was well-positioned to caution his fellow townsmen against being led astray. He was counted among Willington's political elite, based on his election as selectman four times between 1771 and 1784; and he was a member of an exclusive group in town known as the "inner circle of twelve families." He also served as town clerk from 1764 to 1770. And he was a representative to the state's general assembly at least seven times between 1766 and 1785.[138] Holmes also had some experience dealing with medical practitioners of various sorts. His father, David Holmes, had been a respected physician, as well as a political and military leader, in Woodstock, Connecticut.

Men in a position of authority, such as Moses Holmes, kept abreast of the comings and goings of people in town, especially strangers. A foreign quack doctor would not have escaped Holmes's attention. Many traveling practitioners advertised in local newspapers. In October 1785, for example, several months after the exhumations in Willington, the *Connecticut Courant* included a paid advertisement announcing that "Dr. John Newman, of the City of Newport" had "arrived in the City of Hartford." The ad boasted of Newman's "great success in curing of Cancers." Newman also claimed he could cure ulcers, corns, fever sores, and "the King's Evil"—a folk designation for scrofula, a form of tuberculosis that infected the lymphatic nodes, especially of the neck. The ad noted that Newman would be in the city for about fifteen days. For a scold such as Moses Holmes, that would be about two weeks too long.

In early 2023, I received an email from the author of *In the Shadow of Salem: The Andover Witch Hunt of 1692* (2018). Richard Hite wrote,

> I just came to a realization. I know one of the vampire cases you have researched involves Isaac Johnson of Willington, Connecticut. I have reason to believe he is from the Johnson family of Andover, Massachusetts. It looks like he might be a son of Caleb Johnson (born in Andover in 1694), who went first to Windham and then to Willington, Connecticut where he died in 1760. You have probably heard of the exoneration of Elizabeth Johnson of Andover, the last remaining person convicted of witchcraft in 1692. ... In any case, this Caleb Johnson was a first cousin of Elizabeth Johnson [born ca 1670], meaning that Isaac Johnson was her first cousin once removed. So there is a tie between the witch hunt and one of the vampire cases.

I was intrigued by this family's connection to both witchcraft persecutions and vampire exhumations. Genealogical evidence shows that this Johnson family was rife with accused witches. Elizabeth's mother, Elizabeth Dane Johnson (born ca 1641), and two younger siblings, Stephen (born 1679) and Abigail (born 1682), also were accused of witchcraft. Stephen Johnson, Elizabeth, Jr.'s father and Caleb's uncle, died before the witch hunt, as did Stephen's brother, Timothy. But Timothy's wife and daughter (both named Rebecca) were accused and confessed to witchcraft. The elder Rebecca (born 1652) was a niece of Mary Ayer Parker who was hanged for witchcraft in Salem on September 22, 1692. Elizabeth, Jr. was sentenced to be hanged, but the Governor reprieved her. Since her own grandfather described her as "simplish at the best," she probably was developmentally challenged.[139]

What to make of this connection? Scholars have shown that extraordinary events are strong candidates for incorporation into a family's oral history.[140] It thus seems likely that Isaac Johnson, born some forty-three years after the Andover witch hunt, was aware of his family's accused witches. If so, then what bearing, if any, would that knowledge have had on his decision to enact the consumption ritual? While no direct evidence answers this question, intuition leads to speculation regarding Johnson's spiritual disposition, as his family once again faced a seemingly intractable problem. Magical thinking can bridge the possible and impossible. Preternatural aid might transcend the threshold of human capacity. To master nature is both a scientific and a magical quest, for one must travel beyond the limits of science. As Arthur C. Clarke remarked many decades ago: "Any sufficiently advanced technology is indistinguishable from magic."

Around the same time that Hite informed me of Johnson's witchcraft connection, I began communicating with John Blair, Emeritus Professor of European Medieval History and Archaeology at Oxford University, who was writing a book interpreting vampires in the context of social anxiety and female power. One issue we've debated concerns the age and gender of suspected corpses. In an email to me on December 30, 2022, Blair wrote, "In my book I'm putting a strong emphasis on the female walking dead, and especially on young women who die in their teens. The age-group 12-22 for females keeps cropping up: as poltergeist agents, as victims of witch possession, and (in some contexts) as troublesome corpses."

At his request, I created two charts, one (figure 8) showing the gender of exhumed corpses in New England and the other (figure 9) showing the gender and age of both the exhumed corpses as well as those deemed to be troublesome ("vampires"). The data in these charts prompted Blair to wonder "whether (1) the scientific overlay disguises a more traditional scenario in which young females were targeted above the odds, and (2) whether this is in some weird way an inverse-image of the witch-trials of two centuries earlier, occurring in those regions where the (psychological) demons had not already been purged by the witch-mania."

While the charts do indeed show that more females than males were exhumed and identified as troublesome corpses, I have argued these data might reflect differences in gender roles during this period. Males generally worked outside the infected home while females stayed indoors and tended to the ill. So, more females than males contracted consumption. Vital records (when and where they exist) show exactly this pattern of mortality, as more females than males died of consumption. And consumption also infected the young (of both sexes) more than the old. I think this was one reason the disease was deemed unnatural: for young people to be taken before their parents and grandparents seemed to invert the natural order of things. However, it certainly is possible that both discrimination against females as well as gender-related role relations were operating together.

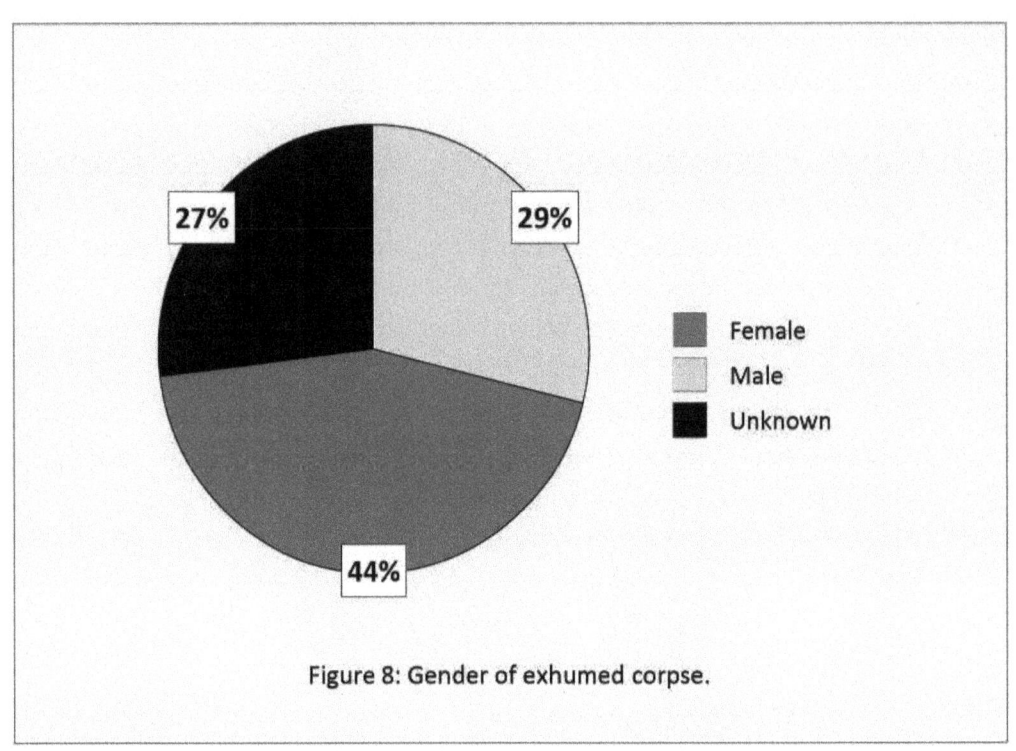

Figure 8: Gender of exhumed corpse.

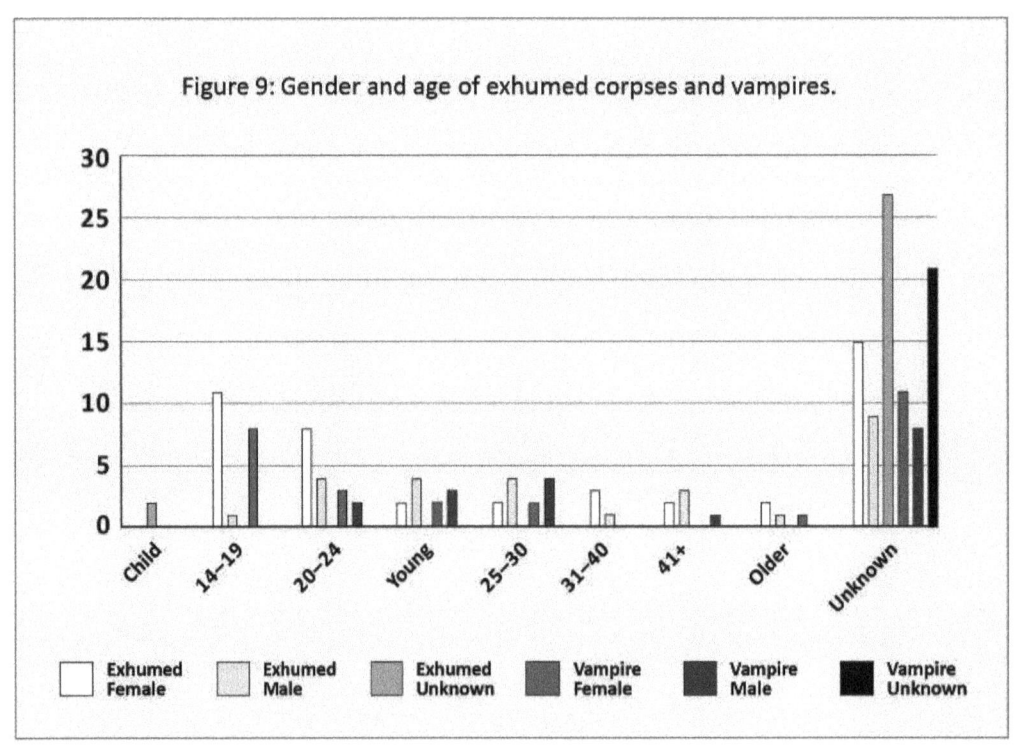

56

***

A town council voted to allow a man to exhume his daughter. Half the population of another town witnessed an exhumation and heart-burning. To these extraordinary public enactments of the consumption ritual, we now add a newspaper notice inviting "all interested" to a family's exhumation. Under the heading, "Bodies to be Exhumed," the *Montrose Democrat* (April 12 & 26, 1871) twice published the following, signed by William B. Tourje of Harford, Pennsylvania:

> The subscriber being the only survivor of a large family of brothers and sisters, all of whom have died of that dread disease, consumption, and whose children are following them by the same disease, has consented that the bodies of some of the first who died may be exhumed to satisfy the belief entertained that it will arrest the further progress of the disease, which is destroying the remaining survivors. The exhumation will take place on Saturday, April 29th, at 10 o'clock A.M., at the burying ground on David Whitney's land, in Lenox township, this county. This notice is given that all interested may be present.
> WM. B. TOURJE
>
> Harford, April 10, 1871[141]

William Tourje was sixty-four years of age when the notice was published. Although this event took place in northeastern Pennsylvania, the federal census of 1850 shows that William and his brother, Jonathan, were born in Rhode Island. Between 1817 and 1819, the family relocated from Rhode Island to Pennsylvania.

May 17, 1871, was a significant day in the annals of New England's consumption tradition, as three separate texts addressing this family's exhumations appeared in the *Democrat*. On the first page, under the heading, "The Lenox Exhuming," the paper's editors published a statement advising readers that William Tourje, himself, did not have faith in the ritual, even though he acquiesced to others in the family who wanted it performed. Some twenty years later, a local newspaper reported that George Brown, of Exeter, Rhode Island, expressed the same rectitude that we see in the following:

> The feeling which pervades the community respecting all such fanatical superstitions as the "consumption tradition" at the present day, makes those who actually believe in it to be objects of wonder and astonishment, and even pity, and strikes the mind of a Christianized community with a sacrilegious horror. We do not here refer to the matter for the purpose of deducing any moral dissertation from the subject, as the number who believe in it are very limited indeed, and we certainly feel to consider them more to be objects of pity than censure, as their mental and moral development must be very limited for these days of education and Christian enlightenment.
>
> We particularly refer to this matter at this time for the purpose of informing our readers of the true position in which Mr. Wm. B. Tourje, the subscriber to the notice which appeared in our columns, stands, as his name is the only one that appears and hence an unwarranted superstitious odium may attach itself to it, and that we may do justice to him we now make mention of the following facts:
>
> When Mr. Tourje presented us the notice for publication, he emphatically stated that he had no faith in it whatever, and that he had for many years been the only opposer in the

matter among the friends and has persistantly [sic] refused until even quite an unpleasant feeling had arisen, and that he now merely consented to gratify their wishes. These facts are corroborated by other parties and altho' Mr. Tourje's name has been made prominent in the matter we think it due him that these facts should be made known.[142]

In the first paragraph, the editors essentially say we do not want to pass judgment on the topic, but those who believe in this horrible superstition are mentally and morally deficient, and we pity them. That others in the community defended Tourje's probity in the matter suggests that this family affair had been a topic of local conversation. The population of Harford, in 1870, was 1,595 (at 1,751, Lenox had slightly more residents). Information and innuendo would have spread quickly.

The day's seminal story, an eyewitness account entitled "The Lenox Exhuming Case," was in the last column of the first page:

> Mr. Editor:—Noticing the article announcing the exhuming of bodies, I attended the opening of the graves at the burying ground, near David Whitney's, in Lenox, which was done at the request and for the satisfaction of the family friends of the deceased, to test the theory of a tradition that the dead were preying upon the living—quite a number of the family having died of consumption—and other surviving ones being affected with the same disease. Having been acquainted with the families from early youth, curiosity led me to see what the result might be.
>
> At the appointed hour quite a large number of people—probably a hundred—were upon the ground, when the work commenced. The grave of the first person being opened, nothing of exciting interest was developed, there only remaining in the coffin the partially decayed bones of the deceased, the death having occurred July 6, 1850, at the age of 60 years and 5 months. They next proceeded to take up the remains of one of his sons, John A. Tourje, Jr., who died, November 22d, 1857, aged 40 years. On the coffin being taken up and opened nothing but the naked bones were presented to view of the spectators, without any of the indications sought for. A sister of the last named was next disinterred, who died, November 10th, 1867, aged 41, being the last who had died, except one, that has been buried only about four months. The graves of all were considerably filled with water and the coffin of the last one being raised, drained of the water and opened, behold! a vine was found!! the flesh was all decayed, but everything else remained as when placed in the ground, as near as could be expected; the silk dress and a bosom pin, in which the deceased was buried remaining entire. The vine found, was taken out, it being found lying in some shavings which were placed in the coffin at the time of burial, beside or near the head of the remains. Drs. Wright and Green being present, the vine was handed them for inspection. As soon as a knowledge of the above facts became known to a sister-in-law of the deceased, wife of the second one named in the foregoing, she stated that a myrtle vine was brought from the yard of Mr. Whitney and placed in the coffin before burial, the corpse being shown in the cemetery before depositing in the grave. Further investigations were made, but I understand that there is some talk of taking up the wife of Mr. Snyder, the last one buried. —There is in all, of the family, 13 in number, children and grand-children, buried in this year, and one buried in the burying ground at the Baptist church near Elder Tower's, and all

but one at least, have died of consumption, the youngest being a grand-child only three years of age, and only one of eight children now living.

    Spectator.[143]

Research supplies some of the missing information. The first grave opened must have been that of John A. Tourje, Sr., who died at the age of sixty years and five months, on July 6, 1850. The second grave opened was John A. Tourje, Jr., who died on November 22, 1857, at the age of forty years. The third grave opened was that of John Jr.'s sister, who died on November 10, 1867, at the age of forty-one years. The Federal census of 1850 indicates that this sister was Elizabeth Tourje, and that William (who placed the exhumation notice in the newspaper) and John, Jr. (whose body was exhumed) were sons of John, Sr. The two brothers were born in Rhode Island; Elizabeth was born in Pennsylvania. Four months prior to the exhumations, in about January of 1871, another family member, the wife of Mr. Snyder, had died. According to the anonymous spectator, her disinterment was then under consideration.

Recovering profiles of these family members was more difficult than I imagined. I encountered multiple spellings of the surname, including Tourgee, Tourjee, Tourje, Taiger, Targer, Tarjee, and Tourja. In the 1850 Federal census, the name was transcribed as Tompos! Moreover, few relevant vital statistics, including cemetery databases, are available. A history of Susquehanna County explains why I could not locate graves for the Tourje family: "In the northwestern part of the township [Lenox] a burial-place was started many years ago, in which were interred many members of the Tourgee and Gardner families. This ground was not appropriately selected, and is not receiving the care the resting-places of the dead deserve at the hands of the community in which they are located."[144] The family was seemingly not sufficiently important to receive more than a mention in county histories. The two physicians placed at the scene by the spectator, however, did make it into the histories of Susquehanna County. Dr. W. N. Green and Dr. Samuel Wright (described as a "botanic physician") lived at Hopbottom, a small community adjacent to Lenox.[145]

The unusual level of public engagement and openness of the proceedings are remarkable. A chain of circumstances has given us insight into a level of local involvement that normally goes unrecorded. We are not informed concerning which ritual procedures, other than exhumation, the participants planned to perform. Since the corpses were decomposed, and the coffins filled with water, the conditions necessary for proceeding with the ritual apparently were not met. A suspicious grave plant appeared in one of the coffins, but its presence was rationalized. In the newspaper, William Tourje's public invitation to the family's exhumations was followed by a clarification by the editors, then a detailed description of the exhumations by a spectator, and, finally, a reader's letter commenting on the consumption tradition.

The latter text appeared in the *Democrat* (May 17, 1871) as "A Letter from Nevada." The writer, Albert Churchill (1836-1906), introduces himself as a subscriber who was raised in Snake Creek valley, some four miles north of Montrose. He subsequently relocated—not to the state of Nevada as the heading implies, but (as my research showed) to Nevada, *Missouri*. He writes that, after becoming a medical doctor and leaving Pennsylvania, he "resided in Louisiana and Texas, about seven years, and about five years in Missouri." As we might expect of an establishment physician, his reaction to Tourje's exhumation notice is acrimonious. For the sake of comparison, Churchill takes us to distant lands, such as "Hindoostan," where "heathen mothers" throw "their children into the rivers" and "widows" bury themselves "with the dead bodies of their husbands." He is "horrified" by the thought of cannibalism. "Who can imagine," he asks, "anything more revolting that eating the ashes of a *corpse*? and *it* the body of a near relative!" He writes that he has "not seen or heard of such a loathsome, detestable practice or belief as that of eating the ashes of a *dead body!*" Churchill is interred in Deepwood Cemetery, in Nevada, Missouri. In what can only be described as an unfortunate error, given his obvious distaste for heathen beliefs, the name engraved on his gravestone is "Churchell."

***

The publicity given the Tourje case in the *Montrose Democrat* prompted a reader to submit an account of a consumption ritual that he had witnessed more than three decades earlier in New Hampshire. The letter, signed by A. O. Warren, ran on April 26, 1871, in the "Local Intelligence" section. Warren's letter to the editor was a welcome exception to the typical consumption-ritual narrative that has me begging for more detail. After delineating the ritual's general pattern, he described a specific event—which he witnessed! He generously provided a time frame (a little before April 1838), a place, the family surname, and the patron's given name. I could not, in good conscience, ask more of a narrator.

By characterizing the consumption ritual as a tradition, Warren implicitly acknowledged that it was handed down through generations by word of mouth and practice: it was an established pattern, not the quirky, one-off event many narrators presume. Two other cases in Cheshire County —New Ipswich ca.1810 (chapter 13) and Keene ca. 1827-30 (chapter 3)—support Warren's inference. Warren has capitalized the word "Consumption" throughout his letter, thus transforming it from a common noun to a proper noun, a special status suggesting that consumption was not a run-of-the-mill disease, but a singularity that people had to confront on its own terms. Because Warren believed that consumption was grounded in heredity, he concluded that nothing—including traditional ritual or, by implication, established medicine—could cure the disease. While he found it "strange" that anyone should turn to "such a tradition," we have seen that desperate people grasp at straws. And it appears that this family had heard stories attesting to the ritual's efficacy. Oral tradition filled a void by giving them agency to act. Testing the tradition's "virtue" through performance recalls the word "experiment" that occasionally was used to characterize the consumption ritual.

Here is Warren's fascinating account:

Mr. Editor: —I saw a notice in your last issue stating that on the 29th day of April, there would be an exhumation of graves in the Cemetery, in West Lenox, near Mr. Whitney's, with a view of arresting the ravages of Consumption in a certain family, &c.

The tradition is this, when a number of the members of any family seem to be failing, one after another from Consumption, it is because the dead are preying upon living, and that the *last* dying does not decay, and return to dust, until the death of another member of the family. Or should it be found on examination that the vitals had decayed, the first member, in the counxion [sic] will be found undecayed, as to its vitals, or there will be found a vegitable [sic] growth in the dust, centering its roots in the vitals, as near as may be.

This plant, or the undecayed vitals, must be removed, and burned to ashes and taken by the living, and then the fearful ravages of Consumption will be arrested.

Now, Mr. Editor, allow me to give you the history of what I once saw myself, and what has been the result.

More than thirty-three years ago, there lived in Jaffrey N. H., a family by the name of John Frost, three members of which had died with Consumption. One other member was but just alive, as was supposed, dying with Consumption, and others were in different stages of that fell disease.

Hearing from different sources reports confirming the truthfulness of the statement in the tradition, above refered [sic] to, they resolved to test its virtue by an exhumation.

Public notice of that fact was given, and probably a hundred people were present to see it. Two graves were to be opened, the first and last, in the line of death. The first grave was that of the grandmother of the family, on the fathers side, dead eighteen years.

On opening that grave, all that it contained of mortality, save dust, was the lower jaw and one large bone in one of the lower limbs.

There was *no vitality* or *form* of it, *either vegetable or animal.*

The second grave contained the last dead of the family, a girl that had been dead about nine months. On opening that grave the body was found not to be decayed but slightly in any part. The same would be true of any one dying with any disease except some putrid disease. Nine months would produce but slight change in a body excluded from the air and five feet in the ground. But the body was opened and the lungs, heart, liver and stomach were taken out and delivered to the family for them to reduce to ashes and take to cure the farther progress of Consumption in their ranks.

Now what was the result? Is the tradition true?

When I was in New Hampshire last fall, by the same cemetery, standing beside these graves and others, I asked a citizen of that place if any other members of the Frost family had died of Consumption since the above described examination? He told me that two, and I think three, had died since that time and others were feeble showing the failure of this infallible remedy. It did not arrest in any degree the ravages of Consumption in that family.

But I have no doubt it did as much for them as it could for any one. In this case and all other similar ones, the tendency of Consumption is inherited and cannot be eradicated any easier than the Ethiopean [sic] can change his skin.

It seems strange to me, that any one should think of receiving medical assistance from any such source or that any one should believe in such a tradition. It has no reason or facts to support it.

After the case above reported, I took pains to inquire as to the correctness of such cases, and I found in every case that, "it was said by some body, that some body said, some one else, some where, had heard of some one who knew of a case over some where, that was so."

Very Respectfully, A. O. Warren.

Warren wrote that the public was informed of the impending exhumation, and he estimated that a hundred people attended. While *formal* prior public notice is unusual, it seems likely that the time and place of an impending ritual would circulate orally among those interested. Warren recorded that it was decided to open two graves, the first and the last to die. In terms of the tradition as outlined by Warren, it makes sense that these two corpses would be exhumed. But the order seems to be reversed. The paternal grandmother, said to have been dead for eighteen years, was the first exhumed. My research indicates that the best match is the *grandfather* of John, Jr., Benjamin Frost, who died April 12, 1819, at age seventy-one—between eighteen and nineteen years prior to exhumation. His grandmother, Rachel Kimball Frost, died at age eighty-five on July 28, 1840—about two years *after* the date of ritual provided by Warren.

Examining the first corpse, the exhumers found "*no vitality* or *form* of it, *either vegetable or animal*." So, they disinterred the last to die. The vegetable growth that was sought seems to have a sympathetic connection to the vital organs. If the organs have decayed, the vine may have become an agency of contagion by incorporating and transmitting the evil once residing in the organs. Warren confirms this mutability, writing that the plant or the "undecayed vitals" should be "burned to ashes and taken by the living." In the consumption ritual, to burn and consume is to transform and heal.

Warren wrote that second grave exhumed contained "the last dead of the family, a girl that had been dead about nine months." Palmyra Frost, daughter of John, Sr., died February 27, 1837. The time frame matches Warren's "more than thirty-three years ago." Nine months after Palmyra's death would have been November 1837—thirty-three years and five months before Warren's narrative. He saw nothing unnatural about the corpse's state of preservation, yet "the body was opened and the lungs, heart, liver and stomach were taken out and delivered to the family for them to reduce to ashes and take to cure farther progress of Consumption in their ranks." Those in the family requiring treatment obviously were not at the cemetery, so the vital organs had to be transported to their home for rendering into medicine.

Near the end of his narrative, Warren presented his examination of the tradition's efficacy. He had returned to his hometown in Fall of 1870, a few months before the notice in the *Montrose Democrat* prompted his account of the Jaffrey exhumation. He presented his case (like the attorney he was), describing how he reached a conclusion. "When I was in New Hampshire last fall, by the same cemetery, standing beside these graves and others, I asked a citizen of that place if any other members of the Frost family had died of Consumption since the above described examination?" Warren seemed intent on prosecuting this event. He found that John Frost, Jr., was still living on "the old homestead" in Jaffrey. (John, Jr., lived to the age of seventy-two, dying in 1900.) Warren wrote that his informant "told me that two, and I think three, had died since that time and others were feeble showing the failure of this infallible remedy. It did not arrest in any degree the ravages of Consumption in that family." Following Palmyra's death in 1837, three more in the family did, indeed, die: her brother, Liberty, in 1836, her grandmother, Rachel, in 1840, and her father, John, in 1847.

Warren's conclusion seemed predetermined by his belief that consumption was hereditary, asserting that the ritual "did as much for them as it could for any one," since "the tendency of

Consumption is inherited and cannot be eradicated." Warren found it "strange" that anyone "should believe in such a tradition," as there were "no reason or facts to support it." Warren closed his argument with a paradigm of oral tradition, a sort of nineteenth-century version of the FOAF of urban legend, the elusive *friend of a friend*. "After the case above reported, I took pains to inquire as to the correctness of such cases, and I found in every case that, 'it was said by some body, that some body said, some one else, some where, had heard of some one who knew of a case over some where, that was so.'"

Andrew Oliver Warren (1817-1895) and John Frost, Jr. (1818-1900) were cohorts, born and raised in a town of about 1300 people. Although we have little information regarding their relationship, it seems likely that their paths crossed during their formative years in a small town. John's father was a deacon of the church and deemed "a worthy citizen." John, Jr., also was known as a "worthy man and highly respected citizen," who enlisted and served in the 14th Regiment of New Hampshire Volunteers throughout the Civil War. His maternal great-grandfather was Deacon Ephraim Adams of the nearby town of New Ipswich, where, as we shall see (chapter 13), a consumption ritual was performed about 1810.

Part III   Newsworkers

Where consumption went, vampires followed. Vampire stories increased apace with consumption cases and, not coincidentally, the number of newspapers in America. These stories were ideally suited for newsworkers seeking to attract an audience, as they combined features of both hard and soft news. Marketable narratives were created by blending natural with supernatural, belief with unbelief, and historical fact with literary convention.

The nineteenth century was primed for vampire stories. Advances in postal service and transportation, followed by the invention of the telegraph (America's first medium of mass communication) greatly widened the scope and reach of reporting.[146] The competitive environment of the newspaper business was heightened by the sheer number of periodicals being published.[147] Newspapers broadened their appeal by offering an increasingly varied content that matched the country's growing cultural diversity. Most of the incidents discussed in this book found their way into popular periodicals at some point during their narrative journey.

Newsworkers mediate reality. To paraphrase social historian Michael Schudson, they select, highlight, frame, shade, and shape the nature of things.[148] Although media scholar Daniel C. Hallin created his three, concentric "spheres of reporting" to depict the changing role of journalists relative to the American political process during the Vietnam war, his scheme is useful in conceptualizing how newsworkers interpreted vampire incidents during the previous century. Hallin imagined "the journalist's world as divided into three regions, each of which is governed by different journalistic standards." The center region is the sphere of consensus, occupied by issues considered noncontroversial by journalists and society at large ("motherhood and apple pie," for example), so they need not be examined and debated.[149] Outside of the center is the sphere of legitimate controversy, where journalists, following the lead of "major established actors of the American political process," balance issues (such as "electoral contests and legislative debates") objectively. The outlying region is the sphere of deviance, where beliefs and attitudes rejected by mainstream society are viewed as hardly worth even commenting upon. "Here neutrality once again falls away," Hallin wrote, "and journalism becomes, to borrow a phrase from Talcott Parsons, a 'boundary-maintaining mechanism'; it plays the role of exposing, condemning, or excluding from the public agenda those who violate or challenge . . . consensus. It marks out and defends the limits of acceptable" views and actions.[150] In this region, Schudson wrote, "coverage of issues or groups goes beyond the reach of normal reportorial obligations of balance and fairness." These topics, when they are addressed at all, "can be ridiculed, marginalized, or trivialized because reporters instinctively realize that mainstream culture treats them with derision and contempt."[151] The outlying sphere of deviance is where most nineteenth century newsworkers consigned consumption rituals.

The split between facticity (news) and poetic license (entertainment) did not develop overnight in American journalism. Newspapers had long experience in exciting the vulgar tastes of readers. Folklorist George S. Carey, in his analysis of folklore in the newspapers of Essex County, Massachusetts, demonstrated that in the nineteenth century: "Without fail they told their subscribers of maudlin murders, of live burials, of suicides, mutinies, and endless piracies. They kept the public abreast of ventriloquism, and made periodic reports on spiritualism and mesmerism. With this emphasis on the sensational, the strange and the occult, it is hardly surprising to find elements of traditional material appearing in these weekly news sheets."[152] Carey found that "nineteenth century newspapers thrived on exchange," which created "widespread plagiarism." And this, he continued, "produced a quantity of very popular stories which moved from newspaper to newspaper without any credit being given for origin. But from what we know about the nature of folklore, it is quite possible that many of these stories had their ultimate beginning orally, and . . . some of these tales doubtless moved fluidly back and forth between the printed page and the folk milieu."[153]

I will add vampires to Carey's list and affirm the easy interchange between folk and newspaper narratives. Newspaper vampire stories range from eyewitness accounts through hearsay and gossip to legend. Sources were mostly unnamed. Even if not derived from oral tradition, many of these tales share attributes characteristic of folk legends, viewed by most folklorists as "traditional prose narratives set in the recent past with humans as the main characters and often regarded as grounded in fact by the people who pass them along, although frequently elaborated in their telling."[154] Legend as cultural debate—or, as folklorists often phrase it, the "dialectic of the legend"—is at least implicit in many of the newspaper narratives we encounter in these chapters.

The tendency of news writers to marginalize both the tales and their tellers—assign them to the sphere of deviance in Hallin's terminology—was not lost on Carey. "There is a decided tongue-in-cheek ridicule appearing in many accounts which reveals a none-of-us-believe-this-but attitude on the part of the writer," Carey wrote. "In fact, certain stories admit of nothing more than castigation of the country bumpkin."[155] Adjectives used to describe the consumption ritual and its participants are condemnatory: monstrous, uncivilized, heathenish, unreasoning, delusional, absurd. Omitting personal information, such as surnames, smoothed the path of dismissal. Masking dehumanizes people as it "others" them. Berating another's superstitions allows one to bask in the glow of communal self-congratulation. Contrast burnishes one side as it dims the other: "Look how far we've come!" Not as far as understanding—much less curing—the scourge in question, of course. The morbid exhumation narrative (which sometimes included cannibalism) acts as a magnet: one end repels and the other attracts. Does scorn offset fascination? Resting on the cusp of credibility, teetering between doubt and veracity, the newspaper vampire story, like a folk legend, is debatable—and is, indeed, debated: possible/impossible; would/would not; did/did not.[156] Anticipating the doubts of their readers, newspapers sometimes added a credibility statement, such as "We have been put in possession of the following facts, for the truthfulness of which we have the most undoubted authority" (Cayuga County, New York) or, more simply, "These are facts" (New Marlborough, Massachusetts).

In their accounts of New England's consumption rituals, newspapers presented little more than the attention-grabbing core elements that could move a weak narrative as far as the next column inch. While writers and editors often omitted details that might lead to identifying the families involved, they were not shy about describing an exhumation's gory details. These lean accounts disclose stability in the ritual's performance throughout the nineteenth century. Variation in its procedures may be partially the result of the process that folklorists term "localization," where communities change a tradition's details to suit their own circumstances.[157] The infrequent, but welcome, occasions when narratives include an account of how a family or community came to know of the ritual speak to the processes of its transmission. By publishing and republishing vampire narratives, newspapers, themselves, disseminated knowledge of the practice to a wider audience.

Many narratives contain enough evidence to allow an evaluation of their accuracy. Establishing that the people identified in a story lived in the place and during the time described in a story lends credibility. Newspaper accounts frequently include an explanation of a ritual's controlling principles—such as associated beliefs, practices, theories, and opinions—sometimes obtained by interviewing people in the community. My research has shown that participants in the ritual were of varied classes, faiths, and professions or occupations. Still, a disdainful fog, formed through estrangement, enveloped them. The fog lifts as we move closer to the community. When we have verbal sketches of the people involved, we see the faces of those who struggled to defeat the unseen evil that was destroying their families and communities.

Gravestone of Salmacious Adams, West Glens Falls Cemetery, Queensbury, New York. *Courtesy of Cyril Place.*

5   Wrath of the Devil's Dark Angel

On April 26, 1790, the following story appeared in the *Albany Gazette*:

> Two children of Mr. Josiah Grant Huit, of Ballstown, being lately attacked with symptoms of a consumption, he was advised, in order to their cure [sic], to dig up and burn the body of a daughter, who died of the fatal disorder, about twenty months ago—Strange to relate! the unfeeling, infatuated father, tore the remains of his child from her coffin, placed them on a pile, and reduced them to ashes![158]

In an era when the spelling of surnames was inconsistent, tracking down Josiah Grant Huit entailed widening the search to include orthographic variations, a strategy that eventually led to a Josiah Grant Hewitt, whose sparse biography matched the article's sketchy details.

Virginia Hewitt Watterson compiled a history of the descendants of Captain Thomas Hewitt of Stonington, Connecticut, which included the following information regarding Josiah Grant Hewitt: he was born in Stonington on May 12, 1742, the son of Nathaniel Hewitt and Rebecca Grant. In 1763, he married Mercy Williams at the First Church of Stonington, where the couple resided until at least the summer of 1787 (when Josiah was named executor of his grandfather's Stonington estate). The first Federal census shows that, by 1790, Josiah had moved with his family to Albany County, New York. (Ballstown was in a section of Saratoga County that became part of Albany County in 1791.) If, as stated in the newspaper text, the exhumed daughter died about "twenty months" prior to the article's publication (April 1790), the unnamed Hewitt girl would have succumbed to consumption in August 1788. The family must have relocated from Stonington to Ballstown sometime between the summer of 1787 and that of 1788, if the daughter died, was buried, and then disinterred in Ballstown. The 1790 federal census, and other genealogical sources, show that Josiah and Mercy had at least three sons and three daughters, all born in Stonington: Desire (born in October 1763); Sarah (christened on April 16, 1769); Thomas Williams (christened on December 2, 1770); Josiah Grant, Jr. (born circa 1773); Palmer (born circa 1775); and Lucy (born in 1779). Lucy was not the daughter exhumed in 1790, for she married Lee Watrous, on November 20, 1797, at the Reformed Dutch Church, in Stone Arabia, Montgomery County, New York. Nor was it Desire or Sarah, as both were still living at the time of the 1790 census. Watterson noted that "by 1800, he [Josiah Hewitt] and the family were living in Palatine, Montgomery Co., NY. They were still in Palatine in 1810, with 2 females still living at home with the parents."[159] The daughter whose corpse was exhumed and incinerated in 1790 must have been one of the unnamed "others in this family" referenced by Watterson.

The concise newspaper text of the Huit/Hewitt exhumation came to my attention through a footnote in an elegy, delivered in 1790, by the librarian at Harvard College, Thaddeus Mason Harris.

Harris was born in Charlestown, Massachusetts, on July 7, 1768. When he graduated from Harvard, in 1787, he was offered the position of private secretary to General George Washington. But, he contracted smallpox and was unable to accept Washington's offer. After teaching school in Worcester, Massachusetts, for a year, during which time he also studied theology, he was appointed librarian at Harvard. In 1793, he accepted the call from the First Unitarian Church at Dorchester, Massachusetts, where he remained as pastor until three years before his death, on April 3, 1842.

Harris's elegy, *The Triumphs of Superstition*, is not a celebration; and the role played by Josiah Grant Hewitt/Huit (or, Hunt, as Harris transcribed his name from the newspaper) is less than heroic:

> Reason and virtue are alike serene;
> Undaunted midst the roaring thunder's smile,
> Nor know one tremor in the midnight scene,
> Whose shades impress grim horror on the vile.
>
> But see, far glittering through the dusky night,
> Yon blaze, whose ruddy streaks illume the sky,
> Shed on the devious path its wanton light,
> And dart its beams upon the indignant eye!
>
> Behold Religion from the spot retire,
> Her mild cheek mantled with a crimson glow,
> While Superstition guards the lighted fire,
> Waves high the torch, and triumphs in her woe.
>
> And there the infatuated father stands—
> Forgot each gentle feeling of his soul—
> Virtue no more his harden'd breast commands,
> Nor pity melts, nor can e'en love control.
>
> To the lone spot by Superstition led,
> (Fondly unkind, and indiscreetly brave)
> Where rest in sacred sleep the silent dead,
> To rob his lovely daughter's peaceful grave.
>
> Ere her chill'd brow had lost its pallid hue,

Ere faded on her cheek the dying smile—

   Torn from the shroud, her form expos'd to view,

Rudely he plac'd upon the kindled pile.

In a footnote to these lines, Harris transcribed the newspaper text and allowed himself to think about, yet not be convinced by, the back story. Engaged and enraged by the act itself, he faulted the father for being seduced by a superstition that could never consummate its promise: "The design was to restore to health two remaining children —Apparently pardonable, but at the same time irrational and cruel. For *no end is attained but by the proper means*, and what connexion is there, can there be, between the means here used and the intended effect?"

Included in the front matter of Harris's published elegy was a summary, penned by an anonymous author who knew more about the exhumation tradition than is contained in either the poem itself or the footnoted newspaper text: "This elegy was an expression of its author's abhorrence of a superstitious practice in which the bodies of those who had died of hectic fever were dug up from their burial sites and removed from their coffins. The bodies were then burned, and the resulting ashes were administered as a remedy to others of the same family also suffering from a hectic fever. This was believed to be a means of curing the illness." Even as early as 1790, this "superstitious practice" was circulating in oral tradition or print media.

When Harris delivered his elegy, it was too soon after the ritual to know if it had, indeed, failed. Two-and-a-half centuries later, we remain in the dark. Nor do we yet know the identity of the daughter whose corpse was exhumed and reduced to ashes. The Hewitt family had deep roots in southeastern Connecticut. Stonington borders Westerly, Rhode Island, which witnessed at least one therapeutic exhumation. And there were two in the nearby Connecticut town of Griswold. Hewitt may have carried knowledge of the exhumation ritual to his new home in New York. But it is equally plausible that it was already there.

***

The southern portion of the Canadian Province of Quebec, known as the Eastern Townships, borders New England and was, in 1791, "an almost unbroken wilderness." To encourage its settlement, the Crown offered land grants to those willing to undertake the arduous task of turning forests and swamps into farmsteads and towns. C. M. Day, in her history of the Eastern Townships (1869), wrote, "a majority of those who came to this section at that time, were from the New England States. As a general thing, they were men of large families and limited means, and were no doubt induced to take this step by the feasibility with which land could be acquired. These New England men were mostly of that energetic, self-reliant, and independent temperament, that gave assurance of success in any enterprise they might undertake." While Day declined to conclude what prompted these New Englanders to face danger and uncertainty, she speculated that "a spirit of ardent enterprise, restless ambition, or hope of gain" may have motivated them.[160]

The backwoodsman, Jethro Smith, was one of those pioneering adventurers who crossed the border into Canada, settling in the newly formed community of Stanbridge East. His plight was recorded nearly one-hundred years later in a history column of the *Waterloo Advertiser* (July 28, 1899):

Lot 5 was settled in 1798 by Jethro Smith of Waterbury, Vt. Smith was known among the early settlers as an experienced hunter and trapper whose spring guns and steel traps skilfully [sic] set at night wrought havoc among the bears and wolves which at that time infested the country. Smith is said to have been a superstitious person who firmly believed in the powers of witchcraft and necromancy and that the souls of men after death possessed unlimited influence over the living, a strange sort of spiritualism not easily defined. The following story which I have from our aged friend E[.] J[.] Briggs of this place, may serve to illustrate to a certain extent the blighting influence of witchcraft which swept over the New England states two hundred and fifty years ago at which time to have disputed the power of disembodied spirits would have brought ruin to the unbeliever and unwary, a belief that still existed at the period of which I am writing. Three of Smith's sons had died lately of consumption. The fourth and last son was stricken with the disease and was living at the point of death. While watching at the bedside of his afflicted son Smith claimed that he had been accosted by a spirit from the unknown world who urged him to exhume the remains of his third son who had lately died and burn that portion of the body from whence the fell disease originated, by which means the wrath of the devil's dark angel would be appeased and the health of his dying son restored. Smith called on George Briggs, who at that time was living near by, and entreated him to be present at the burning. Briggs, fully aware that his neighbor was laboring under a terrible hallucination, strove hard to put him aright, but nothing could be done. That evening at early twilight a log heap standing not far from Smith's residence was discovered to be on fire. Briggs hurried away to the place where a harrowing scene presented itself. The afflicted man bowed down with sorrow and long watching, was walking slowly around the burning heap calling aloud to the angel of death to make good his promise and restore to health again his dying son. On the following morning Smith's fourth, and last son died, leaving the disappointed and hapless man a physical wreck for life.

The *Waterloo Advertiser* was a weekly newspaper that began publication in 1856 and covered many of the English-speaking communities of the Eastern Townships. The Township of Stanbridge was formed in 1801.[161] The Municipality of Stanbridge East, situated in southeastern part of the township, consisted of fifty-six numbered lots of land in four ranges (*Waterloo Advertiser*, July 7, 1899). Jethro Smith's Lot 5 was one of fourteen in the Second Range.

From May to December in 1899, the Stanbridge East column in the *Waterloo Advertiser* included ten historical articles that recounted experiences of its early settlers. These sketches are a valuable resource for unfolding the extraordinary tale of Jethro Smith. Our storyteller, E. J. (Elijah Jenkins) Briggs, was the nephew of George Briggs, the near neighbor who, although declining Smith's request to attend the ritual, arrived at the scene in time to witness its dramatic climax. George Briggs (1775-1851), from Hancock, in western Massachusetts, settled at Stanbridge East in 1809 on Lot 6 in the Second Range, adjacent to Smith's property. He was soon joined by his brother, James, also from Massachusetts. George sold his property in 1811 and relocated to Lot 2 in the same range. Since George was said to be Smith's nearby neighbor, the ritual was likely performed sometime between 1809 and 1811. A son of George's brother James, E. J. Briggs (1808-1900) was ninety-one years of age when his version of the story was published in the *Waterloo Advertiser*. He probably heard about the exhumation as a firsthand account from his uncle and, perhaps, also secondhand from his father and other members of the community.

Another of Jethro Smith's neighbors, also said to be from Waterbury, Vermont, illustrates why the consumption-afflicted father turned to a last resort. Lot 3 was settled in 1802 by William Huckins, who, as the *Waterloo Advertiser* (July 28, 1899) noted with at least a hint of sarcasm, "claimed to be a noted physician. With a medicine bag containing lobelia, dragon root, cowbane and other poisonous nostrams . . . [h]e attacked disease in all its forms a few doses of which taken internally generally relieved the unfortunate patient from further ills both then and forever."

Some unusual, dramatic elements of this story suggest that it may have undergone significant changes during nearly a century of telling:

- drawing a straight line from the witchcraft of the late seventeenth century to the consumption ritual, thereby fusing these two, distinct belief systems.
- Smith's learning of the ritual directly from "a spirit of the unknown world," when it seems more plausible that he learned of it in New England.
- the ritual aimed to "appease the devil's dark angel," thus releasing Smith's son from consumption's grasp.
- Smith's circumambulating the burning pyre while pleading with the "dark angel" to make good on his promise.

Of course, it is possible that Jethro Smith *was* having a psychotic episode ("laboring under a terrible hallucination") and that, therefore, the description of his erratic behavior was essentially accurate—not an implausible scenario given the stress of carving a home out of a wilderness and, at the same time, coping with dead and dying sons. According to the story, the episode left Smith "a physical wreck for life."

Who was Jethro Smith? Finding the answer promised to be a simple task: search the records of Waterbury, Vermont for Jethro Smith, then arrange the resulting puzzle pieces to form a picture. But there were no pieces to arrange because there was no Jethro Smith recorded in Waterbury. Widening the search beyond Waterbury created a new question: Which Jethro Smith is the right one? Of the five candidates, one was born too early: 1730 (in Haddam, Connecticut) would put a seventy-year-old man pioneering in a wilderness with at least one dying son. Three other candidates were not in the right place (Stanbridge East) at the right time (1798). One, born in 1759, would have been in his late forties or early fifties, a reasonable age, when the tragic events climaxed. But census and vital records show that this Jethro Smith remained in Bristol County, Massachusetts his entire life. Another, as recorded in the federal census, was between the age of twenty-six and forty years and residing in Stowe, Vermont with (presumably) his wife, two sons and a daughter (all children being under the age of ten) in 1800. The fourth candidate was born (1780) and married (1804) in Thetford, Vermont. The federal census of 1810 showed him still living in Thetford.

I believe that our Jethro Smith, as recorded in the 1790 federal census, was living in Washington Plantation, York County, Maine as the head of a household that included three males under the age of sixteen years and two females. Presumably, he and his wife had three sons (under the age of sixteen) and a daughter. Perhaps one son had already died of consumption by 1790, or, also likely, his fourth son was born after 1790. The evidence unpacked below strengthens the case that this Jethro Smith is the best fit for the exhumation narrative.

Smith was born circa 1753 in Massachusetts Bay Colony. Records variously describe his place of birth as Massachusetts, New Hampshire, and Maine; however, both Maine and New Hampshire were part of Massachusetts Bay Colony at that time. The recording of Smith's birth does not appear until twenty-two years after the fact, when he was serving in the Revolutionary War. These records list Smith's hometown as Massabesec (also Massabesic and Massabeseck in some sources), York County, Maine, which was part of the town of Sanford. Smith's confusing service record reflects the colonies' muddled approach to raising and maintaining military forces during this time.

Smith first enlisted on October 6, 1775, probably for an eight-month term, the local standard in Sanford at that time. He served in the 10th Massachusetts Bay Provincial Regiment of the Continental Army, which was encamped at Cambridge, Massachusetts.[162] He was still in service on December 13, 1775, as the record indicates that he had put in an order for a "bounty coat" or its equivalent in money.[163] Seven months later, Smith was serving in the 2nd New Hampshire Regiment of the Continental Army. Records show that he had deserted and returned to his unit—twice! His first return was on July 8, 1776, at Chimney Point, Vermont, a strategic site on Lake Champlain. On August 10, 1776, Smith again returned from desertion, this time at Mount Independence, another strategic location on Lake Champlain.[164] A note attached to both of Smith's desertion documents states that this regiment was "organized in compliance with a resolution of Congress of Nov. 4, 1775, which provided that the men be enlisted in the Continental service to continue to Dec. 31, 1776."

Just as Smith's term of enlistment (voluntary or not) was set to expire, Thomas Paine published these memorable words: "These are the times that try men's souls. The summer soldier and the sunshine patriot will, in this crisis, shrink from the service of their country." Did Jethro Smith, whom desertion records describe as five feet, eight inches tall, with dark hair, light-colored eyes, and a "scar on ye face," decide to return to his home in Massabesec? If so, he had lots of company. "Desertion was such a problem during the war that George Washington actually expected the Continental Army to dissolve. When the Revolution began, American soldiers were enthusiastic about fighting for the cause of independence. But as the war dragged on and the hardships of campaigning mounted, once-eager volunteers began to melt away in droves. In fact, prior to the victory at Trenton in late 1776, as many as half of Washington's men had absconded."[165] It is not difficult to imagine Smith weighing his military obligation against the need to care for his budding family.

Between 1778 and 1781, Smith moved about fifteen miles northwest of Massabesec, to Washington Plantation, also in York County. The town's proprietors granted land to persons who had settled there prior to 1782, and Smith received six acres. Like Stanbridge East, where he would relocate in 1798, Washington Plantation was untamed. Its first settlers hunted and trapped the abundant wild game and profited from the hardwood timber they cleared for their homesteads.[166] It was just the sort of remote place where a man's past would not follow him.

The end of Jethro Smith's tale—with his last son dying the morning after the ritual, "leaving the disappointed and hapless man a physical wreck for life"—is, indeed, the end of his trail. There is no gravestone, obituary, or other record to tell us the fate of Jethro Smith.

***

The *Springfield Republican* covered a wide range of small towns throughout the Connecticut River Valley. Beginning in 1824 as a weekly (becoming a daily twenty years later) the *Springfield Republican* regularly published articles on local history that also incorporated folklore, folklife, and oral history. So, I was delighted but not surprised to discover several exhumation accounts in the archives of this fertile source of community narrative. One that was published in 1876, sixty-two years after the reported incident of 1814, named participants. The following story opened several doors to recovering the event, evaluating the narrative's credibility, and sketching in context and meaning:

> The superstitions of New England are as remarkable, considering its modes of enlightenment, as its noble deeds, and much of it, like witchcraft, is traced to our ignorant puttering with the Bible. The delicious pictures drawn by local "school marms" surely did not check it. Among the superstitions of Springfield is remembered the belief that consumption will find relief by the burning of the remains of a relation who died of the same disease. In 1814 the remains of a woman, named Butterfield, were dug up from the old cemetery. It was 4 in the afternoon, and, school being out, all the boys were present. It was on the north side of the Lane (Elm street) and the vitals were carried down to the river bank and burned. Galen Ames of this city was one of the school boys who saw the smoke from this altar of superstition rise from the river bank. William Ames of Dedham, who also visited the place and saw the ashes, makes the bon-mot that the "relative died.[167]

Despite promising details, but true to form, the vampire proved elusive. There is a death notice in the vital records of Springfield for Nabby, wife of Benjamin Butterfield, who died April 6, 1810, at the age of 21.[168] I also found some unsourced family trees (that is, family trees that do not cite supporting documents, such as vital records or census reports) that show a Nabby Fobes, of Springfield, marrying, first, Calvin Bliss, and second, in 1810, Benjamin Butterfield. This Nabby's date of birth is given as January 2, 1788. To date, she is the only candidate for the woman whose body was exhumed.

The star of this narrative is William Ames of Dedham, the only character to have a speaking role. Smoke rising into the air above the riverbank attracted the attention of boys at play and the scene they came upon stuck in the memory of one of them, a naturally curious boy. William, son of the prominent lawyer and legislator, Fisher Ames, was born in Dedham, Massachusetts, in 1800. For many years he made two annual trips to Springfield, timing his visits to coincide with the arrival of shad in the Connecticut River, in the spring, and the celebration of Thanksgiving, in November.[169] The *Republican* would announce his arrival (along with other seasonal residents) in a special column. Did the exhumation occur during William's spring visit or around the time of Thanksgiving? In 1876, the *Republican* noted that Ames, was "locally known as an industrious antiquarian" who had interviewed "old residents." The article mentioned that he had written "some reminiscences" some thirty-one years earlier."[170] What are the odds that Ames omitted the exhumation narrative from his *own* reminiscences? In a story of this kind—possibly damaging to people still alive, yet too compelling to be left untold—there is a collision between the chronicler's need to be discrete and his desire to share a story begging to be shared. One way out of this dilemma is to disguise the identities, but leave enough clues for insiders, who may already know the story, or for others, who may wish to follow the faint trace. Everything points to William Ames as the source of the narrative.

Four years after the Butterfield exhumation narrative was published in the *Republican*, on May 19, 1880, William Ames died in Springfield, near the time that shad were running.

The story reappeared in the same newspaper nearly sixty years later, in 1934, in one of a series of illustrated histories of Springfield, by artist Robert N. Holcomb. Of about ninety installments that ran in the Sunday edition from 1932 to 1935, only one focused on "superstitions" —and you might guess which narrative was featured. Holcomb's first illustrated panel depicts a man in colonial garb shoveling dirt from a grave that contains a coffin. The caption reads: "Just an old Springfield custom—digging up dead bodies." The second panel shows three men, also in colonial garb; one of them is pointing to smoke rising on the horizon. The caption reads: "The smoke of superstition."[171]

***

The opening two paragraphs of an exhumation narrative that appeared on the front page of the *Vermont Watchman and State Gazette* (March 1, 1831) minced no words in complying with the accepted social standards of the time, contrasting "superstition" (the brief title of the article) with Christian civilization:

> It is to be expected that when we visit the benighted corners of the earth, we shall witness the fatal effects of gross ignorance and pagan idolatry; and when we peruse the histories of heathen nations, there we shall discover the natural consequences that arise from ignorance, credulity, and moral blindness. But we are not a little surprised, when we discover the same spirit existing at the present day, in this enlightened portion of the globe, where the blessings of civilization and learning are so generally enjoyed, and where religion has diffused its renovating beams; but such is the fact. Yes, in this christian [sic] land, there are thousands who bow and do homage at the altar of superstition,—who travel back for many centuries and borrow the opinions and principles of heathen nations, and practice them while surrounded by the various and convincing proofs of the fallacy of their course, that may be drawn from learning, civilization and christianity [sic].
>
> A case occurred a few years since, illustrating the above, in the state of Massachusetts, and for its authenticity, I rely upon the statement of an individual of unexceptional character, who was personally acquainted with the affair. It is only one among the many instances that have transpired within a century or two, which go to show how the mind can be brought under the influence and control of the most ridiculous and absurd opinions, and made to submit to the lowest degree of degradation.

The author obviously was aware of other similar events in the region. Plainly, he wanted readers to accept his informant as a reliable witness, who was said to be "personally acquainted with the affair." That he was also present at the ritual is reinforced at the narrative's dramatic climax. As the father stands face-to-face with the abhorrent task he has reluctantly accepted, we get a verbatim monologue enclosed within quotation marks:

> In a town near the centre of that state there resided a Mr. P—. His family consisted of wife and four daughters. He lived with his family in the greatest social harmony, for several years, and a fair prospect was held out, of ending his days in peace. Providence, however, had ordered it otherwise. His wife was attacked with the consumption, and while pining away

under that disease, an incident occurred from which could be dated the ruin of the family. A notorious old woman in the capacity of a fortune-teller paid his wife a visit, and during the interview, related many fearful stories and predicted many direful events that were to befal [sic] the family, and among other things stated, that the disease with which Mr. P's wife was affected, would be the cause of the death of the whole family. The mother was soon to die, and the other members of the family would follow her in quick succession, by the same disease. However, in the midst of this general extinction of the family, a door of hope was opened. She stated, that if on the death of one of the members of the family, the body was opened and the vital parts taken out and burned, the contagion would leave the family. The peculiar situation of the family, and the confidence that was placed in the revelation, as they believed, of the old woman, could not and did not fail of producing considerable excitement. The mother soon died, and one of her daughters was attacked with the same disease, and in the course of a few months followed her mother to the tomb. Another daughter was attacked with apparently the same disease, and before her death she plead for the adoption of the remedy prescribed by their dooms-dame; but for some reason it was neglected and she died. A third fell a victim to the disease and soon followed her mother and sisters. Thus in the short space of a few months, the mother and three daughters were consigned to the tomb.—All this went to prove the truth of the old woman's prediction. Mr. P. during the sickness of his family, could never be made to acknowledge that he believed in the predicted fate of himself and family; but when the present situation of his family was compared with the prediction, his heart failed him, and his faith began to waver. At last the remaining daughter was seized by sickness, exhibiting all the symptoms of her predecessors. The love of life and faith in the prediction of the fortune-teller, made her very importunate for the adoption of the prescribed remedy. She prevailed.

It required much resolution to carry into execution such an undertaking; but to screen, if possible, the only remaining vestage [sic] of a once blooming, happy family, was Mr. P's determination. Accordingly he made his intentions known to several of his friends, and requested them to assist him in his undertaking. The day was set for the accomplishment of the act—it was a dull, cold and gloomy day in December.—He proceeded with his friends to the graves of his wife and children, and disinterred the last buried corpse. "Not a word" says my informant, "was spoken by any one. A death like sadness was seated on every countenance—he unscrewed the lid of the coffin, and with much composure removed the shroud that enveloped the remains of his daughter. He turned himself about, with an imploring look to his friends, and broke the deathlike silence, with 'this is more than I can bear.' No one came to his assistance and he was obliged to go through the painful operation alone. The vitals were removed and burned—the coffin closed and replaced in its bed." All looked forward with anxiety to see the result: but alas, the poor deluded mortal who had clung to life so closely, was doomed to see her hope depart and to take up with an early grave. The feelings of the father were much tried, and when he viewed the graves of his family, he could only consider them the victims of delusion.

Mr. P— lived for many years after the decease of his family, and has often been heard to lament his folly in pursuing the directions of the fortune-teller: and likewise he believes that one principal cause of their death, was the effect the prediction had upon the imaginations. The superstitious belief that a feeble, finite worm of the dust was possessed of the secret of their lives, hastened them to their graves.

A fortuneteller, in the form of a "notorious old woman," appears to prescribe a cure for the ailment that she accurately predicted would descend upon the family, opening "a door of hope." The phrase, "the contagion would leave the family" (*if* the prescribed ritual was performed) confirms that the notion of consumption's contagious nature was circulating among the general population even as the medical establishment was hotly debating contagion versus heredity to explain the disease's etiology. Jack Larkin, longtime Old Sturbridge Village historian, wrote, "Contagion was almost a folk belief, widely held by ordinary people but frequently dismissed as superstitious."[172]

Two motifs in this ritual have established pedigrees in folklore. In some folk rituals, silence is an essential condition.[173] The requirement that a certain task be accomplished without assistance from other people is also well-known in a variety of traditional settings, from folktales to rituals. In many of these cases, however, supernatural assistance may be implicit. And, of course, as we have seen often, turning to the implied supernatural power of the consumption ritual is the last resort of a person urgently hoping to break free of a grim forecast. When the author suggested that the aggrieved father blamed his family's deaths on "the effect the prediction had upon the imaginations," he summons a concept that may have been known but was not labeled "psychosomatic" until about four years after this narrative was published.

Finally, the phrase that heads the article, "for the Watchman and State Gazette," hints that the author, whom we know only as "H," wrote the piece solely for this newspaper. His identity remains a mystery, as does the town near the central part of Massachusetts, where the event occurred.

***

Newspapers cannibalize themselves and each other, and then, down the relentless road of daily offerings, return to the feast as if they never tasted flesh. Occasionally, a lotus-eating journalist lapses into recall. In 1832, newspaper publisher and editor, Seba Smith, wrote about an exhumation he had witnessed at some unspecified time, in a place unnamed, probably Maine (chapter 4, above). He prefaced his description of the event with a general statement of the fundamental theory and practice of "one of the most popular superstitions of New England." If some portion of a consumptive's corpse fails to decay, it is presumed to remain alive by feeding on its relatives until "the whole family one after another sink under the same disease and drop into the tomb." But "if the decaying body is taken up and the living part discovered and consumed by fire, the sick member of the family will recover." The narrative appended to Smith's account, while only mildly cannibalistic, is gruesome in its detail:

> In one of our neighboring towns there is or was a family, the most of whom have died of consumption. -- One, and we believe only one of the family remained, and he was apparently fast following his relatives to that "bourne, whence no traveller e'er returns." Credulous enough to confide in the belief stated above, and anxious probably to do every thing possible to revive his wearing out nature, and thinking that if the living part of the

dead bodies of his friends could be consumed he should recover, he formed and executed the following plan with the assistance of his neighbors, to accomplish his object.

"Armed at all points," that is to say, with all the paraphernalia necessary to the exhumation, dissection, &c. &c. of the dead bodies, his friends and neighbors consisting of a heterogeneous compound of men, women and young ones, to the number of twenty or more, some three or four weeks ago, went to the grave yard - dug up the dead bodies, some of which had been sleeping there some eight or ten years - disrobed them of their cerements, and with a jack knife sharpened for the purpose - at the side of a public road, *cut them up* to find some part in them where the work of decay had not commenced. Having succeeded, after having sliced, upon something less than a half dozen dead bodies, in finding such a part or parts as they looked for, a fire was kindled on the spot, and the whole consumed, the sick man *standing in the smoke*.

Now, it is not improbable that the man may recover his health; but that the burning of any part of the dead bodies, or his standing in the smoke, or that any part of their savage powwow [sic] will have that effect is to us very improbable. FAITH will be a greater efficient in the good work, than all their inhuman ceremonies.[174]

We are informed that the event took place three or four weeks prior to its publication, which would have been in late October 1832. While the narrator avoids addressing who and where, he is not shy about supplying, in vivid detail, the grisly catalogue of how. Standing in (or inhaling) the smoke of a burning vampire (or its life-sustaining body parts) is an established variant of the practice; not ordinary is the exhumation of half a dozen family corpses. Perhaps most extraordinary of all is the large contingent of assistants, including men, women, *and* "young ones." Of course, sensational details ("a jack knife sharpened for the purpose") have always lured consumers of the popular press.

This narrative first appeared in the *Norway* [ME] *Politician*, a weekly newspaper published only in 1832 and 1833. Prior to becoming the *Politician*, it was the weekly *Oxford* [ME] *Observer*, published from 1824 to 1832. Both towns are in Oxford County. Most of the newspapers in which this text appeared had appended it to Seba Smith's eyewitness account. For the sake of context (not to mention added enticement) publishers who reprinted an exhumation narrative from another newspaper might pull a previously published exhumation narrative from their own files, or, less often, run one gleaned from a hitherto unpublished source.

***

As exhumation events get published, and re-published, the chances are that a similar incident—buried in the newspaper's morgue or languishing in some other all-but-forgotten source of historical exotica—will be rescued. An exhumation story from Middletown, Vermont was one of those revivals. The text was appended to an account of a consumption ritual that was performed in Chazy, New York, in 1818 (see chapter 12, below) and was reported in several newspapers the following year. The Middletown narrative, however, did not appear until the Chazy case was reprised, with several updates, in 1875. The *Rutland* [VT] *Daily Globe* (which credited the *Plattsburg Sentinel* for their text of the Chazy incident) attached the following text:

A similar instance took place in Middletown, in this county, in 1839. A lady was rapidly failing with the consumption, and some of her friends heard of the theory of taking the ashes of the lungs of a person who had died of the same disease, as a cure. A few weeks before, a relative had died of the disease, the body was exhumed and the lungs taken out and burned, the sick woman ate the ashes, but she died within the next ten days. In this case only, lungs were cremated, and the body re-interred.[175]

Lacking surnames and similar situating information, this incident is a dead end. Lending an air of authenticity to the narrative is its traditional pattern: friends learn of the consumption cure through oral tradition; cannibalism is explicitly present; ultimately, the ritual failed to revive the health of the ailing individual.

6   The Above Are Facts ... Sickening and Horrible

Tucked into the southeastern corner of Cayuga County, at the southern tip of Skaneateles Lake—central New York's most pristine Finger Lake—Kelloggsville is not named on most maps. But it is conspicuous in the annals of New England consumption rituals, for an exhumation there generated two different but complementary accounts of the incident. The first, apparently provided by an eyewitness, was published only days after the event. The second, narrated by a physician who was at the scene, appeared eighteen years later.

The account first appeared in the *Auburn* [NY] *Advertiser* in August 1851 but was quickly picked up by other newspapers. The following version was published in the *Syracuse Star* (August 30, 1851) under the title "Inhuman Conduct":

> We have been put in possession of the following facts, for the truthfulness of which we have the most undoubted authority. We are surprised, nay astounded, to learn that there are persons in Cayuga county [sic]—a county celebrated for the general intelligence and refinement of her people—who are so superstitious as to believe in the efficacy of such monstrous treatment for the cure of consumption, or who are so uncivilized in their feelings as the account proves them to be.
>
> It is stated that almost two years since a young man died in a neighboring town of consumption, and was buried. Other members of the family are afflicted with the same disease—a brother is not expected to live but a short time.
>
> On Sunday last, some of the brothers and other friends of the deceased proceeded to the grave with shovel and pick—dug up the body—opened the coffin—cut the shroud—and then a *surgeon* (what a surgeon!) was found to cut from that mass of corruption, the heart and lungs, which were in a tolerable state of preservation, considering the length of time since the death of the person. The heart and lungs were laid upon the grass—impregnating the atmosphere with their horrid odor—and wrapped in the pocket handkerchief of one of the brothers, and carried home and burned, while the members of the family inhaling their incense, and afterwards *ate the ashes!*
>
> The above are facts, but they are so sickening and horrible that we felt inclined at first to keep them from the public eye, but a second thought led us to hope that by giving them publicity, steps would be taken to civilize and refine the feelings of those who were participants in the transaction. —*Auburn Adv.*

We do not have to guess what stimulated Dr. William H. Cook (1832-1899) to relate his own version of this ritual nearly twenty years later. He introduced it with a story that appeared in several newspapers in 1869, only weeks prior to Cook publishing his account in the *Physio-Medical Recorder*. After quoting the 1869 narrative as it appeared in the *Great Barrington Courier* (which we will discuss in Chapter 7), Cook provided the following:

A similar case came under our own observation in Cayuga County, New York, in the summer of 1851. During the spring of that year, we several times visited a consumptive patient, in company with a friend practicing in the village of Kelloggsville. The malady was hereditary in the family of the patient, and the patient himself, a young man of about 23 years, was fast passing through the latter stages thereof. Hearing of the superstition mentioned in the above extract, he became possessed with the idea of thus using the heart and liver of a brother who had died of phthisis about two years before. No persuasion could change this desire nor argument shake the faith he had been led to place in the virtues of this treatment. The disinterment of the brother's remains was resolved upon by the family; and, to gratify the impassioned wishes of a man evidently but a short remove from the grave, our professional friend consented to take the liver and heart from the exhumed cadaver. The friends chose a Sabbath afternoon for opening the tomb, and we were present to sustain our brother physician when he made public announcement, to the vast crowd present, that he had merely consented to please a fellow sufferer who must soon stand in the presence of his Maker, but that the thing itself was but a dark and unwarrantable superstition. He and ourself had seriously questioned the professional propriety of taking any part in the proceeding; but had concluded that such a piece of ignorance could best be ended by our publicly announcing its folly, while the assembled crowd would soon be made aware of the utter failure of so barbarous a "cure."

 The remains were found in an unusual state of preservation, every outward fibre seeming to be as perfect as on the day of burial, and the abdominal viscera being but partially decayed. The heart and liver, almost unaffected, were removed and passed over to the friends, who did with them as seemed to them good. The young man died in the following month; and the public felt so horrified at the procedure, and so satisfied at its failure, that the hold of that superstition upon the mind of that community was effectually destroyed.[176]

Cook's report, entitled "A Superstitious Cure for Consumption," exemplifies how one's own personal story can be recalled, even years later, because it resonates with a similar story.

 One difference between these two narratives leaps off the page: the newspaper account includes, in graphic detail, how the extracted vital organs were treated. From their "horrid odor" and wrapping in a handkerchief to be transported home and burned, to inhaling the smoke and ingesting the ashes, the reader is exposed to each gruesome step in the ritual process. Cook skims past these distasteful particulars, writing simply that the friends "did with them [the heart and liver] as seemed to them good." While the newspaper article provides no sense of how many people were present at the grave site, Cook labeled the gathering a "vast crowd." Cook asserted that the "remains were found in an unusual state of preservation," while the anonymous informant said their condition was "tolerable." Both Cook and the newspaper invoked functions of censure and education, Cook to explain why he participated in the ritual and the newspaper to justify publishing the account.

 Reading the newspaper article without knowing what motivated Cook's attendance might lead one to question why a reputable physician would take part in the ritual. Indeed, at least four newspaper iterations inserted just such a reaction: ". . . then a *surgeon* (what a surgeon!) was found to cut from that mass of corruption . . .." Who was this surgeon? According to the 1850 federal census, three physicians were residing in Kelloggsville. William F. Cooper (1802-1881), touted in a county history as the "celebrated physician" of Kelloggsville, practiced there from 1827 until his death in 1881. On August 24, 1851 (the Sunday that the ritual likely was performed), Dr. Cooper had just celebrated his forty-seventh birthday and had been practicing medicine in the hamlet for about twenty-four years.[177] David A. Rupp (b. 1821) was about thirty years of age at the time of the ritual. Federal and New York census reports show that by 1860 he had relocated to the town of Skaneateles, on the north tip of the lake, where he remained until his death, sometime between the Federal census in 1870 and that of 1880. Horatio H. Marsh (b. 1820) practiced in Kelloggsville from 1848 to about 1850, so it seems likely he would not have been present for the exhumation in the summer of 1851.[178] With Marsh out of the running, that leaves either Cooper or Rupp as the attending surgeon. I give the nod to Rupp, since he was much closer in age to Cook, who was just nineteen years of age at the time. How would Cook and Cooper have become friends, with Cooper residing in the small hamlet of Kelloggsville, having settled there when Cook was a six-year-old living in New York City? Rupp, born in Pennsylvania, was living in New York by 1850. Perhaps, like Cook, his medical education was in the state of New York.

All but lost in Cook's narrative is the belief, shared by both the American and British medical establishments at the time, that consumption was hereditary. When we take up the medical context (part VI), we will see that this notion was being rejected on the European Continent by physicians who argued that contagion was the agent responsible for consumption's proliferation. John S. Haller, Jr., in his *Kindly Medicine: Physio-Medicalism in America, 1836-1911* (1997), wrote that Cook believed "germs had nothing to do with the cause of disease and existed only because disease had damaged the tissues and the germs 'find a place to exist, like maggots in bad meat.'" To Cook, the germ theory was merely "wild speculation" that would not long stand.[179]

Although William Henry Cook was not an establishment physician, he was medically, and self-consciously, Yankee and anti-European. Born and educated in New York City, the precocious Cook married a woman seven years his senior a month after his eighteenth birthday; by age twenty, he was teaching chemistry at Syracuse Medical College. His wife, Catherine Ann Masseker, "obtained her early medical training in her father's botanic drugstore" and later "became a leader in the temperance movement."[180] The "we" who several times visited Kelloggsville during the summer of 1851 probably were Cook and his wife. They were pioneering American medical herbalists, employing native plants as a pharmacopeia to make their own prescriptions, an approach known as botanic or Thompsonian medicine. Haller described Cook as "a purist in reform medicine and a convert to physio-medicalism from eclecticism."[181] Physio-medicalism advanced the belief that the human body has an inherent "vital force" that can be used to heal, a notion that was also the mainstay of the Paracelsian medical tradition. In the vampire theory, of course, this "vital force" was a malicious intruder that needed to be destroyed for healing to occur.

Newspaper narratives of the Kelloggsville exhumation demonstrate the ease and relative rapidity with which widely dispersed newspapers were able to access the story. The narrative first appeared in the *Auburn Advertiser*, during the last week of August 1851, but was picked up and run, either in its entirety, or in an edited form, in a score of other newspapers, from New York to California. How was that possible? By 1850, there were 12,000 miles of telegraph line in the United States and, tapping into this unprecedented network of communication, the Associated Press was on its way to becoming the country's most powerful news service.[182] Newspaper editors thus had the technology to review many recently published articles and simply reprint them or add their own, perhaps locally inspired, remarks. The *Cincinnati Enquirer*, for example, appended the following commentary to the Cayuga exhumation narrative: "Is this an instance of superstition or fanaticism, if you please, which deserves such single and special distinction? Wherein does it partake of the marvellous [sic] above Millerism, Mormonism, Spiritual Rappingism, and a dozen other isms that have been crazing the weak brains of some phases of mortality for years back?" When the *Madison Daily Tribune* (September 9, 1851) ran the Cayuga article (citing the *Buffalo Republic* as its source), it also appended the commentary from the *Cincinnati Enquirer*. The reference to "isms" was germane to mid-nineteenth century readers. Both Mormonism and Millerism emerged from Upstate New York's "Burned Over District" (so-called because of the many cults that came and went during the nineteenth century). In 1851, the Mormon church was still a long way from becoming accepted as an established religion, as members had recently settled in Utah after a brief and disastrous stop in Illinois. Millerism was on a decline following the failure of William Miller's prediction—calculated from a passage in the book of Daniel—that the Second Coming of Christ would occur on October 22, 1844. The Fox sisters, also from Upstate New York, entered the "ism" scene in 1848 with their mysterious rappings—which they claimed emanated from spirits dwelling within their house. These initiated a decades-long infatuation with spiritualism in this country, fueled by a steady flow of newspaper articles.

While the appended newspaper commentary makes its point that the consumption ritual, like the "isms" mentioned, was a breach of establishment beliefs and practices, the casual lumping of the two easily dismisses the rather long and widespread acceptance of the exhumation tradition in New England—a tradition, moreover, that had been rooted for centuries in the Old World. Given the details provided by at least one, and likely two, eyewitnesses, the authenticity of the Kelloggsville narrative seems undeniable.

***

Several newspapers that published the Cayuga County (Kelloggsville) exhumation narrative also appended an account from Litchfield County, Connecticut. Most of those newspapers credited the *Albany State Register* for the following narrative:

> Strange as such a revolting superstition may appear, it is not new. A similar case occurred in Litchfield county [sic], Conn., several years ago. — A large family of grown up sons and daughters, seemed all destined to become victims of that fell destroyer, consumption. The disease commenced with the oldest. A few months after his decease, the next in age was taken sick, and in the course of the year died. The next year another; and so on for several successive years, the victims being taken in the order of their ages from oldest, towards the youngest. The living became haunted with the horrid superstition that they were preyed or fed upon by the dead, till finally when another fell sick, the last buried, was disinterred, and his heart, liver and lungs taken out and burned. And although he had been dead several months, it was stoutly affirmed by some of the witnesses of the spectacle, that fresh blood was found in the heart and lungs. The consumption which had commenced in the living subject, however, was not arrested, but went on to its termination. Some of the younger members of the family escaped, and are still living. —*Albany State Register.*[183]

The source of the *Register* account is unknown. Its anonymous author stated that the superstition "is not new," reinforcing a conclusion that, despite frequent exclamations to the contrary, journalists with several years' experience and a good memory knew that they were dealing with a rather long-lived, perhaps even widespread, tradition.

***

In early 1853, the *Livingston* [NY] *Republican* (February 24, 1853) and several other newspapers published an account of a consumption ritual entitled "Heathenish Superstition." Lacking detail, the story still manages to include the commonplace self-righteous condemnation. It is not clear if the vital organs were removed before or after interment:

> A case occurred in this town a short time since that would more properly comport with the customs of the dark ages or heathenism, than those of a land of Bibles, schools, and the civilization of the 19th century. A girl of 15 or 16 years old died of Consumption, and the family under the belief that it would prevent other members of it from dying from the same disease, as several had previously died, had the heart and liver taken from the body and burned. We regret that any man of sufficient intelligence to aspire to the dignity of a physician should in any manner allow himself to be a party to such a transaction—*Mt. Morris Union of Wednesday.*

The *Livingston Union* was a weekly newspaper, published in the town of Mount Morris (located in Livingston County) from 1848 to 1862. The county is in the Finger Lakes region of western New York, south of Rochester and east of Buffalo.

The federal census of 1850 included nearly one hundred females in the town who would have been fifteen or sixteen years of age in late 1852 or early 1853. But a check of a Livingston County cemetery database disclosed no burials for any of the young women during the relevant time frame. A search of local histories and newspaper obituaries also failed to deliver relevant information. An early history of New York's townships, however, did mention that the village of Mount Morris was founded in 1804, "mostly by families from Connecticut."[184] Was the consumption ritual brought to western New York by them? We are informed that a physician participated directly in the excision of the vital organs, but we are left with more questions than answers.

***

Appended exhumation narratives were offered as graphic examples that the superstition presented in the primary narrative was not a singularity. More infrequent is a response that offers an opposing account of the same exhumation. The two reports that follow illustrate the value of having more than one description of the same event.

On May 24, 1854, the *Norwich* [CT] *Weekly Courier* published the following story under the heading, "Strange Superstition":

> A strange and almost incredible tale of superstition has been related to us of a scene recently enacted at Jewett City. It seems that about eight years ago, a citizen of Griswold, named HORACE RAY, died of consumption. Since that time, two of his children--both of them sons, we believe, and grown to man's estate--have sickened and died of the same disease, the last one dying some two years since. Not long ago, the same fatal disease seized upon

another son, whereupon it was determined to exhume the bodies of the two brothers already dead, and *burn them*. And for what reason do our readers imagine? *Because the dead were supposed to feed upon the living*, and that so long as the dead body in the grave remained in a state of decomposition, either wholly or in part, the surviving members of the family must continue to furnish the sustenance on which the dead body fed. Acting under the influence of this strange, and to us hitherto unheard of, superstition, the family and friends of the deceased, accompanied by various others, proceeded to the burial ground at Jewett City, on the 8th inst., dug up the bodies of the deceased brothers, and burned them on the spot. The scene, as described to us, must have been revolting in the extreme; and the idea that it could have grown out of a belief such as we have referred to, tasks human credulity. We seem to have been transported back to the darkest age of unreasoning ignorance and blind superstition, instead of living in the 19th century, and in a State calling itself enlightened and Christian.

About two weeks later, on June 7, the *Worcester* [MA] *Aegis* published another version of this story submitted by one of its correspondents, introducing it with the following: "Under this cover, the Norwich courier of last week treats all lovers of the horrible to such a dish of their favorite dainty as should satiate the most craving appetite. We give the Courier's statement, and below append a confirmation of its almost incredible items, from a correspondent in the town mentioned." After inserting the *Courier*'s text (exactly as above), the article concludes with the correspondent's narrative:

> There is a son of the late Henry B. Ray, who is very low; his disease is consumption; he has lost a father and two brothers within a few years of the same disease. Some of his ignorant and superstitious neighbors have made him believe that there was a part of one of those bodies living and preying upon him, and that he would not only fall a victim, but the said body, or the living part, would continue to grow and flourish, until the whole family were swept from the earth. They therefore suggested the propriety of taking up the bodies and examining them, and should there be any thing found connected with them not in a state of decomposition, to take such part or parts and burn them; stating to him that things of this kind were of frequent occurrence, and that he would regain his health, and together with the rest of the family would remain exempt from this disease. In this way they obtained consent and proceeded to exhume the bodies of his father and brothers. The father and one brother had gone to decay. On opening the heart of the other brother, they claimed there was fresh blood, and forthwith built a large fire in the grave, put the coffin with its mutilated remains into the same, and piled on the fagots until the body that once contained a soul, lay in ashes before them. It is looked upon here as a barbarous proceeding; but public opinion gives to Mr. Ray's advisers the largest share of the odium attached to it.

The text does not disclose if the writer was an actual witness—but he resided in Jewett City, site of the exhumation. It is apparent that he was familiar with both the family and the events surrounding the exhumations.

Other elements in his version lend credibility. Where the *Courier* writer mistakenly identified the father as Horace Ray, the *Aegis* correspondent correctly named the father as Henry B. Ray (who died in 1849). In another discrepancy, the *Courier* article stated that the corpses of both sons were exhumed and burned. But the *Aegis* correspondent wrote that, while the bodies of the father and both sons—Lemuel (who died in 1846) and Elisha (who died in 1851)—were exhumed, only the corpse of one son was burned; which one is not disclosed, but the typical pattern in the New England ritual would have stipulated that the corpse of the last to die, should be burned.

The power of community sentiment, seen in other exhumation accounts, is evident in the *Aegis* narrative. Instead of faulting the remaining members of the Ray family for initiating the exhumations, the correspondent asserts that "public opinion" blames those who advised the family to undertake the ritual. For the ailing son, Henry Nelson Ray, the results were disappointing, for he and his wife and children died not long afterward.

***

On April 13, 1859, the *Glen's Falls Republican* featured the following text, under the heading, *Barbarous Superstition—A Dead Body Disinterred and the Heart taken out to cure a Man of Consumption!!*, on its front page:

> A libel and disgrace upon the intelligence of this community was perpetrated last week at the small settlement of Goodspeedville, some mile or more from this village. It appears that a man by the name of ADAMS, living in Goodspeedville, died and was buried seventeen months ago—leaving a wife to mourn his loss. The widow removed to the West and remained there until a short time since, when she returned to this town. Upon her return she found her deceased husband's brother dying with consumption, and declared that he could be cured only in the following manner, which she said was practiced where she had been living: The body of her husband should be taken up, the heart dissected, and if any blood was found in the heart, it should be *burned*, and the sick man would recover! *This monstrous proposition was immediately acted upon*—the dead body disinterred—a physician (?) called, who took out the heart and lungs, but not enough blood being found to answer the purpose—the body having lain too long—the further prosecution of the infamous project was abandoned.
>
> And this occurred in a country that boasts of its superior attainments in religion and morality—of its free churches, free schools and untrammeled press—under the blazing light of the "progressive" nineteenth century! Comment is unnecessary. The poor-deluded fanatics who were participants in this unheard of outrage upon the sanctity of the grave, are to be pitied more than condemned, and immediate steps should be taken to place them out of harm's way in an Asylum.

The town of Goodspeedville was founded by Stephen Goodspeed (1810-1905) in 1843. Like the widow Adams, Goodspeed went west; unlike her, however, he did not return, and the town subsequently changed its name to West Glens Falls.[185]

To identify the Adams family mentioned in this narrative, I began by searching online cemetery databases. To have a surname, place of residence, narrow time frame, and likely cemetery to investigate was unusually auspicious. When I locate the gravestones of credible candidates, as I did in this case, I visit the town and cemetery to take photographs, get a sense of place, and converse with local inhabitants. Of the more than five-hundred interments in the West Glens Falls cemetery, located in Queensbury, there are only eight with the surname Adams: one died in 2001, but the other seven seem to be related and lived during the time of the exhumation.

The dangerous Adams corpse was exhumed about seventeen months after death, which would have been November 1856. One of the Adams interred in the cemetery, Salmacious, died on October 26, 1856, at the age of twenty-five—a nearly perfect match. His brother, Marcellus, died on June 26, 1858, about two months after the exhumation. A likely scenario is that Marcellus was the dying brother for whom the aborted ritual was to be performed. The family connections calculated based on gravestone inscriptions are verified by census records. In the 1850 federal census, Zophar, a fifty-three-year-old shoemaker, was living with his forty-eight-year-old wife, Sally, and their three children: Salmatius (incorrectly transcribed from Salmacious), a nineteen-year-old sawyer; Noratius, sixteen year of age and, like his father, a shoemaker; and Marcellius, thirteen years of age. The cemetery inscriptions indicate that Sally died in 1851, and Zophar, a veteran of the War of 1812, survived until 1881. Since Noratius is not buried in this cemetery, he may have survived the family consumption epidemic long enough to relocate. One interred member of the Adams family, Medory, died as an infant in 1856.

Two other aspects of this narrative are noteworthy. There was a physician at the scene. And, while we do not know the identity of Salmacious Adams' wife, she seems to have acted as the belief specialist in this case, having learned of the ritual wherever in the "West" she had briefly relocated. If it seems odd that she would encounter the ritual beyond New England, we should remember that people living in New England had been moving westward for decades in search of better farmland and other opportunities. They did not leave their traditions behind. Additionally, newspapers had long been circulating vampire narratives from New England.

***

An appended newspaper narrative that appeared in the *Lowell Daily Citizen and News* (May 17, 1858) is a solitary enigma. Of the handful of newspapers that published the Adams story from West Glens Falls, only the Lowell, Massachusetts newspaper included the following account, which was tacked on with no contextual information:

> In Coventry R. I., is a cemetery in which we have had pointed out to us a long row of graves belonging to a single family, and a clump of pine bushes on a neighboring knoll. Some 20 years ago, the father and two other children died of consumption and two of the survivors were apparently soon to follow. The doctrine was then broached and generally believed, that if the most recent body should be exhumed the heart would be found in perfect preservation, and that if the heart, with its contents, should be burnt to ashes, the disease would be stayed. Accordingly, late one Saturday night, a party of the friends of the family, took up the body and removing it to the group of pines abovementioned, extracted the heart, which was found as expected, quite undecayed and filled with blood. The organ was burnt and the ashes carefully replaced with the rest of the body in the grave, and at sunrise the repulsive task was completed. Of course the spell did not succeed; the sick soon died, and one by one the survivors followed them, until at the time of our visit, the family was extinct, sixteen white tombstones in a row, marking their resting-place. The superstition alluded to by our contemporary, therefore, is not an unheard of thing, but it is or *has been*, quite prevalent in some of the less enlightened sections of even our own New England. At the present day, however, Coventry can with no degree of truth be placed in this latter category whatever it was long years ago.

The narrator obviously was aware that the ritual was not unusual, writing that it had been "quite prevalent in some of the less enlightened sections of even our own New England." The stipulation that the ceremony had to be enacted before sunrise is a folkloric motif that, while not absent in the New England tradition, is not typical. Even more unusual is that the burned heart was placed back into the grave. A close variant is when Justice Forward, of Belchertown, Massachusetts, had the burned organs of his daughter placed into a box and reburied above her coffin (see chapter 2).

Identifying this narrative's family was daunting; there are nearly two hundred cemeteries in the town of Coventry, located in western Rhode Island, and the number of interments in each range from one to more than three thousand. Nearly three hundred surnames in these cemeteries have ten or more interments represented. One would have to be inordinately obsessed to wade into this find-a-grave quagmire. Even then, the probability of success would be close to zero. So, this incident remains suspended in uncertainty.

***

In an odd reversal of fortune, the man who wrote, "I thrive best on solitude," eventually succumbed to a communicable disease. Perhaps less unforeseen if you lean toward consumption as more hereditary than contagious. Henry David Thoreau (1817-1862) had ample reason to lean that way: his grandfather, John, died of consumption, as did half of John's eight children—three of them before the age of thirty. Thoreau's older brother (also John) also contracted consumption just as he and Henry were beginning to thrive with their recently established school in Concord, Massachusetts. But before consumption claimed him, John died of "lockjaw" (tetanus) following a "slight accident."[186] In early May 1861, on the advice of a physician, Henry traveled to Minnesota.[187] But any hope for a recovery was short-lived. After two months away, Henry returned to Concord, "little benefited by his long journey. … The following January he became confined to his room; he grew steadily worse, and died on May 6."[188] After more than fifteen trying years, Henry David Thoreau acquiesced to consumption at the age of forty-four.[189]

Three years prior to his premature death, he must have been keeping a watchful eye for possible cures, for on September 26, 1859, Thoreau made the following entry in his journal: "I have just read of a family in Vermont who, several of the members having died of consumption, just burned the lungs, heart and liver of the last deceased, in order to prevent any more from having it."[190] There can be little doubt where Thoreau obtained his information about this event. Ten days before Thoreau's journal entry, the following narrative, under the heading, "A Strange Proceeding," was published in the *Bellows Falls* [VT] *Times* (September 16, 1859):

> Mrs. Prescott Lawrence, of Winhall, died on Wednesday of last week, of consumption, and as a member of the family had previously died of the same disease, the family went through the superstitious farce of burning the lungs, heart and liver of the deceased *to prevent any more from dying of the same disease*! As strange as it may appear we are assured from perfectly reliable authority that this is a fact, and that it can be substantiated by any number of witnesses.

The timing is right, and the texts are nearly identical, so it seems virtually unassailable that Thoreau found his narrative in one of the fifty newspapers that published this article, with minor variations, following its first appearance in the *Bellows Falls Times*.

Mrs. Prescott Lawrence was born Clarissa Huntting (sometimes spelled Clarisa Hunting; spelling and transcription variations are common in older records) on February 19, 1831, in Londonderry, Windham County, Vermont. She was the first of seven children of Newell Huntting (1802-1897) and Lucinda Carkin Huntting (1809-1861). When Newell died at age ninety-five, on November 28, 1897, he was the oldest resident of Londonderry, where he had been living with his son, Jonas. His lengthy obituary in the *Londonderry Sifter* (December 3, 1897) called him "a worthy town man of good moral and Christian character." Born in Marlboro, New Hampshire, on August 21, 1802, Newell was four years of age when his father, Deacon Jesse Huntting, moved the family to Londonderry. Newell's young life followed a path common to many of the region's early settlers, who carved a living out of a seeming wilderness: they encountered wild animals, felled trees for pasture and agriculture, and built rude log houses. At age twenty-five, Newell relocated to Boston, where he met and married Lucinda Carkin, a native of Dracut, Massachusetts. The newlyweds returned to Londonderry, purchased a forest tract, built a farm, and began raising a family. In 1847, the Hunttings moved a few miles west to the adjoining community of Winhall, in Bennington County. Their children, in order of birth, were:

1. Clarissa M. (February 19, 1831)
2. Adeline L. (November 26, 1833)
3. Hannah M. (December 5, 1835)
4. Sarah J. (October 21, 1840)
5. Jonas N. (January 21, 1842)
6. Mary Agnes (April 4, 1846)
7. Viola E. (November 24, 1847)

Clarissa married Prescott (Prescot) Lawrence and, in 1857, they had a son, Charles N. Lawrence. I have discovered no records documenting these two events, but the Vermont Vital Records show that Charles N. Lawrence married Sarah A. Tenney in 1881. This record lists Prescott Lawrence as his father, Clarissa as his mother, and estimates his birth year as 1857.[191] In the meantime, two of Clarissa's sisters had died: Hannah M. Huntting, on July 1, 1853, at age eighteen,[192] and Sarah J. Huntting, in 1855, at age fifteen.[193] Certified copies of their death certificates (made in 1920) unfortunately omit cause of death. However, an entry in the Winhall Town Records (1857-1976) shows that Clarissa died of pulmonary consumption, on September 7, 1859 (a Wednesday, as noted in the *Bellows Falls Times* account) at age twenty-eight years and six months.[194]

Clarissa's husband, Prescott Thomas Lawrence, was born July 24, 1824, in Peru, Vermont, a town that abuts Winhall to the north. The 1850 federal census identifies Prescott Lawrence as a twenty-six-year-old laborer born in Vermont. He and his sixteen-year-old sister, Hannah Lawrence, were living with the Thomas French family in Peru. In the 1860 census, Prescott—by then a cooper, still living in Peru—was in the household of S. B. Russel (Samuel B. Russell) and his wife, Hannah, listed as twenty-seven years of age. In the ten-year span between these two census reports, Prescott married, had a son, then became a widower—and Hannah had married Russell. Also residing in the Russell household in 1860 was Anna Lawrence, a sixty-one-year-old widow, and three-year old

Charles N. Lawrence. Anna is Anna Abbott Lawrence (1799-1866), Prescott's mother, who probably was taking care of her grandson while her son made barrels.

If Clarissa's vital organs were removed and burned "to prevent any more from dying of the disease," the ritual might be judged at least partially successful, as her siblings Adeline (1833-1918), Jonas (1841-1909), Mary Agnes (1846-1915), and Viola (1847-1895) all lived well into mature adulthood. However, Clarissa's mother, Lucinda Carkin Huntting, died in Winhall on June 2, 1861, at the age of fifty-two years and ten months. Her death certificate lists pulmonary consumption as the cause of death.[195]

The obituary of Clarissa's father, Newell Huntting, notes that, when he was a lad in the early 1800s, he was "often sent on horseback with grain to mill at Manchester, a distance of 12 miles, the road being a path following marked trees." The town of Manchester is where, in 1793, Rachel Harris Burton's lungs, heart, and liver were removed from her corpse and burned to ashes on Jacob Mead's blacksmith forge to cure the new Mrs. Burton. It was reported that from five hundred to a thousand people were present to view this event (see chapter 11). Did some of these townsfolk delight young Newell with this story as he waited for the grain to be milled? This version of the consumption ritual had been in the vicinity for at least sixty-six years by the time it was performed again, in Winhall, before a "number of witnesses," to stop the deaths in the family of Newell Huntting.

***

Medical science in the mid-nineteenth century wavered between two possible explanations for the spread of consumption, both of which seemed able to account for observations long associated with the disease: it appeared to run in families, so many believed that there had to be a hereditary component; yet it also seemed to be transmitted from one person to another, regardless of kinship, so there must be a contagious aspect, as well. An account from Grafton, New York, an 1867, weds these two explanations, arguing, in effect, that the same human logic underlies both folk and scientific approaches to problem solving. Is an assertion that "the decaying vitals of that body are continually emitting the elements of that disease" based on superstition or science? Science has proved that disease-causing microbes can survive in dead organisms. Is burning the infected organs superstition or science? Science tells us that harmful bacteria can be rendered inert through incineration. In this instance, superstition and science seem to have arrived at the same place.

The following narrative, from the *Troy* [NY] *Daily Whig* (January 16, 1867), gives little information that leads to external corroborating evidence, yet it provides insight into the logic of this "strange transaction":

> Superstition in this county. —A correspondent of a New York paper, writing from this county, gives the following account of a strange transaction which recently occurred in the town of Grafton. The writer says:
>
>> I acquaint you with the particulars of an affair which occurred on the mountains of Grafton, a town in this county, a few days since. There is an old superstition, (which dates back even beyond the witchcraft of our own New England) that when any member of a family falls a victim to consumption, the decaying vitals of that body are continually emitting the elements of that disease, which the surviving relatives will receive, and thereby generation after generation will be followed by that dreadful scourge, prevails. According to

the superstition, there is but one means of eradicating this tubercular diathesis [i.e., tendency] from the family. That is to remove the vitals from the corpse and burn them. Accordingly, a family which has lost three of its members from that disease—one a year since, the other two within a few days—adopted the barbarous remedy. The bodies were disinterred, and from those which were in a sufficient state of preservation, the heart and lungs were removed and burned, after which the ashes with the bodies were returned to their graves and buried. The scene was witnessed by a large number of the relatives and friends of the deceased, who returned to their homes conscious of having done a good work.

Grafton has always been a source of annoyance to our county politicians, especially to those running for office, and the proverbial looseness of its political morals can only be accounted for on the theory, so well illustrated in the above transaction, that "where ignorance is bliss, 'tis folly to be wise."

Close variants of this text appeared in at least nine different newspapers from mid-January to early February of 1867. All headed it "barbarous remedy," but omitted the belittlement of a near neighbor the *Troy Whig* used as a tag.

These rituals were not clandestine. Frequently, many relatives and friends attended. This narrative's author was identified in several newspapers as "a correspondent of the *Litchfield Sentinel*, writing from Eagle Mills, Rensselaer county [sic], N.Y." I wonder what he had in mind when he wrote that the superstition "dates back even beyond the witchcraft of our own New England." Was he making a connection to European beliefs? Stories about European witchcraft were plentiful enough in nineteenth-century American newspapers. In 1869, a Bulgarian witchcraft story appeared in newspapers across the country. The following text is from the *Boston Daily Advertiser* (March 17, 1869): "Superstition lately led to a horrible outrage in the commons of Jabalca, in Bulgaria. The body of a woman, suspected of witchcraft, was disinterred, and the heart having been cut out, was burned to ashes." The *Jackson* [MI] *Citizen* (April 6, 1869) inserted in its version, "only last month—not two or three hundred years ago, as might be supposed." Substitute the name of a New England town for "Jabalca, in Bulgaria," and "causing consumption in her family" for "witchcraft," and you have a typical New England exhumation narrative. The significance of the dichotomy between consumption and witchcraft will come to the fore when we consider possible sources of the New England tradition.

7 Eliza Heard and Believed these Stories

Newspaper stories lacking researchable clues are discouraging. This one from the Connecticut River Valley had potential. Puzzle-lovers will appreciate the challenges—and rewards—of this tale, published in the *Springfield Republican* (September 3, 1869) as "Strange Superstition in Berkshire":

> A correspondent of the Great Barrington Courier reports this strange case of superstition and delusion in southern Berkshire: In the southern part of the town of New Marlborough, near the Connecticut line, lives a family by the name of F——. It seems this family is predisposed to consumption. About the first of last January one of the family—Emily, a girl 18 years old,—died of this disease and was buried at Cornwall Hollow, Ct. The mother and a brother Charles, and a sister, Eliza, remain. Charles is now wasting away with the same disease. During the summer a man by the name of Case, who lives in a town to the east, was in this section hiring cattle pasture [that is, he was looking to lease land for pasturing his cattle]. He met this family of F—— and told them some awful stories of how persons nearly dead with consumption had been cured by the digging up of some relative who had died with the same disease, taking out the liver, lungs, heart, etc., where fresh blood would be found, and burning them, after this the sick ones would improve until health was restored. The reason assigned was that there was a sort of vital current existing between the living and dead—that those organs in the dead body that contained fresh blood and appeared to be alive, would continue to live until the vitality of the living subject was exhausted, unless said organs were taken out and consumed by fire. Eliza heard and believed these stories, and if they ever appeared absurd they soon became matters of fact to her. Dr. S. of an adjoining town, was importuned to do the dissecting—no rest could be obtained until the thing was accomplished. Strange, the doctor consented. On the 10th of last August the doctor and half a dozen friends dug up the body of Emily, cut out the liver and a portion of the lungs, took them some distance from the grave and burned them. The heart and a portion of the lungs were found decayed—that portion of the lungs supposed to be destroyed before death—and fresh blood was found in the liver, enough to besmear the doctor's hands. These are facts, and they say Charles's health has begun to improve.

Newspaper vampire stories reveal little and cloak much. This one provided names, but first and last were disconnected. The F family apparently had a foot in both Massachusetts and Connecticut. They were living in the southern part of the small town of New Marlborough, situated in southern Berkshire County, Massachusetts, when Eliza was told the story. But her sister, Emily, was buried in Cornwall Hollow, in northern Litchfield County, Connecticut, less than twenty-fives miles south of New Marlborough. A cemetery search for an Emily F, who was born about 1851, and died near the first of January in 1869, might reveal the family's surname; finding her siblings, Charles and Eliza, would strengthen the connection. Thanks to Connecticut's monumental cemetery database, the Hale Collection, now digitized and accessible online, I launched my initial search. Funded by the Works Progress Administration in the early 1930s, Charles R. Hale recorded vital information taken from headstone inscriptions from 2,269 Connecticut cemeteries.[196] I hoped the

Hale Collection might yield more complete information than an on-site visit. Since the time of Hale's documentation, some inscriptions have become illegible, and several gravestones have been damaged, destroyed, or simply disappeared. On the other hand, as I also discovered, gravestone inscriptions and transcriptions can contain inaccurate information.

In the Cornwall Hollow Cemetery, there was just one surname—Ford—that corresponded to the published text; this family's deaths clustered between 1855 and 1871. Vital records and census reports provided bits and pieces, but gaps and inaccurate gravestone transcriptions impeded my investigation. What I gleaned from my visit to the cemetery in September 2022, when combined with archival data, led to the following narrative: When Amos Ford, the head of a hardscrabble family of subsistence farmers, died in 1862, his widow, Rachel, attempted to continue farming and keep her family afloat. But two sons, Harvey (1855) and Manson (1858), had already died and three other offspring—Watson, William, and Emily—were showing signs of consumption. After the spring of 1867, with Watson (1865) and William (1867) now dead, Rachel sold the small farm in Cornwall Hollow, Connecticut, and relocated just across the state line to Sandisfield, Massachusetts. While she kept house, her sons, Charles and Arthur, worked on neighboring farms. Her young adult daughter, Eliza, found work a few miles away in Connecticut as a live-in servant for the family of Charles Barnes, neighbors of Rachel's relatives. Then Emily died and Charles became too ill to work.

The Ford family's weather-beaten sandstone gravestones trace a precipitous decline. Harvey, Manson A., and William A. have individual gravestones. The inscription on the stone of Manson, not yet five years old when he died, is especially heartrending: "Should tears and prayers have held him here/He beed not passed away." After the father, Amos, died in 1862, the struggling family had to share a single stone. On one side are Amos (1814—1862), his wife Rachel Parmalee (1815—), his son Watson (1849—1867), and daughter Emily J. (1851—1869, but rendered inaccurately on the stone as 1849—1867). The other side of this stone has son Arthur H. (1855—), his wife Jennie E. Demarcia (1854—1887), and their daughter Jennie E. Ford (1887). Did Arthur's wife die in childbirth? Evidence on the gravestones indicates that their inscriptions were last updated after 1887 (when Arthur's wife and daughter died) but before 1893 (when his mother, Rachel, died). Had the stones been brought up to date, surely Rachel's death date would not have been left open.

The remnants of the Ford family were desperate when a stranger arrived, in the summer of 1869, with tales about a remedy that Eliza hoped would halt the deaths in the family. Enter "a man by the name of Case" from "a town to the east," who "was in that section" to lease land for pasturing his cattle. He met the Ford family "and told them awful stories of how persons nearly dead with consumption had been cured by the digging up of some relatives who had died with the same disease. . .. Eliza heard and believed these stories, and if they ever appeared absurd, they soon became matters of fact to her." Who was Case? A search through vital records of the towns just east of New Marlborough showed that the surname Case was most conspicuous in the town of Otis. Someone from Otis would, indeed, have reason to seek additional pastureland in New Marlborough. In his *History of Western Massachusetts* (1855), Josiah Gilbert Holland observed that "most of the inhabitants of Otis are farmers, and depend on their dairies and the raising of stock for a living."[197] Other sources affirmed that fattening beef cattle was one of the major uses of natural resources in the Connecticut River Valley of Massachusetts.[198]

A Campbell Case, born in Otis in 1809, was almost certainly the vampire storyteller. He appears in both the 1850 and 1860 federal census as a farmer living in Tolland, Hampden County, Massachusetts—less than fifteen miles east of New Marlborough. By 1860, he obviously had prospered as a farmer, for that year's federal census lists his real estate value as nearly $6,000 and his personal estate worth more than $4,000. The following year, according to the *Springfield Republican* ("Farms and Farmers" column, January 21, 1861), "Campbell Case of Tolland" was "one of the largest farmers in Hampden county [sic]." Over the years, as his estate continued to grow, Case acquired additional pastureland beyond his own farm in Tolland. Case's wide-ranging, ongoing search for pastureland would have given him an opportunity to both hear and tell stories about exhumations. In 1871 (according to the column reporting news items from Western Massachusetts, in the *Springfield Republican*, October 17, 1871), "Constable Billings . . . seized a gallon of gin, recently, from Campbell Case of North Blandford." With a bit of gin, the well-traveled Case might have been inclined to loosen his talktape, as James Joyce would say, and relate "some awful stories" that he had heard about how some people in the southwestern portion of the Connecticut Valley cured consumption.

I concluded that "Dr. S. of an adjoining town" was residing in a town next to Cornwall Hollow, since "the doctor and half a dozen, more or less friends dug up the body of Emily." Would all of them have spent a long day traveling from New Marlborough to Cornwall Hollow for the exhumation? The family's stronger ties to the Cornwall vicinity reinforce the conclusion that the attendants and the doctor were from the Connecticut side. Even so, I searched a variety of sources in both Berkshire County, Massachusetts and Litchfield County, Connecticut and found no likely candidates for Dr. S. The narrative's anonymous author wrote that it was "strange" that "the doctor consented" to "do the dissecting." Perhaps his motive was not so strange. If simple curiosity was not sufficient, the rare opportunity to dissect a human body may have been the deciding motivation.

Finally, we must ask if the cure worked. Eliza, the sister "who heard and believed" the stories, survived the family's consumption plague, living to the age of eighty-seven years. But Charles, whose health had "begun to improve" did not fare well. A death record for Charles W. Ford, a "single farmer," shows that he did not recover, dying of consumption in Sandisfield, Massachusetts (near New Marlborough) on March 3, 1871, at age twenty-four. I have not located his gravestone. Given the desperate financial situation of the family, I would not be surprised if Charles was buried without an inscribed gravestone.

Anonymous "correspondents," such as the one who contributed the exhumation narrative to the *Great Barrington* [MA] *Courier*, either participated in or were close observers of the events they described; they provided intimate, sensational details and on-the-scene coverage with a cachet of credibility— a heady mix that newspaper editors found impossible to resist. A bonus for the correspondent's initial publisher was the wide circulation of its newspaper's name. To that end, the editors of the *Great Barrington Courier* (cited by numerous newspapers as their source) ran the Ford family story twice, separated by two months, under different headlines: "Strange Superstition in Berkshire County" in September and "A Human Hyena" in November. Searching through newspaper archives, I found at least fifty versions of this narrative, published in newspapers across the country from September through November 1869.

Several newspaper editors used the Ford family story as a bridge to make a particular topical point. A few added other exhumation accounts, presumably to suggest that the incident from Cornwall Hollow was not singular. The *Lawrence* [MA] *American and Andover Advertiser* (September 10, 1869) appended a synopsis of a case from Plymouth County, Massachusetts (which we take up in chapter 18); and the *Rutland County* [VT] *Herald* (September 30, 1869)—noting "this is no new story" —summarized a consumption ritual that took place three decades earlier in Rutland County, Vermont (discussed in chapter 5). Some newspapers placed the Ford family exhumation narrative into a larger context of general superstition, ranging from the well-worn story of the Salem witch trials to a peculiar and horrific enactment of what might be termed an urban legend from Spain. Henry D. Jencken, an English attorney working in Spain for several local clients in the late summer of 1869, was assaulted by a throng of people with stones, clubs, and knives; they apparently believed him to be one of the "Tios del Sain," that is, child-stealers who were suspected of catching and butchering young children for the purpose of using the fat of their entrails to repair telegraph wires. After his story ran in several newspapers in England and the United States, appended to the Cornwall Hollow narrative, Jencken, himself, provided the *London Times* (August 14, 1869) with his own detailed narrative of the assault, which he somehow managed to survive, although not without some grievous injuries. Conspiracy theories, such as "Tios del Sain," were not new in the nineteenth century and, in fact, have continued into the present. Their credence, fueled by rumor and legend, now relies more on electronic media than word-of-mouth. But their essential characteristics, and sometimes dreadful outcomes, remain consistent.

***

One of the best explanations for how an interred corpse could harm the living is included in a story from Jay Hill, located on the Androscoggin River in southern Franklin County, Maine. During September and October in 1869, several newspapers published variants of the following text (this one was in the "State Items" column of the *Daily Eastern Argus*, September 7, 1869), attributed to the Farmington (Maine) correspondent of the *Boston Herald*:

> Some time during the last winter a young man named Henry Cole, a resident of Jay Hill, died of consumption, being the second one in the same family to fall a victim to this great scourge of New England. Another member of the family, Frederick, a brother of Henry, is now supposed to be in the last stages of consumption, and in order to stay the ravages of the disease the body of Henry has been disinterred for the purpose of removing and burning the heart, this being supposed to be a sovereign remedy.

Since Farmington, the seat of Franklin County, is only a few miles north of Jay Hill, it seems likely that the correspondent had some ties to the community, if not the family, about whom he wrote. The 1860 federal census has the entire family living together in the Livermore Falls section of Jay. The father, Otis Cole (1809-1893), was listed as a farmer with real estate valued at $1600 and a personal estate of $923. In addition to his wife, Agatha (1810-1892), the household includes three sons: Frederick G. Cole (born in 1840) and twins Henry and Horace Cole (born about 1842). Genealogical research, combined with gravestone records from Jay Hill Cemetery, where these family members are interred, indicate that Henry died March 20, 1868, Horace died April 10, 1869, and Frederick died January 27, 1870. Matching genealogical information with that in the newspaper article yields a familiar sad story: Henry was the first son to die, followed a year later by Horace; by then, Frederick was showing the dreadful signs of consumption, so the body of the first son to die, Henry, was disinterred and his heart burned to save Frederick. The "sovereign remedy" apparently was unsuccessful, for Frederick died about four months later.

The narrative, as it appeared in the *Cincinnati Daily Gazette* (September 29, 1869), includes a tag not found in the other newspaper texts I accessed: "the popular theory being that there exists between blood relations a sympathetic link which death does not entirely sever, and which, unless interrupted, oftentimes works injury to the living and sometimes results in death." The "sympathetic link" mentioned seems to anticipate Sir James George Frazer's notion of "sympathetic magic." Both echo the theoretical underpinnings of the Paracelsian approach to medicine. Anthropologist Karen Gordon-Grube noted that the ancient ideas of "like cures like" and "sympathetic action" (or "action at a distance") were fundamentally important to Paracelsians who "saw the universe as a 'living unit' in which the occult or magical influences of the World Soul were everywhere at work; every part of the universe bore a sympathetic relationship to the rest."[199] Each person was a "microcosm" and "had within himself all the forms of external nature," thus creating "a sympathetic attraction" between the "microcosmic man" and the "corresponding element" in the macrocosmic universe. Paracelsian physicians healed by channeling the restorative influences of the "World Soul" into the ailing organ of the patient.[200] For example, Edward Taylor of Westfield, Massachusetts, Puritan poet and town physician, "approved of the ingestion of 'mummy,' which was medicinally prepared human flesh."[201] Gordon-Grube termed mummy the "sovereign remedy" of the Paracelsian.[202] New England's consumption ritual, however, posited a malign, rather than a benevolent, "sympathetic link" between the living and the dead. By the end of the nineteenth century, Frazer had explicated the foundations of this theory, writing:

> Side by side with the view of the world as pervaded by spiritual forces, primitive man has another conception in which we may detect a germ of the modern notion of natural law or the view of nature as a series of events occurring in invariable order without the intervention of personal agency. The germ of which I speak is involved in that sympathetic magic, as it may be called, which plays a large part in most systems of superstition. One of the principles of sympathetic magic is that any effect may be produced by imitating it.[203]

From September 25, 1869, to February 17, 1870, a similar story appeared in at least eight newspapers. It may be a case of mistaken location. If the Cole surname were excised, and New Hampshire substituted for Jay Hill, the following short description, which was published in the *Commercial* [NYC] *Advertiser* (September 25, 1869) would be a suitable match: "The body of a New Hampshire youth has been exhumed and his heart cut out and burned, from a superstition that this alone would save the life of his brother, who is dying with the same disease—consumption." If that is, indeed, the case, it exemplifies how newspapers simply copied pieces verbatim from other papers or off the telegraph wire, with little or no vetting. Errors or omissions were thus likely repeated serially. Complicating matters, an article in the Manchester (NH) *Mirror and Farmer* (October 2, 1869) quoted the brief narrative (attributing it to the *Boston Post*) then added: "Not much; that intelligent action was performed in Hampshire County, State of Massachusetts." This New Hampshire newspaper did not want its state to own the incident, so they simply transferred it to Massachusetts based on a shared name? But, if this is *not* the Jay Hill event, then we can add it to the "information lacking" folder.

***

On Friday, May 29, 1874, the *Lewiston* [ME] *Evening Journal* published the following notice:

> A correspondent informs us that in one of our cemeteries on Thursday, a body was exhumed and placed face downward, in accordance with the directions of a person in the clairvoyant state. The family has lost seven members by consumption, and there is an ancient superstition that to stop the ravages of consumption in the family, it is only necessary to bury the last victim face downwards. We have heard of many specifics for consumption, but this one has ghastly priority and novelty.

While this specific for consumption may have been ghastly, it certainly was not novel. Between June 1 and July 9, at least thirteen newspapers in nine states, from New York and Minnesota to Texas and Louisiana, published an abbreviated variant of the text in which the "person in a clairvoyant state" becomes, simply, "a clairvoyant." The text as it appeared in the *Albany Evening Journal* (June 6, 1874) is typical of this variant: "In Maine, recently, by the advice of a clairvoyant, a body was exhumed and reburied with the face downward, as a means of staying the ravages of consumption in the corpse's family."

Addressing vampire hunters and slayers, Bruce A. McClelland wrote, "The presence of a humanlike threat that is either not visible or generally not detectable naturally brings about the need to locate an individual who possesses the unique power to identify and hence accuse the duplicitous aggressor."[204] In most of New England's consumption rituals, there is no need for such a specialist. Even in this narrative, the last victim is the default scapegoat. So, why the need for clairvoyance? Knowing that, according to contemporary usage, a "clairvoyant" is "a person who is able to perceive matters beyond the range of ordinary perception,"[205] only begins to illuminate this case. To tell the fuller story, we should construe the term—combining the French words for "clear" and "vision"—as journalists might have in the mid-nineteenth century. A search of the word in newspapers, beginning in the nineteenth century, shows that "clairvoyant" appears in the early 1820s, often associated with animal magnetism, spirit-rapping, and Mesmerism—newly emerging notions with scientific pretensions. Over the following decades, con artists and the credulous began to form a partnership, creating a flow of money from the latter to the former. The *Boston Evening Transcript* (July 13, 1843) included the following article: "'Clairvoyant' female. In Paris there are said to be about one hundred clairvoyant females who pick up a living by pretending to see diseases—for which they prescribe. The Medical and Surgical Journal says that there are one or two only in Boston—but they enjoy a fine revenue for their impositions and the credulity of those who consult them." One of the two may have been the self-anointed clairvoyant, Mrs. W. Fergus, whose paid advertisements appeared in Boston newspapers that year.

In 1874, Dr. O. Fitzgerald, another self-proclaimed "wonderful clairvoyant, physician and surgeon," was traveling between Portland, Maine, and Lowell, Massachusetts. He would take a room in a hotel and advertise in the papers that he could see clients for a period of two to four days only. This was a good strategy, allowing time for disappointed clients to cool off or give up, while the "doctor" was in another city. Then he would return "by popular demand" (if the ads are to be believed) for another few days. Dr. Fitzgerald was advertising a stay in Portland from the twelfth to the sixteenth of May. Lewiston—where, on Thursday, May 28, 1874, the consumptive corpse was exhumed under clairvoyant advice—is only about thirty-five miles north of Portland. But distance was not necessarily a barrier to consultation; some clairvoyants advertised that, for a certain fee, they would answer questions by mail.[206]

We cannot be certain what the correspondent who wrote the article about the exhumation in Lewiston had in mind when he used the phrase "person in a clairvoyant state." If he was, indeed, referencing this category of professional con artists, then using the face-down variant of the vampire ritual makes sense. Inverting a corpse is gentler than mutilating it, or, even more distasteful, cannibalizing it. These more gruesome folk rituals may be suitable when kin, friends, and neighbors argue for it—but a professional, who also is a stranger, would be on more tenuous ground if he prescribed such actions. Prone (that is, face-down) burial, moreover, has an established pedigree in world tradition.[207] Sir James George Frazer hypothesized that face down burial might have been an outgrowth of the belief in reincarnation. "When several children have died in a family, the last of them is placed in the coffin face downward and is buried in that posture, . . . to prevent its soul from entering again into its mother's womb and being born again into the world."[208] In the New England consumption tradition, a reappearing spirit is never welcome.[209]

***

Although he would firmly disagree on philosophical grounds, it seems almost preordained that Benjamin Franklin Underwood (1839-1914) would eventually edit *The Index*, a weekly periodical "devoted to free religion" (as its banner proclaimed). Descended from "good Rhode Island stock" on his father's side, and educated in Westerly, Rhode Island, Underwood was a self-proclaimed skeptic by age twelve. Two years later, upon reading *The Theological Works of Thomas Paine*, he had lost faith in the Bible. In his first public speech, delivered in New Haven at the age of eighteen, Underwood espoused the cause of materialism,[210] a philosophy that scorned "the supernatural in general (e.g., ghosts and magic), and religion in particular (e.g., immortal souls and divine intervention)."[211] In 1883, when the following entry was published, Underwood had recently assumed co-editorship of *The Index* (with its founder, William J. Potter):

> That superstition is deadly appears from what recently happened on the Island of Vinal Haven, Me., now regularly visited by summer boarders on account of its rare coolness. A young American broke open the coffin of a friend, who had died of consumption, in order to turn over the corpse, by which it was hoped that the life of a member of the same family, similarly afflicted, might be saved. The effluvium has so injured the believer in vampires that his own death is near.[212]

We can understand why this narrative, in which contagion retaliates against a vampire hunter, would have resonated with Underwood: it is a small-scale exemplar of science defeating superstition, mirroring the results of Robert Koch's announcement, a year earlier, that he had discovered the tuberculosis germ, a scientific triumph that eventually replaced vampire hunters with microbe hunters. Underwood's use of the word *vampires* reflects its increasing popularity for characterizing New England's consumption rituals. The first instance of using the word to characterize the New England ritual (as far as I have been able to discover) was in 1842. By the 1880s, the *vampire* appellation was well established among outside commentators.

Regarding the "young American" on Vinal Haven (Vinalhaven, since 1924), or the source of his story, we remain in the dark. Since at least three of the five instances of turning a corpse face down occur in Maine (two in Oxford County), one wonders if this is a regional variant of the tradition.

***

In March and early April 1875, at least ten newspapers published the following text: "A singular rite of superstition was recently conducted at Newport, R. I. The relatives of a person who had died of consumption, and who believed that cremation of the intestines of the deceased would prevent the disease from attacking the survivors, exhumed the body, and burned those parts, but another member of the family has since died of the disease." The article first appeared in the *Springfield* [MA] *Republican* (March 6, 1875). It seems odd that only one Rhode Island newspaper, *Bristol Phoenix* (March 13, 1875), published the account. In Indiana, the *Fort Wayne Daily News* (March 10, 1875) ran the article in its "News Gatherings" column. In the other seven newspapers, all in the Midwest, the article was in a column entitled "Incidents and Accidents." The newspapers that ran the story exemplify the state of small-town Midwestern papers who, in the latter half of the nineteenth century, were receiving most of their news reports from a news service, predominantly the Associated Press.[213] Unfortunately, the narrative's lack of detail makes it a dead end.

***

On June 8, 1890, the *Providence Sunday Telegram*, buried the following article, untitled, at the bottom of page four: "Superstition has not yet wholly died out in Rhode Island. In one of the country towns recently a young man who was threatened with consumption was seriously advised to have exhumed the body of some near relative who died of the disease, and to have the heart thereof cremated. He followed this suggestion, but our informant does not say what success attended the process."

A good argument can be made that the "informant" was either reporting what he had witnessed firsthand or heard circulating orally within this unidentified country town. The identical text was published five days later in the *Narragansett* (RI) *Times* (June 13, 1890), crediting the *Providence Telegram*. Both texts closed with a derisive tag that ushered readers into the sphere of deviance: "This is as curious a superstition as the one where children fear to throw away any extracted tooth, because some dog might get hold of it and then a dog's tooth would grow in the cavity."

***

Thunder booms from a sunny sky the moment the grave of a dead consumptive is disturbed by his two brothers, on a family-approved mission to save a fourth brother, dying of consumption. Also at the grave, from which the corpse must be disinterred, are a Universalist minister and a medical doctor. The clumsy doctor accidentally pricks the swollen heart he has excised from the exhumed corpse; spurting blood is construed as a sign of life. The corpse is reburied, but the heart is carried back to the family home, where it is burned to ashes in the cook stove. After the ashes are administered to the dying brother (we are not informed how), he recovers.

This story has nearly everything one expects to find in a New England vampire narrative. Unfortunately, the missing elements are the who, where, and when that journalists are expected to include in their stories. The full text of this story appeared in the *Boston Sunday Herald* (August 13, 1899), embedded in an article headlined, "Lingering Belief in 'Hoodooism'." Three subheads roughed out the arc of the anonymous author's narrative, a reproof already commonplace when this article was published:

Medievalism in New England Not Wholly Eradicated—Signs by Which Credulous Live.

Wrecks of Old Religions.

Queer Illustrations That Are Drawn from the Fertile Field of Rural Superstition and Ignorance.

Setting the exhumation scene for the unnamed characters—who resided "somewhere in New England, at a time within living memory"—the author opened his article with an esoteric literary reference: "In his essay on 'Politics,' the Yankee Plato remarks, in his customary fashion, that 'although we think our civilization near its meridian, we are as yet only at the cock-crowing and the morning star.' Ignoring for the present numerous arguments that might disprove that assertion, it is the object of this paper to call attention to some facts that tend to support it." The author continued in the same vein by citing Oliver Wendell Holmes's designation of Boston as "the Hub," where "New England has laid to its good soul the flattering unction of intellectual supremacy." Despite the "constant drainage of good blood toward the West," who can rightly claim Boston's status is undeserved or that people have ceased to "look to the East for the appearance of wise men?"

The story, itself, published in the *Boston Sunday Herald* (August 13, 1899), is free of strained analogies or literary allusions:

To be concise. New England is still tinctured with mediaeval and antemediaeval superstition, from which few of the people are entirely free, and in some localities beliefs are prevalent in things akin to sorcery, magic and various occult influences. As an illustration, the following incident will serve:

Within five hours' ride of Boston, in a village only five miles distant from a city of about 20,000 inhabitants, a man died from consumption, and, some months later, his brother fell sick. After a family consultation, two other brothers, accompanied by a physician and a Universalist minister, proceeded to the cemetery, opened the grave, removed the body of the dead man, took from it the heart, restored the corpse to its place, and then, burning the heart to ashes in the family cook stove, administered a portion of the ashes to the sick man, who promptly recovered. Of these things there is no doubt, for they were matters of common knowledge in the place at the time.

But, in addition, one of the brothers who assisted in the disinterment and the subsequent grewsome surgery, has often said that when they first began to disturb the earth at the grave they were startled by a terrific peal of thunder from a sunny sky. Later, when the heart had been removed, the physician carelessly cut it, whereupon a jet of bright, crimson blood spurted out. The narrator's explanation was that some life had miraculously remained in the body for the benefit of the other brother. Names and dates are at hand, but, as the parties are still living, it would not be fitting to publish them.

The author conceded that "this is an unusual case," but maintained that more mundane practices, such as "nailing up a horseshoe for good luck" or transferring a wart to some unsuspecting passerby, are no less superstitious. He showed familiarity with the contemporary theory of cultural evolution by asserting that all such "lingering superstitions" are merely survivals, "the wrecks of old religious beliefs." While the "reverence or fear" manifested in these superstitions survived, he argued, the comprehension ("if ever there was any") had been lost. His knowledge of geology was evident when he likened these survivals to stratigraphy: "There seems to be a more or less thin layer of paganism underlying, even New England Christianity, especially among the inhabitants of the remote country districts, and this layer now and again appears on the surface, just as the deeper strata of granite or slate crop out on hilltop or river bluff where the overlaying formation has been sliced away by glacial planing or water erosion." His assertion that paganism lingered longer in "remote country districts" also suggests that the author was aware of the notion that folklorists and linguists refer to as "the archaism of the fringe," according to which older cultural forms are retained in areas removed from the centers of innovation and change (such as "the Hub").

Incorporated into this archaic "layer of paganism" was the belief in good and evil spirits that "every man . . . watched carefully for indications of the good will or anger of these spirits. "All diseases," the author continued, "were traced to the work of some demon." Therefore, the "occult evil influences" had to be "driven out and kept at bay by the wearing of amulets, talismans and charms." I suspect that the writer had read George R. Stetson's treatise on New England's vampires, published in 1896 (which we inspect in chapter 16). The author strayed from Western superstitions when he described a West African practice employed to divine the identity of a thief; one wonders if this is the "hoodooism" that appears in the article's headline, "Lingering Belief in 'Hoodooism'," for nowhere in the article is there a reference to African American beliefs or practices that conventionally have been subsumed under the term *hoodoo*. The brief detour to Africa is followed by a discussion of familiar American folk beliefs ("superstitions," if you must), ranging from scheduling tasks in accordance with phases of the moon to divining the future and reading signs of good and evil.

Despite his focus on rural examples, the author did allow that "the city is hardly less fruitful of similar instances, although of a different nature," citing card readers, fortune tellers, astrologers, palmists, mediums, and purveyors of quack devices, such as galvanic belts. And he also acknowledged that "other parts of the United States"—indeed, "other countries in the civilized world"—might be just as superstitious as New England. The author closed his article by identifying the essayist he quoted in his opening sentence: "It may be, therefore, safely asserted that, whether or not Emerson was right in placing the 19th century civilization at the cock crowing and the morning star, the people have not as yet, even in these last stages of the century, and in New England, come into the light and warmth of an educational noonday."

This article poses unanswered questions, some of which lend themselves to speculation. First, let us consider its noteworthy elements:

- The **family consultation** is a common theme in New England's exhumation narratives. It serves to validate the need to perform the ritual, relegating responsibility to a collective kinship unit and thereby relieving a single individual of whatever accountability, blame, or shame might ensue.

- The **presence of a minister** (in this case, a Universalist) at the scene—unusual but not unknown (see, for example, the cases from Belchertown and Chazy)—extends the validation beyond the family unit, suggesting that a larger community (and established religion) gives at least tacit approval to the proceedings.

- A **physician on site** is found in more than a dozen of New England's exhumation narratives; in about half of these cases, the physician is present only as an observer, not actively participating in the ritual. In this case, his role was to perform the autopsy. Was his cutting into the heart a matter of being careless (as the narrative states), or was it intentional (which concurs with the tradition of checking for "fresh" blood in the heart)?

- The dual nature of vampirism is evident in finding **"fresh" blood in heart**: Was it a curse (as is the usual interpretation), or was it a miracle (as in this narrative)? Depending on the context of interpretation, it could be either, or both.

- This narrative incorporates the most common variant of **cannibalism** in the New England tradition: burning the heart to ashes, which are then ingested by the ailing family member. Some form of endocannibalism is present in more than fifteen percent of the New England cases. A rare twist here is burning the heart in the family's "cook stove."

- An unusual outcome of the ritual is that the **patient recovered**, a result seen in less than twelve percent of New England's vampire cases.

- The peal of **thunder in a clear sky**, which was said to coincide with the first spadeful of earth, is a magical motif that we will encounter (chapter 15) in a narrative from Woodstock, Vermont, published in the *Vermont Standard* in 1890, more than sixty years after the exhumation.[214]

- The report that thunder was heard when the exhumation first began was attributed to one of the two brothers who attended the exhumation; the way the statement was framed by the author suggests that this **brother had shared his exhumation narrative frequently** (he "has often said"). Was he the "narrator" that the article's author referenced, or was there another, third-party narrator? Depending on the answer to this question, the narrative could be a personal-experience story or memorate (if told by a brother who was there) or a community-based legend (if it had become common knowledge).

- The author had names and dates but would not furnish them because the parties involved were still living. That information at least provides a **temporal framework**: if the people involved were still living in 1899, we might guess that the oldest participant would be between seventy and eighty years old and, thus, would have been born between 1810 and 1820. If the brothers were in their twenties when the exhumation took place, that would place it about 1830 to 1850.

- We also get a broad sense of **where the exhumation occurred**: "Within five hours' ride of Boston, in a village only five miles distant from a city of about 20,000 inhabitants." Of course, it makes a huge difference if the rider is traveling by horse or by train. On horseback, he would be able to go about twenty miles, at most, providing the roads or trails were on level terrain. But since Boston's South Station opened, coincidentally, on January 1, 1899, I will guess that we are dealing with a five-hour train ride. The New England Limited, which was the top of the passenger line in 1891, made the 214-mile Boston to New York trip in six hours. So, a five-hour radius from Boston would be about 180 miles, which incorporates most of New England, including: New York well west of Albany; almost as far north as Burlington, Vermont; into the portion of Maine that includes its most populous areas; well into Connecticut; most of New Hampshire; and all of Rhode Island. In short, a five-hour rail radius does not exclude many populated areas of New England. Additionally, the number of New England cities that had a population of at least 20,000 around the turn of the century also is very large.

Redactions of the article, published in the *Cincinnati Enquirer* (December 30, 1899) and the *Elmira* [NY] *Daily Gazette and Free Press* (January 20, 1900), exemplify the popular press's pull toward a broad appeal and, by way of contrast, highlights the narrative's urbane origin. The term "hoodooism"—which appeared in the *Herald* headline—was conspicuously absent in the Elmira newspaper. In its place were the headlines: "Superstitious Yanks" and "Witchcraft." The narrator of the redactions pointedly denies the relevance of the expurgated headline: "You won't find voodooism, but you will find a belief in almost everything else that is uncanny." These revisions reinforce my hunch that the first rendition of the article, which was included in the "Special Features" section of the Sunday edition of the *Boston Herald*, began as an orally delivered paper or, perhaps, an essay previously published in a literary periodical or professional journal. The article's beginning and ending, both of which pay homage to the thoughts of Emerson, also were deleted from the Elmira version, as were the writer's personal recollections of interacting with the local folks in New Hampshire and Maine. The exclusion of "high culture" literary citations broadens the appeal of the article, while the elimination of the writer's personal-experience narratives lends a more generic quality.

Unmistakable traces of the Boston Brahmins—the clannish, hereditary elite culture of New England characterized not only by good breeding and manners, but also by superior education and understated wealth—are layered into the *Herald* narrative, with the at-least implicit presence of Ralph Waldo Emerson, James Russell Lowell, Oliver Wendell Holmes, Sr., and Edward Everett Hale. On the nineteenth anniversary of Emerson's birth, Hale identified Lowell as the person who dubbed Emerson "The Yankee Plato." And it was Holmes who first referred to Boston as "The Hub." The writer obviously was fluent in the world of the Boston Brahmins (an appellation also coined by Holmes). His status in Boston must have been secure financially as well as culturally and socially, for he was able to travel to Maine for hunting and vacation in New Hampshire and Vermont. The writer's dialectical treatment of the native speech of those in outlying areas gives voice to an obvious divergence in status. Who but the privileged could breakfast in Boston and sup in Down-East Maine? When he refers to himself as "the writer" and terms his article a "paper," he (perhaps unintentionally) reveals that he probably was not a news writer, but rather a professional in some other field—one where "paper" is construed to be an essay, article, or dissertation on a particular topic. Perhaps tellingly, Hale and his family often spent the summer in South Kingstown, Rhode Island, a region where consumption rituals and their narratives were notably prevalent.

8   My Sisters Were Drawing My Life Away

How can a newspaper article answer the basic questions of who, what, when, where, why, and how, yet remain unapproachable? The usual impediment to recovering a plausible narrative is lack of information. In this case, however, *inaccurate* information was the culprit.

The false trail begins in 1879, in Moncure Daniel Conway's *Demonology and Devil-Lore*: "In 1874, according to the *Providence Journal*, in the village of Peacedale, Rhode Island, U.S., Mr. William Rose dug up the body of his own daughter, and burned her heart, under the belief that she was wasting away the lives of other members of his family."[215] Conway's misspelling of Peace Dale as Peacedale (not unusual, even in Rhode Island) was picked up by subsequent authors. In 1914, Dudley Wright, in *The Book of Vampires*, substituted Placedale for Peacedale.[216] And that misspelling was, in turn, carried on by others, including Montague Summers in *The Vampire in Europe* (1929). But Conway's "*Providence Journal* of 1874" citation was retained.[217] Although I searched painstakingly through reel after reel of microfilm, I failed to find the cited article. Moving on to evidence from genealogy and gravestones, I concluded that the exhumed girl's father was William G. Rose (1821-1911), who is interred in the George Rose Lot (South Kingstown Historical Cemetery #10), located on Mooresfield Road, in South Kingstown. I also concluded that the daughter whose body was exhumed probably was Ruth Ellen Rose, born in 1859 to William and his first wife (who died in 1863). The gravestones of both Rose and his second wife, Mary G. Tillinghast, are in the Rose Lot. But I failed to find the grave of Ruth Ellen, who died on May 12, 1874, according to her obituary in the *Narragansett Times* (May 15, 1874). The missing grave and the missing newspaper article, combined with the apparent fact that Rose seems never to have lived in the village of Peace Dale, cast a fog of uncertainty over my conclusions.

Nearly five years after *Food for the Dead* was first published, a research librarian lifted the fog. With access to extensive, digitized databases that had powerful search capabilities, he was able to locate, in a matter of minutes, the newspaper article that I had fruitlessly pursued. That article apparently first appeared in the *Providence Herald* in 1872—not in the *Providence Journal* in 1874, as reported by Conway and subsequent authors.

Yet, as I plowed deeper into the electronic newspaper databases, I found more complications. My librarian benefactor had sent me, not the article from the *Providence Herald*, but one from the *New York Evangelist* that closed with the credit, "*Providence Herald, Sep.* 5." Was the *Evangelist* text, published under the headline, "A Strange Story of Superstition," on September 19, 1872, the same as that in the *Herald*? A key-word search of newspaper databases yielded nearly two dozen texts of the Rose narrative. Five newspapers ran the story under the heading: "A Strange Story of Superstition." Nine others led with this headline (or close variant), situating the event in Connecticut: "Horrible Scene. Burning of Hearts in Connecticut—Absurd and Brutal Superstition." Almost all these newspapers cited their source as the *Providence Herald* (usually with September 5 as the date). Given that the source newspaper is in Providence, and that every village, town, and other geographical feature cited in the article is in the state of Rhode Island, one wonders why the headlines situate the exhumations in another (albeit, adjoining) state. As in the chain of repeated errors in the vampire literature, from Conway to Summers, one newspaper routinely copied another without fact checking.

Since none of the databases I searched included relevant issues of the *Providence Herald*, I still had not found the cited parent article. But my search uncovered an article from the *Providence Evening Press* that was published on September 4, 1872—one day *before* the supposed original text! This article's author, in his first two words, lays claim to village residency, insinuating an insider's viewpoint. The following narrative, published under the heading of "The Latest Sensation," in the "Peacedale" (not the correct "Peace Dale") section of Rhode Island state news, includes an unusual amount of detailed information:

> Our village was considerably aroused on Thursday last, by the announcement that two graves had been dug up at the north end of the village, near the Saugatucket river. Upon investigation, we learned the following circumstances: It seems that the family of Mr. William Rose, who reside on the shore of Narragansett Bay, at what is called Saunderstown, are subject to consumption, several of the family having died of it, and one of them being quite low just now. At his urgent request, the father, who was assisted by Charles Harrington of North Kingstown, repaired to the family burying ground, which is located near Watson's Corner on the Saugatucket river, and after building a fire, first dug up the grave of his son, who had been buried twelve years for the purpose of taking out the heart and liver, which were to be placed in the fire to be consumed in order to carry out the old and we might add foolish saying that the consumptive dead draw nourishment from the living. But as the body was entirely decayed except a few bones, it was covered up and the body of his daughter, who had been dead seven years, was taken out of the grave. This was also found to be nearly decayed except the liver and heart, which were in a perfect state of preservation. The coffin was nearly perfect. After the heart and liver were taken out of the body, it was placed to the fire until consumed after which the remnants were interred again, the fire put out, and two departed to their respective homes. Only a few spectators were present to witness as they say the foolish and superstitious act. The news of the above spread like wild fire and no little excitement was caused in both our village and others surrounding. It seems that this is not the first time that graves have been dug up where consumption was prevalent in the family and the vital parts burned in order to save the sick. A few years ago the same act was performed in the village of Moorsfield [sic] and also in the town of North Kingstown. Of course there was a failure to effect the cures intended by the operation in both cases.

When I finally gained access to the Providence *Morning Herald*, I found there was no exhumation article in the issue of September 5, 1872. Extending my search to adjoining issues, I discovered that the article was published on September 4, the same day that it appeared in the *Press*. The small differences between these two texts reflect the *Press*'s insider viewpoint vs. the *Herald*'s outsider viewpoint. The *Press* has, "Our village was considerably aroused on Thursday last, by the announcement that two graves had been dug up at the north end of the village," where the *Herald* has, "The village of Peacedale was thrown into quite a state of excitement on Thursday last, by the report that two graves had been dug up near Watson's Corner." In the *Press*, Charles Harrington, who assisted Rose in the exhumations, resides in "North Kingstown." In the *Herald*, and subsequent newspaper texts crediting the *Herald*, Harrington lives in "North Kingston"—a subtle mistake often made by Rhode Islanders *not* living in South County. Kingston is a village in the town of South Kingstown, whereas North Kingstown is a town adjacent to South Kingstown. The following sentence in the *Press* text is absent in the *Herald* text: "The news of the above spread like wild fire and no little excitement was caused in both our village and others surrounding." Still unresolved is how these two significantly different texts, describing the same event, were published on the same day. That the *Morning Herald* was published by Noah D. Payne, and the *Evening Press* was published by Hiram H. Thomas & Co., suggests they were not sharing resources. Finally, in the *Herald*, ". . . witness the horrible scene" is, in the *Press*, ". . . witness as they say the foolish and superstitious act." If the *Evening Press* narrative was written by an insider, as the syntax attests, the empathy one might expect from a fellow villager is not apparent.

Empathy, or at least polite discretion, *is* apparent in the text as it appeared, on September 6, in the *Narragansett Times*, South County's local newspaper, published in Wakefield, South Kingstown. The identities of the participants (living and dead) were masked:

> Exhuming dead bodies and burning the vitals in conformity with the practice of antiquity. —It has been reported to us that a Mr. R. who has a son sick with consumption, and who believed in the old doctoring that the consumptive dead draw nourishment from the living, exhumed, on last Thursday, two bodies of the R. family who were buried near Watson's corner, a girl and a boy. Of the boy, who had been buried twelve years, there was nothing left but the bones, the flesh having returned to dust from whence it came; but the body of the girl, who had been buried only seven years, was to some extent in a state of preservation. It is said that the vitals were still in a perfect state of preservation. Mr. R. having heard that if the vitals of the dead of the family who died of consumption were taken out and burned, the living of the family would be freed from the destroying disease, accordingly Mr. R., assisted by an Indian, built a fire near the grave and burned the vitals of the dead girl. Having replaced the remains in the grave, Mr. R. and his assistant returned home, believing that they had saved another son from death by consumption.

I wonder if "the old doctoring" is supposed to be "the old doctrine." Or, perhaps, the anonymous author does mean "doctoring," in the sense of applying remedies to an ailing patient. In any event, unlike the other newspaper narratives, there is no outward scolding or marginalizing of the participants. The opening phrase, "it has been reported to us," strikes me as referring to a personal communication (oral or written) rather than an already published narrative (additionally, the *Times*' usual practice was to source other newspapers when applicable). The *Narragansett Times* was a weekly, published on Fridays. The event was reported to have occurred "last Thursday," which, if taken literally, would have been the day before its publication in the *Times*—probably too late to appear the next day in a weekly. Moreover, we know that the story first appeared on Wednesday, September 4. So, the exhumation would have occurred on Thursday, August 29.

The information contained in the newspaper stories is, with a few exceptions, corroborated by historical evidence, cemetery records, census data, and genealogical sources. The son and daughter were said to have been dead, respectively, for twelve years (since 1860) and seven years (since 1865). The father almost certainly was William R. Rose, who was born about 1824, in Jamestown, located on Conanicut Island (directly opposite Saunderstown, North Kingstown, across the west passage of Narragansett Bay). The vital records of Rhode Island show that William and

Phebe Ann Carr (1829-1885) were married in North Kingstown, by Reverend Edwin Stillman, on November 22, 1843. William and Phebe had at least five children: John R. Rose, born about 1846, died March 21, 1865; Benjamin Carr Rose, born 1851, died 1926; Phebe Ann Rose, born 1852, died 1923; Maria D. Rose, born 1854, died November 5, 1868; and Henry (as he appears in the 1880 federal census) or Thomas W. (in the 1870 federal census) Rose, born 1867, was living with his parents, attending school, in 1880. The *Rhode Island Atlas* for 1870 states that "W. R. Rose lives in the southern part of Saunderstown, North Kingstown near Watson's and the South Ferry," and the accompanying map pinpoints his residence in the South Ferry District of North Kingstown, matching exactly the newspaper narrative.

In the 1870 federal census Rose is identified as a stone mason. Their nineteen-year-old son, Benjamin, is listed on the same page as his parents, but he is living, as a laborer, in the household of Thomas Gould, a farmer. Benjamin married Georgianna Crandall on December 24, 1874; the Rhode Island Cemetery Database shows a Benjamin C. Rose, born April 29, 1851, died August 20, 1926, and buried in North Kingstown's large Elm Grove Cemetery (NK026), where many of this branch of the Rose family are interred. Henry Rose also is interred in this cemetery, but the cemetery database has "unknown" for both his birth and death. So, we can eliminate both Benjamin and Henry (or Thomas) as the son, and Phebe as the daughter, whose bodies were exhumed in 1872. That leaves John and Maria, both of whom had died prior to the 1872 exhumation date. The only discrepancy—not enough to doubt the overall veracity of this narrative—is the number of years they were interred prior to exhumation: John had been dead for seven years, not twelve; Maria had been dead for four years, not seven. If it was twenty-one-year-old Benjamin whose life the ritual was performed to save, then it might have been judged successful, as he lived for another fifty-four years; if it was Henry (or Thomas), we know that he lived until at least 1885.

Locating the family cemetery— "near Watson's Corner on the Saugatucket river" —was trying. The best candidate, the Ebenezer Adams Lot (SK031), just south of Watson's Corner, is on private property, accessible only with the owner's permission. When James N. Arnold (1844-1927), Rhode Island's preeminent recorder of family data, visited this cemetery in 1880, he noted that it contained the remains of seventeen members of the Rose and Adams families. My visit to the cemetery in 2008 confirmed Arnold's observation that the Rose lot was overgrown with weeds and briars, and, unfortunately, contained no inscribed stones. Research showed that Ebenezer Adams was the grandfather of William R. Rose, so it is plausible that Rose and his two children, John and Maria, were interred in this plot.

Two Charles Harringtons contend for William Rose's assistant. The elusive one was born in 1831 and at the time of the 1880 federal census was living in Exeter, listed as a single, white male. The "occupation" space was blank, like other aspects of this man's existence. The other Charles Harrington, born 1839, is buried in the Seth Harrington plot in South Kingstown. His residence in South Kingstown puts him in closer proximity to North Kingstown and William Rose. There is a possible occupational link between the families. William's occupation in the 1870 federal census is stone mason; his son, Benjamin, was listed as a stone mason in the 1900 federal census. Both Charles's father, Seth, and his brother, Daniel, were stone masons, according to the 1870 federal census. The *Narragansett Times* text omits Harrington's name, identifying him simply as "an Indian," a designation that probably would not have been apparent to outsiders. After three centuries of "submersion in American society," the Native Americans of South County (mostly of the Narragansett Tribe of Indians) came to "resemble neighboring non-Indian Americans more than . . . their seventeenth-century ancestors." With the exception of "a continuous tradition of beliefs and legends," William S. Simmons wrote in his study of New England Indian history and folklore, if one looked at "economic and political activities, . . . language, churches, homes, and appearances, one sees in most cases little that can be attributed to the indigenous sources of New England Indian culture."[218] So, even if Charles Harrington was known in his local community as an "Indian," that designation probably would have escaped the casual glance of an outsider. Census records are not much better in that regard. Significantly, the Narragansett Indians were (and are) renowned for their skill as stone masons.

The consumption ritual, like the disease it aimed to destroy, cut across social, economic, educational, and other cultural boundaries. Although the ritual was performed by people who, in good conscience, could not be marginalized as uneducated, backward, and superstitious, as the nineteenth century wore on, those performing the ritual risked ridicule. A degree of insouciance would have been insulating. Evidence shows that both William Rose and Charles Harrington could engage in behavior that was beyond the bounds of probity, even illegal. According to the *Narragansett Times* (April 26, 1861), "A brown mare with an express wagon attached, belonging to Mr. James H. Sweet, of Kingston, was carried off from the street . . . while he was attending" the circus. The horse and wagon later "were found on the railroad track, near the High House, Warwick"—an area now called Greenwood, more than twenty miles from Kingston. The *Narragansett Times* (May 3, 1861) continued this report:

> The horse was much fatigued with hard driving, and was left on the track in a position to be struck by the locomotive when it came along. Some persons in the neighborhood saw the horse . . . and removed him from the track a few minutes before the arrival of the train. Charles Harrington and Charles Streeter were arrested on Monday, on suspicion of being concerned in the affair, and were examined on Tuesday before Wilkins Updike, Esq. They plead guilty and were required to recognize the sum of $400 for their appearance for trial at the next term of the Court. Seth Harrington became surety for them, and they are now at large.

When the Court of Common Pleas for Washington County convened the next Monday, the grand jury returned a bill of indictment against Harrington and Streeter for stealing Sweet's horse and wagon. The *Narragansett Times* (May 17, 1861) reported that the pair was arraigned, and both pled guilty. Harrington was about twenty-two years of age at the time.

The granddaughter of William and Phebe Rose, Deda Belle Macdonald, contributed the following piece of family history concerning her mother, Phebe Rose Caswell, to the Rose Genealogy Newsletter in 1953:

> Phebe's parents were no doubt poor, as people with large families were in those days. Phebe was a well developed and pretty miss at the tender age of twelve. About this time, 1865, the family lived in Saunderstown, R.I. Nearby, a James Gardiner, aged about 73, who was obviously much older than Phebe, needed a wife, as he had a large home to care for. It was an old time country house, with a huge chimney and cupboards etc. all around it. However, this man offered to pay the father, William Rose, $100 if Rose could get Phebe to marry

him. Rose took the money and the daughter went to marry the old man, of whom she was afraid, and sent under protest. When it was time for the marriage to take place, Phebe could not be found. Of course the old man was "put out" and wanted the money back. Rose reluctantly returned the money. Phebe had hidden in the back of the old chimney. . . . Later, under cover of darkness, she returned to her folks and lived a normal life.

The letter went on to describe Phebe's subsequent marriage, at age 15, to James Caswell, a sea-faring man, and the birth of their five children, including Deda Belle, the narrator. The editor of the Rose Genealogy newsletter, in which this anecdote appeared in 1953, added the following cryptic comment: "There were some unpleasant notes about William Rose, but there seems no point in noting them here." Given Rose's apparent wild streak, the "unpleasant notes" could refer to any number of events. I wonder if the exhumation in 1872 is one.[219]

Almost twenty-five years after the exhumations, Benjamin C. Rose makes another appearance in the vampire chronicles. Again, we must assume the role of detective to fill in blanks. In his seminal article on the New England vampire tradition, published in 1896, George R. Stetson (discussed in chapter 16) presented the following text:

In the same village resides Mr ———, an intelligent man, by trade a mason, who is a living witness of the superstition and of the efficacy of the treatment of the dead which it prescribes. He informed me that he had lost two brothers by consumption. Upon the attack of the second brother his father was advised by Mr ———, the head of the family before mentioned, to take up the first body and burn its heart, but the brother attacked objected to the sacrilege and in consequence subsequently died. When he was attacked by the disease in his turn, ———'s advice prevailed, and the body of the brother last dead was accordingly exhumed, and, "living" blood being found in the heart and in circulation, it was cremated, and the sufferer began immediately to mend and stood before me a hale, hearty, and vigorous man of fifty years. When questioned as to his understanding of the miraculous influence, he could suggest nothing and did not recognize the superstition even by name. He remembered that the doctors did not believe in its efficacy, but he and many others did. His father saw the brother's body and the arterial blood.[220]

This account almost certainly is that of the family of William R. Rose of Saunderstown, and Stetson's informant must have been Benjamin C. Rose, the second son of William, born April 29, 1851. If it was twenty-one-year-old Benjamin whose life the ritual was performed to save in 1872, then it might have been judged successful, for he lived for another fifty-four years (dying August 20, 1926, at the age of 75). And, like his father, Benjamin was a stone mason (as noted in federal census data).

Stetson's account includes a few details at odds with the record. Benjamin would have been forty-four, not fifty, years of age when Stetson interviewed him in the summer of 1895. Contemporary newspapers stories relate that a son *and* daughter were exhumed—not just a son. When the son's corpse was deemed too decomposed, Maria's heart was cremated. Which version is accurate? On one hand, we have a newspaper article, based on an informant's account (who may even have been an eyewitness), published less than a week after the event; on the other, we have a son who, in 1872, was 21 years of age (was he at the exhumation?) and probably suffering from consumption, relying on his memory of an event twenty-three years in the past. Another possibility is that Stetson, himself, did not accurately record these details. He cited no publications regarding this case, so it is not known if Stetson had read, or even was aware of, the contemporary newspaper articles. Benjamin's father, William Rose, died in October 1894, about a year before Stetson arrived in South County for his fieldwork.

***

Every cloud has a silver lining, and the false trail that for several years frustrated my attempts to document the William Rose event presented an unexpected gift. On April 25, 1916, the New London (CT) *Day*, ran the following story, headlined: "Old superstition is still believed. People think dead feed upon living—New London County case is cited."

> A Hartford physician had among his patients a young Sicilian woman, a housemaid, who had pulmonary tuberculosis. The disease had advanced somewhat, but had not reached the stage where it would be pronounced incurable. The case was responsive to treatment, principally hygienic and dietetic, and the advisability of a change of climate was taken under consideration. Correspondence was had with Albuquerque, New Mexico, and he was fortunate enough to be able to secure the patient a position in one of the hotels in that city. The arrangements were in progress when she put them in disorder by deciding that she would prefer to return to her homeland. This was not considered advisable, as it was in midwinter, and an ocean voyage at that season and in her condition would have been hazardous.
>
> > "I know that," the girl said, "but I want to go."
> >
> > She had formed a plan of "telling all about America" before she "was taken sick."
> >
> > "But you will not be able to do anything when you reach there," she was told.
> >
> > "I know it," she said, "but in a little while I will be better. I know that!"
> >
> > In this she persisted.
>
> In the course of the conversation it came out that she had lost two sisters in Sicily with the same disease, and that her native town was an unhealthy, consumptive place. Admitting this, her insistence was inexplicable. After a little while she stopped talking about the trip, and spoke of having written to her father in Sicily. A month or more passed and then she came to the doctor.
>
> > "I will go to New Mexico," she said sententiously. "I shall get well now, and that climate will be best."
> >
> > The physician was congratulating himself on having persuaded her, when she said:
> >
> > "Father, as soon as he got my letter, and knew what ails me, went and did what will cure me."
> >
> > "What was that?"
> >
> > With no reluctance she explained.
> >
> > "I was afraid my sisters had been buried so long that he would have to burn the whole bodies. He would have done it, for he loves me so much. But he had the priest write me this letter, which shows what he did."

She handed the doctor a letter written in Sicily.

It read as follows:

"Your father, on having the graves opened, found the lungs of both girls and burned them on a charcoal stove at the graveside."

"What does that mean?" the doctor asked.

"Why, don't you know?" the girl said. "You see I began to realize that my sisters were drawing my life away, and I remembered that it used to be said that consunzione (consumption) can always be stopped in others by having the lungs of the dead taken up and burned. So I wrote father what ails me, and he did what the priest writes for him."

The physician, surprised at such a belief, asked, "And do you think the priest believes that?"

"O, sure!" was the reply. "He knows it is so."

The following week she left for Albuquerque, and has written several letters from there, telling that she is better and corroborating it by a local physician's case records.

"Doctor," she wrote, "if you find any other consumptive cases have had near relatives who have it, just you get them to have the lungs burned, and that will mean that the sick ones will live."

The physician was relating this story to some professional friends recently, when an old physician from New London county said:

"That is no new remedy! And it is not Sicilian superstition only! Seventy years ago, in 1846, Horace Ray, a well-known citizen, died of pulmonary tuberculosis in his home in Griswold, this state. Afterwards two of his sons, grown-up young fellows, died of the same disease, the last one in 1852. About two years later a third son developed the same disease. Where, or how he got the idea I cannot say, but he made up his mind to exhume the bodies of the two brothers and burn them, because, he said, the dead were supposed to feed upon the living. He declared a belief that the bodies were not decomposed, and that as long as they so remained, wholly or in part, he and other surviving members of the family would continue to furnish substance on which the dead could feed, and did feed. Acting under his advice, the family opened the graves, took up both bodies and burned them on the spot. That was on June 8, 1854, in what is now Jewett City, and I have a copy of the Norwich Courier of that week, relating the occurrence."

"Did the young man recover?" asked the Hartford doctor. "Did he—eh?—go to New Mexico forthwith and get well?"

"Results were not given," the New London man said. "I heard the story many times when I was young, and, as I have said, I have a Norwich paper telling about it."

"Let me put in a word," said a younger physician, who was in the company. "I think it was in 1875, when Dr. Dyer of Chicago, known to all as an eminent practitioner, reported in one of the medical journals a case occurring within his personal knowledge, where the body of a woman who had died from consumption was exhumed, and her lungs burned, under the belief that she was drawing after her into the grave some of her surviving relatives."

"Now one more from me!" the old New Londoner said. "I have at home, preserved with that Norwich Courier, a copy of the Providence Journal. At this moment I am unable to give its exact date, but I know it was in 1874. It is an account of one William Rose of Placedale, Rhode Island, digging up the body of his own daughter and burning her heart, under the belief that she was wasting away the lives of other members of the family!"

"Well," continued the Hartford doctor, "plainly southeastern Connecticut, and the contiguous little Rhody, and Chicago, share with Sicily in superstitions!"

So, the old physician from New London has a copy of the *Norwich Courier* that says the Ray family exhumations occurred on June 8, 1854? Well, that newspaper published its account of the event on May 24, 1854, stating that the exhumations took place on the previous eighth—which would have been *May* 8. Then the young physician from Hartford shares the account of Dr. Dyer of Chicago, sourcing a medical journal. The old physician follows up with the William Rose story, which he locates in *Placedale* (instead of the correct Peace Dale), claiming to have a copy of the *Providence Journal* article from 1874 that describes this event. We now know that the Rose story was first reported in the *Providence Evening Press* on September 4, 1872. Dudley Wright is the nail in this narrative's coffin. Wright's *Vampires and Vampirism* was published just two years before the exchange reported above. Wright's mistake-ridden narrative is matched by the New London *Day* account:

> Even America is not free from the belief in the vampire. In one of the issues of the *Norwich* (U.S.A.) *Courier* for 1854, there is the account of an incident that occurred at Jewett, a city in that vicinity. About eight years previously, Horace Ray of Griswold had died of consumption. Afterwards, two of his children—grown-up sons—died of the same disease, the last one dying about 1852. Not long before the date of the newspaper the same fatal disease had seized another son, whereupon it was determined to exhume the bodies of the two brothers and burn them, because the dead were supposed to feed upon the living ; and so long as the dead body in the grave remained undecomposed, either wholly or in part, the surviving members of the family must continue to furnish substance on which the dead body could feed. Acting under the influence of this strange superstition, the family and friends of the deceased proceeded to the burial-ground on June 8th, 1854, dug up the bodies of the deceased brothers, and burned them on the spot. Dr Dyer, an eminent physician of Chicago, also reported in 1875 a case occurring within his own personal knowledge, where the body of a woman who had died of consumption was taken from her grave and her lungs burned, under the belief that she was drawing after her into the grave some of her surviving relatives. In 1874, according to the *Providence Journal*, in the village of Placedale, Rhode Island, Mr William Rose dug up the body of his own daughter and burned her heart, under the belief that she was wasting away the lives of other members of the family.[221]

This refuting evidence casts doubt on the entire newspaper article, including the beguiling narrative of the consumptive Sicilian housemaid. It appears that the article's anonymous author used Wright's paragraph on American vampires as a platform to spin his tale within a tale.

But could the housemaid's story be authentic even though its context was fabricated? It's no stretch to believe that she could have been consumptive. Laura C. Rudolph, writing about Sicilians who immigrated to America during the early twentieth century, observed, "Sicilians were especially vulnerable to tuberculosis and many of them returned to Sicily gravely ill. During the mass migration, there were so many immigrants returning to die that several villages in Sicily set up sanitariums to receive them." As a strategic crossroads in the Mediterranean Sea, Sicily received traditions from many cultures, notably those of Europe, North Africa, and the Middle East. It is not inconceivable that, over the centuries, some version of the vampire belief would have found its way to the largest island in the Mediterranean. Rudolph described a folk culture that was primed for such practices, noting that many Sicilian religious beliefs "were based on a mixture of Catholicism, paganism, and superstition" that "for centuries" had been "virtually untouched and unchallenged from the unrest provoked in other countries during the Protestant Reformation." "Helpless and vulnerable against the elements," Rudolph wrote, "the superstitions, saints, and magic their folk religion provided them with helped to ease the uncertainties and anxieties of their rural lifestyle."[222]

Medical doctor Giuseppe Pitrè (1841-1916) documented and published an enormous corpus of folklore from his native Sicily (25 volumes between 1871 and 1913). He devoted an entire volume to folk medical practices, and included personal narratives and other supporting testaments, as well as relevant proverbs and sayings. Pitrè wrote, "When an illness persists and fails to respond favorably to ordinary expedients, the suspicion arises that more powerful causes are at work than normally present. These may be the work of evil spirits, of those who cast spells, or magicians and, equally bad, of imaginary supernatural and marvellous beings which are at work producing obstinate illness."[223] In my reading of this detailed collection, I found no mention of exhuming deceased relatives and burning their lungs (or entire corpses) to cure consumption—or any other ailment. Surely Pitrè would have encountered and recorded such practices. That he did not leads me to conclude that the entire 1916 narrative published in the *Day* was fictional.

If it was, indeed, fabricated, the obvious question is: Why? Simply an entertaining story? Or did its anonymous author have a more strategic intention? If the story's function was to underscore the folly of using folk instead of establishment medical practices, it subverts its own purpose: the housemaid apparently recovered; and the results of the American rituals were inconclusive. Perhaps the narrative aimed to contrast Hartford's mainstream society to the city's substantial and growing Italian (including Sicilian) population. An article published just eight months earlier in the *Hartford Courant* (August 29, 1915) appears to negate such intention. A full-page spread with photos in the Sunday feature section, devoted to Hartford's "Little Italy," led with a subhead that summarized the article's tone: "There are approximately 12,000 people of Italian birth or descent in our city and they are assessed for $10,000,000 on a list of $110,000,000. Their ability and industry have made them a very valuable part of the city's life." The opening sentence raised, then destroyed, a negative stereotype: "The conventional Italian of the vaudeville stage, with his air of good-natured poverty and his lack of intelligence, was not copied from the Italians of Hartford." Another, perhaps more remote, possibility lies in a perceived pecking order of Italian immigrants: northerners first, southerners second, Sicilians last.[224] Was the housemaid's story constructed to show that Sicilians were comparably more "uneducated" and "culturally backward" than their mainland neighbors?

Intent is an elusive behavior. Maybe it does just come down to an artful headline-grabbing story. But, whatever its purpose, it is fiction.

Part IV   Family and Media Talk about Mercy Brown

The family stories of Lewis Everett Peck and Reuben Brown are based on the exhumation of New England's most well-known vampire, Mercy Brown. Versions of her story have been communicated in every possible medium, from spoken, written, and printed, to film, video, and the Internet (a search of which yields millions of pages for Mercy Brown). She has been expanded and twisted into every imaginable undead shape, even though the facts of her plight are fixed in credible sources.

On March 17, 1892, three corpses—the wife and two daughters of George T. Brown—were disinterred and examined in the Chestnut Hill Cemetery, in Exeter, Rhode Island. The wife, Mary Eliza, had died of consumption on December 8, 1883. On June 6, 1884, Brown's eldest daughter, Mary Olive, succumbed at the age of twenty. Within a few years, Brown's only son, Edwin, began to show signs of consumption. Edwin traveled to Colorado Springs with his wife, Hortense (Hortie) Frances Himes, hoping that a change in environment would effect a cure. By this time, Brown's nineteen-year-old daughter, Mercy Lena, had contracted consumption. She died on January 19, 1892, and her corpse was placed in a stone crypt in the cemetery, awaiting the spring thaw for burial. As Edwin continued to decline, he and Hortie returned to Rhode Island to be with family and friends.

Hoping to save Edwin and spare the rest of the family, George Brown yielded to pressure from kinsmen and neighbors and consented to the consumption ritual. The medical examiner for the towns of Exeter and North Kingstown, Dr. Harold Metcalf, reluctantly agreed to oversee the proceedings. George Brown and Edwin were notably absent. When the body of Mrs. Brown was exhumed, some of the flesh appeared to be mummified, but there was no blood in her heart. Mary Olive's corpse had been reduced to a skeleton, with only a thick growth of hair remaining. But Mercy's corpse, reposing in an above-ground tomb for two winter months, seemed fresh. The heart and liver were removed. The heart showed clotted blood, but the liver, though deemed well-preserved, revealed no blood. Despite Dr. Metcalf's assurance that the condition of Mercy's corpse was unremarkable, attendants at the scene kindled a fire and burned her heart and liver to ashes. As a cure, Edwin was to drink the ashes in water, but no one could (or would) say if he did.

Sadly, the hoped-for cure did not work. Edwin died on May 2, a few weeks shy of his of twenty-fifth birthday. And three of the remaining Brown children also died of consumption. Annie Laura Brown, who had married Charles Edward Taylor in 1886, died August 9, 1895, at the age of twenty-five. Jennie Adeline Brown, married to William Henry Edwards, died on October 2, 1895; she was just eighteen years of age. Myra Frances Brown, who had married Arthur Caswell on May 25, 1898, died only thirteen months later, on June 25, 1899, also at the age of eighteen. Hattie May Brown was the only sibling to survive the family's consumption epidemic, living to the ripe old age of seventy-nine. Their father, George, also lived long, dying at age eighty, at Hattie's home, on November 17, 1922. Edwin's widow remarried and lived to the age of sixty-five.

The next two chapters, focused on varied stories about what happened to Mercy Brown, exemplify the dynamic nature of storytelling events as they are enacted over time in different contexts, including:

- How complex interactions between local knowledge and outsider viewpoints shape narratives.
- The reconstruction of personal stories, showing that even a story that has been told countless times by a single individual over a lifetime is never complete nor set in stone.
- Storytellers as cultural commentators.
- Agency and ownership of stories.
- The significance of role relationships between performer and audience.
- Social uses and cultural functions of stories.
- The legend process.

Crypt where the corpse of Mercy Lena Brown was kept prior to her exhumation, Chestnut Hill Cemetery, Exeter, Rhode Island. *Photo courtesy of Cyril Place.*

9 She Had Turned Over in the Grave

The recounting of family stories about Mercy Brown began soon after the event and continues into the present. Lewis Everett Peck (known as Everett in his community) was my initial link to the family narrative of Mercy's exhumation. I first interviewed Peck, related to Mercy on his mother's side (the Arnold family), on November 18, 1981, at his home on Sodom Trail, in the town of Exeter. At that time, I was directing a two-year survey of folk traditions in South County (officially, Washington County), Rhode Island. A self-described "Swamp Yankee" and jack-of-all-trades, Peck's knowledge of Exeter's history and folklife was broad and deep, so our interview was wide ranging. But it was the story of Mercy Brown that most stirred my curiosity. Peck said that he had heard the story many times from his older relatives, notably his grandmother, mother, and an uncle (who, Peck said, was a small child when the event took place). Following is a transcription (with my commentary in brackets) of relevant portions of that interview:

PECK. Mercy Brown was a relative. I can't tell you right now how we are related, but we are related. My grandmother was a Brown. And it was told to me . . . as a kid. [*explaining the purpose of a cemetery visit*] Decoration Day [May 30] was one of the big days, and Children's Day [at the end of the school year, in June] was one of our big days around here. You didn't go very far around here, you know. And when we went to the cemetery, there was Mercy Brown. . . . And they'd say, "Well, don't go running over there. Don't touch the stone, because of this awful thing that took place years ago." Now, what they do here [*pointing to a newspaper clipping from the* Narragansett Times, *dated October 25, 1979*], they change this around as if I believe in vampires. Now, that ain't what I'm sayin'. I'm just revealing what *they* believed and how they had to handle their own problems, see? So, anyway, . . . there'd been several in the family, they had come down with some disease. Young and old! All of a sudden! And anything they did didn't seem to stop it. Even those that didn't even live here, as far away as Ohio.

BELL. In the same family?

PECK. In the same family. Brother! Was coming down with the same sickness. So, there was twelve men, as it's told to me, of the family that was left. They got together and they figured it was all their turn. "This is it!" And they got together and they took a vote, what to do. And they dug up one grave, *not* several. They dug up *Mercy*. [*softly*] For some reason they picked her, because there was something there that led it to that. Then, they dug her up and she had [*emphasizes each word*] turned over in the grave. [*almost as an aside, in a hushed voice with a tone suggesting understatement*] Well, right away, there's a lot of problems there. So they took her out and [*emphatically*] they cut her heart out. There was blood in the heart. Well, they decided they had to kill it, so they started a fire . . . not far from the grave, and they burnt the heart, took the ashes and done something with 'em. [*softly, as an aside*] . . . I don't remember that stuff there, . . . And, [*matter of factly*] it seems as if that's what took care of it. [*as an explanatory aside*] You know, years ago you didn't have medicines, you . . . had to figure out your own. They were self-independent people, everybody that lived here. There was no such thing as relying on somebody. You did it yourself! [*signaling closure*] And that's in general . . . [*abrupt shift, referencing the newspaper articles*] Now they have gone further, some of these, and hunted updates and different things that I wasn't told . . . but, anyway. [*returning to closure*] And that's in general . . . what happened.

BELL. Did they call them vampires in the story, when you heard it?

PECK. No! . . . It wasn't designated as a vampire. It was just, you don't go over there handling it, and you leave it alone, because this awful thing that had to be done. And there's an old man living today who can remember one of the twelve. He's in his nineties. And his name is Brown, Reuben Brown.

Peck said that, when a local television news station interviewed him, he told them about Reuben Brown. He suggested that they call him before going out to interview him because he might just shoo them away.

PECK. So . . . they call him and he says, "Well, come in!" So they went down and . . . they asked him if he *believed* in it and he said, "'tain't what I believed in, it's what *they* did." You know, that was his answer.

Later in the interview I asked Peck if he knew about other, similar events in Exeter:

PECK. I know that there was another. There was another family that experienced this same thing. Which would be about, I would say, close to a mile east of where this other one took place. There was another one, I've been told. Which I didn't know it at the time. I've heard of this since these paper clippings. 'Cause I get letters from everywhere. Oh, yeah, I've been in places, people say, "Here comes the vampire man" and they go all through it, and "ha, ha, ha." It doesn't bother me because I don't ask them to believe it. I don't care if they do or not.

BELL. When did they start calling it "vampire," anyway?

PECK. I don't even know that. No . . . no . . . my grandmother and my mother never mentioned the word. I'm fifty-two years old and I never heard anybody talk about vampire until we got into this stuff here [*referencing the newspaper clippings*].

BELL. I wonder if it was the newspaper people?

PECK. Well, you never know. I don't know. . . . I couldn't answer that. Nope. [*once again, signaling closure*] So, that's the story on that one there![225]

Peck began his story with "orientation elements" that, according to folklorist Gary R. Butler, provide "the information concerning background that he [the narrator] considers necessary to orient listeners so they can understand what is to follow."[226] Peck briefly explains his connection to the family and how he came to know the story of Mercy Brown, which unfolded as a natural consequence of two generations of the family meeting in the cemetery to decorate graves on what was then known as Decoration Day (now Memorial Day) or coming together at the Grange (which abuts the cemetery) on Children's Day to celebrate the end of the school year. Told to a younger generation by their elders, the story explains the prohibition against touching "the stone [apparently the one upon which Mercy's heart was burned, but perhaps her gravestone], because of this awful thing that took place years ago." Peck's pointing to the newspaper article and taking exception to its suggestion that he personally believed in vampires illustrates another of Butler's structural elements in traditional narrative discourse: "When it becomes necessary, additional orienting information may be inserted into the body of the narrative at a convenient juncture."[227] In his summarizing statement—"you did it yourself"—Peck places the exhumation into historical and cultural contexts, reminding contemporary listeners that today's taken-for-granted medical treatments did not exist during Mercy's era and that, since the medical establishment at that time could not provide an effective treatment for consumption ("some sickness," in Peck's words), it fell to the people, themselves, to find a solution. Peck's final tag ("So, that's the story on that one there!") functions, in Butler's terms, as a "frame-out strategy": Peck signals that he has reached the end of his narrative. Returning to the conversational mode, he invites listeners to comment on, or ask questions about, the preceding narrative.

Peck's family story includes elements that are commonplace in narratives about New England exhumation rituals: a series of deaths prompts the men of the family to confer and reach a consensus (in this case, they "took a vote") regarding whether an exhumation was in order; finding blood in the corpse's heart, they proceed to burn the heart to ashes. Peck said he did not remember what was done with the ashes. Peck concludes that the ritual was effective in ending the string of family deaths ("it seems as if that's what took care of it"), and he avoids passing judgment on his family's course of action. Late in the interview, Peck returned to the question (in some cases, the assumption) about his own belief in vampires, which often arose during newspaper interviews—or even during interactions within his own community: "Do *I* believe in vampire [sic]? No, I don't believe in that. I'm not sure *they* did, but they had to come for an answer.... And some of them old people probably died with that in their mind, that they did the right thing."

Recognized as the unofficial spokesperson for the Mercy Brown event, Peck has been interviewed by the media many times over the years. With a few notable exceptions, his story has remained fundamentally consistent with the version he related to me in 1981. Three years later, however, in an interview with Karen Lee Ziner, in the *Providence Journal*,[228] Peck included some additional orienting information in his narrative. The "some sickness" he mentioned in our interview became "this vampirish disease"—about which Peck elaborated: "I imagine the disease was tuberculosis. But they didn't know much about tuberculosis then." Peck's additions may be the result of a feedback loop involving him and the media who publish/broadcast their interviews with him: when he relates his family story, newspaper and television commentators invariably identify the Brown family's disease as tuberculosis (or consumption) and connect the ritual to vampires. Perhaps Peck is addressing topics he believes are important to media workers, using terminology that makes sense to them and their audience. When media workers interview Peck for an assignment, Peck is the storyteller and they are the audience; but media workers become storytellers when they use his interview as a basis for narrating versions that they create for *their* audiences—newspaper readers, TV viewers, or online searchers. Media workers bring a shadow audience with them to their interviews with Peck, who is also aware of this audience. He may be framing his story with them in mind.

On several occasions Peck described an extraordinary personal experience that he did not introduce during our first interview. Some fifteen years later, in a subsequent interview with me and a documentary film crew from Connecticut Public Television, Peck recalled seeing a bright light shining over Mercy's grave. He related this corpse light incident in 1979, as documented in the *Narragansett Times*,[229] and again in 1988, for the *Boston Sunday Herald* (October 30, 1988). Perhaps his most detailed variant of this encounter appeared in Ziner's *Providence Journal* article of 1984:

> Though Peck says he doesn't believe in ghosts or roaming vampires, and though he insists this is all nonsense, he does admit he saw something strange one night, years ago, near Mercy's grave.
>
> That was when he was a young man out roaming with his brother and they drove up near that hill framed by restless trees, containing the supposedly restless spirit named Mercy.
>
> "I was about 18 or 19 years old when this thing took place. We had a Model A . . . and I went up in the back of the Chestnut Hill Church with my brother David.
>
> "And by God, we looked and we saw a great big ball of light, so bright it was blue." It hovered in the vicinity of the four or five graves where the Brown family members, including Mercy, are buried.
>
> "It was a bright light. It was round. God she was bright, that's the part that stuck in you. I have no idea what it was.
>
> "And to answer you how it went out. I don't know. We didn't stay," he says with a nervous grin that indicates he thinks he and his brother barely escaped an unfriendly encounter.
>
> The brothers drove down the road to a neighbor, also a member of the Brown family. He said of the glowing orb, "Sonny, we've seen it before." "And then he laughed," says Peck.
>
> "Then we talked to someone from the other side of the family, and she'd seen it, too," Peck says, the memory of his boyhood fright driving the glint in his eye.[230]

That Peck's extraordinary encounter was experienced by others in his family is not surprising —in the context of folk tradition, at least. The motif of lights around graves appears in several traditional narrative forms, including folktales, legends, and ballads.[231] In British balladry, the "light-soul" has been interpreted in the context of "blood superstitions" where "the blood of the slain may be transformed into a light or . . . a miraculous light may burn over the buried bodies." Lowry Charles Wimberly cites an Icelandic version of the traditional ballad "Babylon" where "a miraculous light burned over the place where the [murdered] maids had been buried."[232] Paul Barber connects such "corpse lights" to vampires and other reanimated human cadavers:

> Sometimes the revenant is discovered because his grave is visible, usually by either a blue fire or a blue glow. In *Grettir's Saga*, for example, Grettir is attracted to the mound of Kar the Old by fire on the moor. The blue glow, in European tradition, is frequently interpreted as the soul, and it is seen as an indicator of buried treasure through much of Europe, apparently because it shows where a body is buried, and bodies were frequently buried with valuable grave goods.[233]

Romanian *Strigele* (witches or their restless dead spirits) "are seen as little points of light floating in the air."[234] Bram Stoker, not surprisingly, used this motif in *Dracula*, to foreshadow impending horror: "a faint flickering blue flame" appears at the side of the road to frighten Harker and his stagecoach driver as they travel to Dracula's castle.

An early eighteenth-century appearance of a corpse light in the town of Narragansett, Rhode Island—adjacent to the town of Exeter—suggests the possibility of an extant local tradition long before Peck saw the blue light above Mercy's grave. The *American Weekly Mercury* published the following eyewitness account in April 1722; note the archaic spelling with 'f' standing in for the modern 's.':

> Newport, Rhode-Ifland, March 30.
>
> There has lately a furprizing Appearance been feen at Narraganfet, which is the Occafion of much Difcourfe here, and is varioufly reprefented; but for the fubftance of it, it is Matter of Fact beyond Difpute, it having been feen by Abundance of People, and one Night by about 20 Perfons at the fame time, who came together for the Purpofe. The Truth, as near as we can gather from the Relations of feveral Perfons, is as follows.
>
> This laft Winter there was a Woman died at Narraganfet of the Small Pox. and fince fhe was buried there has appeared upon her Grave chiefly, and in various other Places, a bright Light as the Appearance of Fire. This Appearance commonly begins about 9 or 10 of the Clock at Night, and fometimes as foon as it was dark. It appears varioufly as to Time, Place, Shape and Magnitude, but commonly on or about the Grave, and fometimes about and upon the Barn and Trees adjacent; fometimes in feveral Parts, but commonly in one intire Body. The firft Appearance is commonly fmall, but encreafes to a great Bignefs and Brightnefs, fo that in a dark Night they can fee the Grafs and Barque of the Trees very plainly; and when it is at the Heighth they can fee Sparks fly from the Appearance like of Sparks of Fire, and the likenefs of a Perfon in the midft wrapt in a Sheet with its Arms folded. This Appearance moves with incredible Swiftnefs, fometimes the Diftance of Half a Mile from one Place to another in the twinkling of an Eye. It commonly appears every Night, and continues till Break of Day. A Woman in that Neighbourhood fays fhe has feen it every Night for thefe fix Weeks paft.[235]

That both women died of a contagious disease—the Narragansett woman of smallpox and Mercy of consumption—might strengthen a supposition that there was a local corpse-light tradition. If so, it probably had European origins. The association of such lights with virulent diseases was noted by Jacob Grimm, who wrote that a tradition from Voigtland "makes the plague come on as a blue vapour, shaped like a cloud," and "a plague that raged in the Odenwald shewed itself in the shape of a little blue flame in the sacristy of the town-church at Erbach."[236] Viewed in a larger folkloric context, such lights form the basis of a number of different belief tales, including the *ignis fatuus*, better known in English-speaking traditions as the Jack-o'-lantern or Will-o'-the-wisp. This tradition is based on the widespread European belief that "souls which have not attained heavenly peace roam at night like bewildered birds, *in fiery shape*, on field and meadow."[237]

Everett Peck was devoted to collecting items of local history and lore from a variety of sources. Was he aware of the story from Narragansett? During our initial interview, in 1981, Peck projected a rational, down-to-earth view of his family's story and the events that brought it to life; he was disdainful of attempts to discover supernatural elements attached to Mercy's grave, such as whispering voices (reported in a newspaper article in 1979).[238] Was he being contextually sensitive in not mentioning his own apparently supernatural experience with the blue light? Knowing that I was engaged in a long-term documentation project, funded by a federal agency, to preserve his community's traditions, did Peck fear damaging his credibility or defaming his community if he related that story during our recorded interview? Did he view media workers as a different sort of audience, one more accepting of—if not decidedly on a quest for—such "mysterious" and sensational experiences? While such questions remain open for now—and, as we are about to discover, move to the foreground in Reuben Brown's version of the family story—Peck did answer when I asked directly, many years later (1995), why he did not tell me about his experience with the light around Mercy's grave during our first interview: "You didn't ask me," he said.

***

Everett Peck, keenly aware that some people considered his story of Mercy to be "farfetched," advised skeptics to talk to Reuben Brown—perhaps the last living, narrative link to the twelve men of the family who sanctioned the exhumations. Reuben Brown was born in Exeter in 1897, just five years after his uncle, George T. Brown, had consented to the exhumations of Reuben's aunt and two cousins. Peck warned that Brown should be contacted in advance, "because he may put you off the place. Maybe he won't talk to you." So, I was not terribly surprised (though disappointed) when Brown did not respond to my interview requests. However, in 1984, shortly before his death, Brown consented to a newspaper interview with Ziner (for the same *Providence Journal* article that included her interview with Peck). Brown's account, as quoted in the newspaper, includes not only Peck's eerie motifs, but also additional imagery from legend and mass culture. Here is the published portion of Ziner's encounter with Brown (suspension points in original):

> There are those from the Brown family who still care to tell the tale, and perhaps they know it best.
>
> Reuben Brown lives in the woods of Exeter, in a house ancient and creaky and alive with the soft gonging and ticking of an old clock.
>
> Brown is 87, hard of hearing and a mite creaky himself. Still, he's full of wit and he loves to tell stories.
>
> One of those is the legend of Mercy Brown.
>
> For this tale, Reuben Brown leans back in his worn brown leather chair, rests his feet on a wooden stool, and clutches his cane for emphatic, here-and-there taps on the floor. In the faded, sunlit living room, white-haired, 92-year-old Marion Brown sits on a couch and interrupts her husband now and then with laughter or correction.
>
> The whole fearful matter started with unexplained deaths, says Reuben Brown. Young girls, six or seven on one side of the Brown family, pined away and died. All of them "had a mark on their throats."
>
> "People figured they'd been bit by a vampire . . . they all had that mark on them and nobody knows who made it," says Brown.
>
> Some folks were sure that Mercy—already gone to her grave—was the vampire.

A dozen people got together—members of Mercy's family and others in the town—and decided to open the grave and pull Mercy's body into the sunlight to perform a terrible task.

Reuben Brown had a friend who was there.

"I used to know a man who saw them when they unearthed her. He said he saw them cut her heart out and burn it on the rock. . . it appeared that Mercy had moved in the grave. She wasn't the way she was put in there . . . .

"But he said there were no more deaths after that. That's what he said."

Reuben Brown adds this footnote: "My father believed she was a vampire. He said all those girls had the mark on their throat when they died."

Reuben Brown's footnote is suspicious, given his proximity not only to the nascent family *story*, but to the *family*, itself. Did Brown's father—S. Everett Brown, a younger brother of George T. Brown—really believe that his niece, Mercy, was a vampire? Or that all the dead girls (not just two, but six or seven?) had apparent bite marks on their necks? Brown's newspaper story raises many questions. If his tale accurately portrayed the events, why were striking details such as marks on the throats not made public in 1892? Why was this motif absent in Peck's story? Did Reuben Brown relate the story as it was told to him? Did his memory fail? Did he insert details he could not recall? Which elements in his narrative were received tradition and which were shaped, over repeated tellings, to correspond with his audience's expectations?

Interpreting Brown's interview with Ziner as a storytelling event suggests possible answers. In the newspaper article, we have a narrative version of Mercy Brown's exhumation, told by Reuben Brown, then edited by Ziner (the newspaper reporter) who, along with Brown's wife, Marion, were the story's listeners. Mrs. Brown also played a teller's role at some points during the storytelling event, as Ziner writes, "Marion Brown sits on a couch and interrupts her husband now and then with laughter or correction." If Mrs. Brown was able to "correct" her husband's narrative, we can presume that she knew it well enough to have formed some notion of a correct (or proper) text. The newspaper version of Reuben's narrative was therefore a text that emerged through interactions between him and his wife, as well as Ziner. One can presume that the same kind of "formative" and "plastic" interactive storytelling process was in play every time Reuben Brown narrated the Mercy Brown event, so we cannot know how the text has changed throughout many years of telling—a crucial point to keep in mind when we consider some of its striking motifs.[239] Ziner's omission of the elements that were corrected is an unfortunate obstacle to exploring the story's range of acceptable (socially approved) variation, at least during that particular storytelling event. One wonders, also in vain, which, if any, aspects of this "fearful matter" stimulated laughter. Since Ziner characterized Brown as "full of wit" and fond of telling stories, perhaps during this event he related other narratives—some of which furnished a basis for expecting more humor than would flow from stories of digging up and mutilating the bodies of one's own kin.

Ziner's newspaper write-up, describing a storytelling event—her interview with Reuben and Marion Brown—is a text that she created according to the premises of a narrative context quite distinct from that of the interview, itself. Newspapers and other popular periodicals lack the face-to-face directness and immediacy of an oral storytelling event. But both contexts share some fundamental processes: storytelling establishes role relationships— most importantly, tellers (narrators) and listeners (audiences)—and feedback is built into enacting these roles. In face-to-face storytelling contexts, feedback is contiguous with narrating: Mrs. Brown can interject a "correction" the moment she hears something that she believes needs correcting; Ziner can ask questions or, perhaps, use unspoken elements (a frown or shrug of the shoulders) to communicate uncertainty or a need for clarification. Even though the narrative is an emergent, dynamic expression, and each rendition is unique, it nonetheless remains stable enough for a community of listeners to identify it as a distinct story. The recursive feedback loop takes a longer, more indirect track in newspaper contexts, with additional roles and associated relationships that involve narrators, editors, and audiences. Yet, in both oral and newspaper contexts, processes such as role expectations and feedback continually shape and reshape narratives. Texts, or variant representations of the verbal (spoken or written) portion of these narratives, are deployed (we will see that "appropriated" might be a more suitable verb in some cases) by a variety of cultural systems for several reasons (social uses and cultural functions, in a more formal sense).

Five years before Ziner's article was published in the *Providence Journal*, Reuben Brown had been interviewed by Tom Burns, one of the team of four professional folklorists, who, in 1979, documented folk cultural activities throughout Rhode Island for the American Folklife Center at the Library of Congress. Burns did not tape this interview, but his detailed fieldnotes flesh out Ziner's sparse description of Reuben Brown. His residence, described by Ziner as a "house ancient and creaky," was a farmhouse converted from an earlier stagecoach inn. The farm, on Widow Sweets Road, is about one mile west of the Exeter Grange, Chestnut Hill Baptist Church, and Exeter Cemetery (Exeter Historical Cemetery #022, where the Brown family are interred), which are clustered on Ten Rod Road (Route 102).

According to Burns' fieldnotes, Brown's degree in mathematics, with honors, from the University of Rhode Island, led to several business offers. Rather than relocate to New York City, however, Brown chose to remain in Exeter and work on a dairy farm—not surprising, given his heritage. As the long-time editor of *Yankee* magazine, Judson Hale, wrote, "New Englanders are stay-put people. They are reluctant to move to another house, and they certainly don't want to leave the region."[240] By the time of his interview with Burns, Brown owned the farm, He told Burns several stories, including some witty ones alluded to by Ziner, but none concerned Mercy Brown. The humorous stories told by both Reuben Brown and Everett Peck are part of a localized variant of a regional Yankee tradition shared by the two men. Piece by piece, throughout the rest of our interview, Peck (just as Brown did) used both his own experiences as well as the memories and stories of older relatives to transform the geographical space of Exeter into a cultural place defined by social relations— "a sense of communality"[241]—that stretched back to the brink of collective memory.

The particular Swamp Yankee image that Peck and Brown cobbled together was rooted in subsistence. The short definition of a Swamp Yankee is "a rural New England dweller who abides today as a steadfast rustic and who is of Yankee stock that has endured in the New England area since the colonial days."[242] Other characteristics associated with this local variant of a Yankee include rural, agriculture-based subsistence; an intimate knowledge and respect for nature and the natural life; simple, homespun tastes; forthright and plain-spoken, yet also clever (as the stories of both Peck and Brown illustrated). Depending on context—who is using it and to whom it is addressed—the Swamp Yankee assignation can be variously a source of pride or an opportunity for ridicule. Brown and Peck are not only *aware* of the stereotypical traits associated with Swamp Yankees, they obviously delighted in playing to them and with them: perhaps affirming this trait on one hand but undermining that one on the other. Clever self-parody is a feature often ascribed to Yankees in general. In Seba Smith's fictional Maine Down Easter, Major Jack Downing, we saw another localized Yankee variant who became a model for fictional Yankees (chapter 4). The stories of Brown and Peck suggest a shared regional culture that harks back to the shrewd Yankee: both Downing, the Down Easter, and Brown and Peck, the Swamp Yankees, play off images that, while stereotypical, are rooted in an authentic self-perception. The veracity of these humorous stories is secondary to the self-conscious image that the men, themselves, advance in their narration.

***

When Tom Burns interviewed Reuben Brown and I interviewed Everett Peck, we were engaged in "cultural conservation," which at that time consisted of documenting (mainly through sound recordings and photography) traditional expressive culture for archival preservation. Although presentation and interpretation were viewed as likely future options, they took a backseat to basic fieldwork collecting.[243] Everyone involved in the fieldwork process was, to some extent, enacting the cultural-broker role, as both interviewers and interviewees attempted to bridge gaps in knowledge and understanding. Making sense, as a basic impulse, should not be underestimated when considering why people agree to share their local knowledge, such as family narratives, with a wider audience. Everett Peck is clear in his urge to explain why his not-so-distant ancestors exhumed the corpse of Mercy Brown. By grounding his narrative in a past that, although recent, was qualitatively unlike today, Peck enacted the role of culture broker. Emphasizing that Mercy's family and neighbors were part of a cohesive but independent community, living in a time that lacked the medical knowledge and treatment now taken for granted—a context that pressed them into figuring out for themselves how to cope with a "mysterious" and lethal epidemic—Peck asserted the reasonableness of behaviors deemed "superstitious" by outsiders distanced by culture or time.

When members of a close community, such as Peck and Brown, consent to have their expressive traditions documented, they de facto control the content—at least to the extent that they choose what elements to share and how to communicate them. They have a stake, and thus should have a voice, in the disposition of their traditions. Folklorists, by training (and in accordance with professional guidelines), determine both what expressive behaviors are suitable to document as well as which interpretive modes are appropriate. But informants also make these decisions, even if less self-consciously. Whether these conditions are agreed upon explicitly or, in most cases, only partially or implicitly, the "broker" (narrator, in this case) ultimately is the one who shapes the story and, to some extent, at least, frames it by suggesting contexts for interpretation and by providing clues about how the narrative should be made sensible. Informants are encouraged to view their local knowledge—often in the form of expressive\aesthetic cultural forms—as a valued aspect of a unique tradition. In the specific case of the folklife survey when I interviewed Peck, the sound recordings, photographs, notes, and other materials gathered, were deposited in an archive as a "permanent" record of a sampling of folklife from that time and place. Peck's wide range of expressive culture incorporated materials from many sources—including oral, printed texts and maps from newspapers, periodicals, pamphlets, and books, and copies of town records and similar documents—stretching back into the nineteenth century.

Peck was cognizant of his role as a cultural broker; it was obvious when he commented about outsiders who come to Exeter and act as if they are local-culture experts after a short amount of time of casual contact with a few people in the community:

> See, we don't have no true history of the town of Exeter. Out in the safe I probably have got much as anyone, and I won't hand it out. . . .. for a lot of reasons. . . .. It's against my grain to have someone come into town, be here six months and know the whole town and then . . . when you go to explain something they treat you . . . Who's this guy? Who's that fat old guy, he don't know nothin'. You know, and that, that don't go good. And, so, . . . I decided when they come here I didn't tell everything, just enough to let 'em know we did know, and that's it.

Peck was acutely aware of the appropriation process and took an active approach to controlling it. "Appropriation" is the relevant concept when the relationship can be viewed as, in Diane Goldstein's terms, "cultural imperialism" —which is an asymmetrical relationship where, in this case, those who seek local knowledge have a perceived (or assumed) privileged position above those who create and possess it.[244] Peck's comment about the obituary of a local history author, which appeared in the newspaper the day of our interview in 1981, was telling: "Fact, this mornin's paper, the writer—found her dead. . . . I says, 'Well, I don't wish anybody dead but I ain't gonna cry over that one'."[245] At that time, I took Peck's comment as a challenge for me to engage in self-reflection. Now, with four decades of additional folklife work behind me—though that challenge still obtains—I believe that Peck and I share at least one goal regarding how his family's narrative should be framed and transmitted: to counter the bias of looking at the past only through the window of today. As a scholar, I have been able to gather, and muster into this cause, many additional narratives.

Following publication of my interpretation of the Mercy Brown event (and other consumption ritual narratives) in *Food for the Dead* (2001), I noticed that Peck began referring media workers, especially those involved with television and film documentaries, to my book, or to me for interviews, regarding Mercy Brown. When I asked him why he decided to stop granting interviews, he said that, since I had explained everything in my book, there was nothing more he needed to add to that conversation. Media workers still would contact me, asking if I thought Peck might consent to be interviewed. I would pass on what Peck had told me, yet many asked and he consistently declined. So, I was surprised when, in 2015, a college journalism student, Stephanie Ressler, who had interviewed me for her thesis on New England vampires, was granted an interview by Peck. After Ressler posted her thesis online, I asked her how she managed to get an interview. She emailed me:

> I found his number, called him, and he invited me to visit his house. As I described in the story he was very weary [wary?] of sharing his story and what my intentions were with the interview. I had to do some convincing to assure him I would be respecting his family and not improperly labeling Mercy as a "blood sucking vampire." Most questions I asked him he would refer me to your book but he eventually opened up and agreed to retell some of the stories he told you like about some Halloween nights and the story with the TV crew at the gravesite. As he would tell me these stories he would have little recollections that he would

share with me along the way. It wasn't an easy task though, but I was grateful for every tidbit I got from him and also for the extensive research I was able to gain from your research.[246]

In her thesis, Ressler wrote that "Everett was extremely satisfied with how Bell told the story with respect to him and the Brown family history, so he promised himself it would be the last time he would share the details of the story to someone new. Any stranger lucky enough to hear the story out of his mouth is always questioned for their intentions."[247] My good feeling that Peck was satisfied with how I told his family's story is tempered by the unshakable feeling that I have "appropriated" his family narrative and am now its "broker." Peck uses the narrative, as it is currently embalmed in print between the covers of my book, as "the text" on the occasions when he directs inquisitors to read it if they want the "real" story of what happened to Mercy and her family. The publication of his narrative lets Peck, now in his nineties, off the hook, which, in practical terms, means that he has an excuse for not telling the story. Appropriation of cultural property was an unintended, but perhaps inevitable and foreseeable, consequence of my interactions with Everett Peck. Although Peck willingly relinquished his narrative (making the transaction seem less than a hostile takeover), I am still uneasy that the "burden" of storytelling shifted from Peck to me.[248]

The family stories related by Lewis Everett Peck and Reuben Brown reveal that the source of Brown's narrative—his friend who attended the exhumations—was, according to Peck, one of the twelve men who decided that George Brown should consent to the ritual. Brown received the story from a man who was not just an eyewitness, but also a principal actor in the ritual, beginning with the decision-making stage and continuing, perhaps, to the heart burning. A reasonable inference is that this unnamed participant began relating his experiences to people in his social circle very soon after the event. Unfortunately, we do not have the story as it was formulated and communicated by him. But we do have an 1892 account by a "local correspondent" for the *Providence Journal* who interviewed eyewitnesses. In the next chapter, we interpret his narrative, which straddles the line between insider (esoteric) and outsider (exoteric) viewpoints.

10   What Really Happened to Mercy Brown?

Two days after three members of George T. Brown's family were exhumed in Exeter, Rhode Island, the *Providence Journal* published an article describing the scene, beneath these lurid headlines:

EXHUMED THE BODIES.

Testing a Horrible Superstition in the Town of Exeter.

BODIES OF DEAD RELATIVES TAKEN FROM THEIR GRAVES.

They Had All Died of Consumption, and

the Belief Was That Live Flesh and

Blood Would be Found That Fed Upon

the Bodies of the Living.

The brief narrative that followed failed to live up to the promised melodrama, ending with, "Mr. Brown has the sympathy of the community."[249] Two days later, on March 21, 1892, the newspaper atoned for its initial hyperbole by publishing a much longer article that included two different narratives, one embedded within the other. The embedded narrative was contributed by a "local correspondent." The newspaper's editors used quotation marks to separate it from their own commentary, which framed the following text by the local correspondent:

"It seems that Dr. Metcalf attended Mercy Lena Brown during her last illness, and that a short time prior to her death he informed her father that further medical aid was useless, as the daughter, a girl of 18 or 19, was in the last stages of consumption. The doctor had heard nothing further from the family until about a year ago, when a man called on him and stated that Edwin A. Brown, a son, was in a dying condition from the same disease, and that several friends and neighbors fully believed the only way in which his life could be saved was to have the bodies of the mother and the two daughters exhumed in order to ascertain if the heart in any of the bodies still contained blood, as these friends were fully convinced that if such were the case the dead body was living on the living tissue and blood of Edwin. The doctor sent the young man back, telling him the belief was absurd. Last Wednesday the man returned and told the doctor that Mr. Brown, the father, though not believing in the superstition himself, desired him to come up to satisfy the neighbors and make an autopsy of the bodies.

"On Wednesday [note other sources cite Thursday] morning, therefore, the doctor went as desired to what is known as Shrub Hill Cemetery, in Exeter, and found four men who had unearthed the remains of Mrs. Brown, who had been interred four years [she died in December 1883]. Some of the muscles and flesh still existed in a mummified state, but there were no signs of blood in the heart. The body of the first daughter, [Mary] Olive, was then taken out of the grave, but only a skeleton, with a thick growth of hair, remained.

"Finally the body of [Mercy] Lena, the second daughter, was removed from the tomb, where it had been placed till spring. The body was in a fairly well preserved state. It had been buried two months. The heart and liver were removed, and in cutting open the heart, clotted and decomposed blood was found, which was what might be expected at that stage of decomposition. The liver showed no blood, though it was in a well preserved state. These two organs were removed, and a fire being kindled in the cemetery, they were reduced to ashes, and the attendants seemed satisfied. The lungs showed diffuse tuberculous germs.

"The old superstition of the natives of Exeter, and also believed in other farming communities, is either a vestige of the black art, or, as the people living here say, is a tradition of the Indians. And the belief is that, so long as the heart contains blood, so long will any of the immediate family who are suffering from consumption continue to grow worse; but, if the heart is burned that the patient will get better. And to make the cure certain the ashes of the heart and liver should be eaten by the person afflicted. In this case the doctor does not know if this latter remedy was resorted to or not, and he only knows from hearsay how ill the son Edwin is, never having been called to attend him."[250]

We saw in the last chapter that Everett Peck claimed Mercy "had turned over in the grave" and Reuben Brown said, "it appeared that Mercy had moved in the grave. She wasn't the way she was put in there." Brown later added, "My father believed she was a vampire. He said all those girls had the mark on their throat when they died." These two motifs were not mentioned by the local correspondent, nor did they appear in other contemporary sources, which leads me to conclude that these motifs were added to the family stories after the incident was reported in the *Providence Journal* and other Rhode Island newspapers in 1892.

The popular vampire image—well established by the late nineteenth century—was evident in a response to the local correspondent's narrative by the editors of the *Providence Journal*: "All mention of 'the vampire' is omitted from this account of the exhuming, but this signifies nothing. The correspondent simply failed to get to the bottom of the superstition."[251] The editors rectified his perceived inadequacy by offering the *Century Dictionary*'s definition of *vampire*: "A kind of spectral being or ghost still possessing a human body, which, according to a superstition existing among the Slavic and other races of the lower Danube, leaves the grave during the night, and maintains a semblance of life by sucking the warm blood of men and women while they are asleep." Had the family narrators also yielded to the European-derived vampire image of popular media?

J. A. MacCulloch (in *Hastings Encyclopaeida of Religion and Ethics*) identified two distinct vampire types: one was simply "the spirit of a dead person" and the other was a "corpse, re-animated by his own spirit or by a demon, returning to sap the life of the living, by depriving them of blood or of some essential organ, in order to augment its own vitality."[252] The European vampire, and certainly its literary offspring, is characteristically the second type; the New England vampire is not so easily pigeonholed. Its corpse, remaining in the grave, seems possessed by an unnamed, incorporeal evil. Yet, by Halloween of 1981, the bite-mark motif attached to Mercy Brown was back in the *Providence Journal*: "the local story was that she had died of a bite in the neck."[253] The source of this "local story" is unidentified. Was it Reuben Brown?

We know little about the local correspondent whose account was published in the *Providence Journal* four days after the exhumations. He apparently interviewed Harold Metcalf and others close to the event. His detailed description of Metcalf's autopsy of Mercy's heart and liver suggests that he had direct access to that information. Was he a witness, or did he receive those details from Metcalf, either through conversation or by way of a yet undiscovered written autopsy report? The correspondent's narrative hints that he was at the cemetery, but his situating the event on Wednesday rather than Thursday argues the contrary. Without additional evidence, I suspect that the correspondent did not attend the ritual but received pertinent information from Metcalf.

The correspondent shows empathy, recording that George Brown did not believe in the theory but acquiesced to the wishes of his friends and neighbors. That the correspondent did not mention *vampires* hints that his informants were local. Viewing the incident through the lens of newspapers that served the Exeter community sharpens the insider focus.

***

In close-knit rural communities, such as Exeter in the late nineteenth century, people hesitate to share information with outsiders that might reflect negatively on their neighbors and kin. Much local information travels informally, by word-of-mouth, without media involvement. People interacting face-to-face can incorporate feedback and adjust their narratives as they see fit. Newspapers cannot instantaneously regulate the flow of their narratives. And there are few sanctions against—indeed, there are often rewards for—publishing sensational stories. Exhumations, corpse mutilation, and cannibalism enthrall many while defaming few. For insiders, derogatory commentary accompanying disagreeable descriptions added insult to injury. Local newspapers that reported news of South County (including Exeter) counterbalanced the outsider newspaper stories.

The general rift between insiders and outsiders was intensified by long-standing social, political, and economic conflicts between urban Providence and rural Rhode Island. The Brown family exhumations received widespread scrutiny in part because they occurred at the end of a century that had experienced great changes in science and technology. A vampire exhumation in 1892 was a startling nullification of "progress" and "advances in civilization." In the vocabulary of the era's social scientists, it was a "survival from a lower stage of cultural evolution" (we examine this notion in Part VII). But the battle over Exeter's consumption ritual also reflected in-state animosities that had been in place since at least the Revolutionary War. The large debt accumulated during that war strained relations between farmers and urbanized commercial interests in Rhode Island. Farmers had borne the brunt of this financial burden since revenue from real estate taxes had to offset revenues lost due to the disruption of trade and the British occupation of commercial port towns.[254] Farmers argued successfully for the issuance of paper money, allowing them to pay their debts to reluctant creditors with highly inflated currency. Even as Rhode Island was becoming the first urban, industrialized state in the Union, the state legislature continued to be dominated by rural interests, which steadfastly maintained a tax structure and voting requirements that ensured the ongoing disenfranchising of the burgeoning cities in the north. This disparity led to an armed confrontation in the 1842 Dorr Rebellion, which created greater representation for Rhode Island's northern industrial towns. As the momentum continued to swing in the direction of cities during the nineteenth century, farming communities lost their dominance. These hostilities were still palpable in 1892.

The local point of view was framed in a letter to the *Pawtuxet Valley Gleaner* (March 25, 1892) from a self-described neighbor of Mercy's father, George T. Brown:

Mr. Editor, as considerable notoriety has resulted from the exhuming of three bodies in Exeter cemetery on the 17th inst., I will give the main facts as I have received them for the benefit of such of your readers as "have not taken the papers" containing the same. To begin, we will say that our neighbor, a good and respectable citizen, George T. Brown, has been bereft of his wife and two grown-up daughters by consumption, the wife and mother about eight years ago, and the eldest daughter, Olive, two years or so later, while the other daughter, Mercy Lena, died about two months since, after nearly one year's illness from the same dread disease, and about two years ago Mr. Brown's only son Edwin A., a young married man of good habits, began to give evidence of lung trouble, which increased, until in hopes of checking and curing the same, he was induced to visit the famous Colorado Springs, where his wife followed him later on and though for a time he seemed to improve, it soon became evident that there was no real benefit derived, and this coupled with a strong desire on the part of both husband and wife to see their Rhode Island friends decided them to return east after an absence of about 18 months and are staying with Mrs. Brown's parents, Mr. and Mrs. Willet Himes. We are sorry to say that Eddie's health is not encouraging at this time. And now comes in the queer part, viz: The revival of a pagan or other superstitions regarding the feeding of the dead upon a living relative where consumption was the cause of death and so bringing the living person soon into a similar condition, etc., and to avoid this result, according to the same high authority, the "vampire" in question which is said to inhabit the heart of a dead consumptive while any blood remains in that organ, must be cremated and the ashes carefully preserved and administered in some form to the living victim, when a speedy cure may (un)reasonably be expected. I will here say that the husband and father of the deceased ones has, from the first, disclaimed any faith at all in the vampire theory but being urged, he allowed other if not wiser, counsel to prevail, and on the 17th inst., as before stated the three bodies alluded to were exhumed and then examined by Doctor Metcalf of Wickford, (under protest, as it were being an unbeliever.) The two bodies longest buried were found decayed and bloodless, while the last one who has been only about two months buried showed some blood in the heart as a matter of course, and as the doctor expected but to carry out what was a foregone conclusion the heart and lungs of the last named (M. Lena) were then and there duly cremated, but deponent saith not how the ashes were disposed of. Not many persons were present, Mr. Brown being among the absent ones. While we do not blame any one for there [sic] proceedings as they were intended without doubt to relive [sic] the anxiety of the living, still, it seems incredible that any one can attach the least importance to the subject, being so entirely incompatible with reason and conflicts also with scripture, which requires us "to give a reason for the hope that is in us," or the why and wherefore which certainly cannot be done as applied to the foregoing.[255]

At the time of the exhumations, Edwin Brown was living with his wife, Hortie, in the village of Wickford, North Kingstown, the town just east of Exeter—near the home of Dr. Metcalf. In its first issue following the exhumations, the *Wickford Standard* (March 25, 1892) ran this short statement under the "Exeter" column: "There has been considerable said about the exhuming of the remains of a young lady recently, under the old theory that a vampire in her body was sucking the life blood of her surviving brother who is ill. The incident did occur, and many there are who are loudest in its denunciation would have caused the same thing to be done, were they in the place of the bereaved father."[256]

Following Edwin's death, on May 3, 1892, the same newspaper published a lengthy piece, entitled "A Rhode Island Country Town." While there was no explicit reference to either Edwin's death or his family's exhumations, the following excerpts draw their significance from that context. The article begins by extolling the unique virtues of Exeter, "a quiet rural town . . . where the sun rises earlier and where its golden beams linger later than in most other places in Southern Rhode Island." Not only is Exeter naturally illuminated longer than its neighbors, it is also free of modern corrupting influences, for "no travelling theatrical troupe or circus company ever stops over night . . . to poison the minds of the young." Residents of Exeter, "are honest, industrious, amicably-inclined and . . . are generous and social in their relations with each other." Misfortune brings out the best in "these kind-hearted country people," for "in cases of sickness and death . . . their neighborly hands will linger long and lovingly in their efforts to perform . . . the highest service which human power can render." Outsiders, however, such as "young, ambitious city journalists" looking for "something sensational and unreal, . . . have allowed their imaginations to picture a ghoul in every deserted fireplace." "Unaccustomed to the conditions of country life," these writers "regard every man, the cut of whose clothes showed independence in dress, as a deluded believer in some superstitious impossibility." The article concludes by asserting that, relative to other communities, the people of Exeter "possess all the virtues and fewer of the vices, . . . and it is equally true that the average education, general intelligence and freedom from superstition will compare favorably with any other town in the State, although it bears the name of 'Deserted Exeter'."[257]

Local newspaper articles assume an audience whose shared experiences can be invoked without explication: those who need to understand *will* understand because they have access to knowledge not available to outsiders.[258] The esoteric viewpoint expressed in a letter to the *Pawtuxet Valley Gleaner*, responding to the exoteric label "Deserted Exeter," is plain: "We think that someone in 'Deserted Exeter,' who can afford it, should contribute an old shoe from which a medal can be made and suitably inscribed (never mind the expense we say) and then with appropriate ceremony have the said medal presented to the very (?) gentlemanly reporter whose graphic and herculean efforts recently appeared in the Providence Journal. 'Nuff said.'"[259] Although we do not know the identity of the *Journal*'s local correspondent, we might have some empathy for his plight, caught between two competing cultural systems. He is taken to task by his editors for failing to get to the bottom of the "vampire" story, and then raked over the coals by locals for disregarding the integrity of their community for the sake of personal gain.

***

Newspaper articles that appear around Halloween highlight eerie decorations and playful interactions with supernatural phenomena. Halloween was not always the cheerful diversion it has become. A contraction of hallowed (or holy) evening, Halloween originated in the ancient Celtic celebration of summer's end (Samhain). The last day of October was a night of gloom, a grieving for the decline of the sun and the shortening of the days. On this evening, the Lord of Death summoned souls of the recently dead to pass judgment on them. These roaming spirits were believed to have extraordinary power to work harm on the living, and prudent people stayed home to avoid places where the dead might assemble. Halloween is currently an immensely popular celebration for both children and adults, as the once genuine fear of the dead has transformed into an entertaining quest for spooky things and chilling experiences.[260]

Media workers now serve up Mercy Brown as fear-free fright. The headline of Karen Ziner's story— "The Cryptic Tale of Mercy Brown\Was She a Victim . . . or a Vampire?"[261]—combines an aura of mystery with playful punning, a typical approach as exemplified by the following headlines:

- France Carrardo Bolderson, "Horrible History at the Boston Boo-Centennial," *Providence Journal-Bulletin*, October 11, 1975
- Paul Eno, "They Burned her Heart . . . Was Mercy Brown a Vampire?" *Narragansett Times*, October 25, 1979
- C. Eugene Emery, Jr., "Did They Hear the Vampire Whisper?" *Providence Sunday Journal Magazine*, October 28, 1979
- Bruce Fellman, "What Really Happened to Mercy Brown?" *Providence Sunday Journal Magazine*, October 28, 1979
- Bruce Fellman, "Things Still Go Bump in South County Night," *Providence Journal*, October 31, 1980
- "Rhody Town Was Heartless When It Came to Vampires," *Springfield Republican*, November 2, 1980
- Roy Bongartz, "When the Winds Howl and the Trees Moan . . .. But They're Just the Stuff of Old Stories—So We're Told," *Providence Sunday Journal Magazine*, October 25, 1981
- Les Daniels, "Grave Undertakings \ Rhode Island's Stake in Vampires," *Providence Eagle*, October 29, 1981
- Gerry Goldstein, "Here May Rest a Vampire," *Providence Journal-Bulletin*, October 31, 1989
- C. Eugene Emery, Jr., "Century Ago, R.I. Staked Claim in Vampire Lore \ Tale of Family Curse Spurred Neighbors of Exeter Woman to Unearth Her Body — and Legend," *Providence Sunday Journal*, January 12, 1992

Both students and bearers of local traditions may be understandably annoyed when media trivialize community lore. Even so, the relationship between mass culture and local legend is not entirely antagonistic, at least concerning vitality, if not veracity. Well before the advent of the Internet, folklorists recognized the often-symbiotic relationship between oral tradition and mass media. Legend scholar Ronald L. Baker remarked that "there is plenty of evidence suggesting that mass culture nourishes legendry—providing it with fresh subject matter and speeding its dissemination."[262] The Mercy Brown case, in particular, gives substance to Baker's observation that "television, radio, newspaper, and other mass media have engulfed and spread a number of legendary themes."[263] Donald Allport Bird even argues that the media sometimes assume the role of folk narrators, becoming active bearers of tradition.[264] As Norine Dresser remarked in her book on American vampires, "It is the media—radio, TV, movies, newspapers, magazines—that have largely shaped our folklore about the vampire today, altering the historical and Old World roots of belief."[265]

Mass media accounts during recent decades have incorporated elements from the popular conception of vampires, nurtured and sustained by numbing repetition in film, television, novels, and the Internet. In 1970, the author of a *Yankee* magazine article asserted that "the family and friends" of Edwin Brown "unanimously agreed that it must be a vampire that was sucking his blood and causing his loss of strength."[266] Five years later, a *Providence Journal-Bulletin* (October 11, 1975) article, "Horrible History at the Boston Boo-Centennial," reinforced the blood-sucking motif: "It was a sudden deteriorating illness of Edwin A. Brown many years ago that caused his family and friends to believe a vampire must have sucked his blood, causing a once husky and healthy man to lose all his strength."[267] By 1980, it was a "ghoulish curse"—not consumption—that had stricken the Brown family, and Mercy had acquired the attributes of a typical, corporeal pop culture vampire: "Surely the fiend rested there and stole forth from the crypt to drink the spirit of her brother."[268] Just one year later, again in a *Providence Sunday Journal* (October 25, 1981) article, Mercy Brown, like all successful pop culture vampires, had achieved immortality: "The mother, Mary Brown, had died on Sept. 8, 1883, apparently of tuberculosis, but the local story was that she had died of a bite in the neck. The following year, on June 6, her daughter, Mary Olive, aged 20, followed her in death—and again a mysterious bloodsucker was blamed." The article relates that "a doctor prescribed a dose of ashes of Mercy's heart dissolved in tonic" to "exorcise the demon" from her brother, Edwin. The narrative ends by leaving the vampire door ajar: "That vampire still roams free in Chestnut Hill Cemetery, and is said to rage around that midnight fire that glows anew every Halloween in Exeter."[269]

When media stories suggest that Mercy still haunts Chestnut Hill Cemetery, legend trippers descend on the place. Some left sound recorders at her grave to run during the night. Several claimed to hear whispering from the grave when the recordings were played back.[270] Mass media, now firmly entrenched in the legend process as active bearers of tradition, have become spirit belief specialists who interpret and shape, for many people, the supernatural nature of events. Yet, where outsiders look to mass media for interpretations of a community and its beliefs, insiders continue to evoke established folk tradition. Everett Peck's reaction to leaving tape recorders at Mercy's grave harmonizes with the down-to-earth tone of the family legend: "Well, they got some kind of soundin' devices to hear, uh, they say they hear things. Well, now, damn, you ain't gonna hear nothin'. . . . There may not even be *bones* left."[271]

As the popular narrative moved away from the local story, the focus of conflict shifted from a family's effort to stop the onslaught of consumption—a folk medical practice—to combating a pop culture vampire. The contrast between insider and outsider versions of the narrative is deeper than text, plot, or other content. *Function* changed along with substance, as a belief legend grounded in a local community transformed into a media-based fabulate, a story in which matters of belief are subordinated to artistic performance and entertainment.[272] These vampires, as Harry A. Senn noted, "belong to a different intellectual and social context" than the vampires of folk tradition and history.[273] The insider and outsider ways of seeing—call them what you may: paradigms, language games, communication sources, cultural systems—are vying for the same territory. And, significantly, it is outsiders who have imposed, and continue to impose, the *vampire* on these cases.

By the turn of the twentieth century, vampires ceased to be the threat envisioned just a few years earlier in New England, and mass media communicators changed the vampire paradigm.

Vampire stories followed the general trend of blending news and entertainment. "Infotainment" (a neologism combining "information" with "entertainment") blurs distinctions between "hard" news—timely and important information about serious matters that people should have to be knowledgeable citizens—and "soft" news—"interesting matters" that deal with "human foibles" and the "texture of our human life."[274] Because soft news stories do not become obsolete within hours, they fill Sunday papers and the content of infotainment programs, presentations that can be assembled well before they are distributed, broadcast, or posted. These are narrative-driven stories that construct "a shared world with readers emotionally." They are, as sociologist Michael Schudson noted, "part of a process of producing collective meanings rather than as a process of transmitting information."[275]

In 1892, the editors of the *Providence Journal* treated the Mercy Brown event as hard news. They and other outside commentators did not hesitate to hammer home the lessons of this timely, factual, and important incident. As the story lost its luster as news, it was transformed from "testing a horrible superstition" into "things still go bump in South County night." Narrators who sublimate facts for the sake of "color" create good stories at the expense of good history. Those familiar with popular culture vampires (who isn't?) can probably plot the arc of contemporary oral tradition regarding Mercy Brown. Her father, George Brown, comes to the foreground as not only the assenter for her exhumation, but also the actual exhumer and heart extractor, believing firmly that his daughter was, indeed, a vampire. Mercy's gravely ill brother, Edwin, appears at the cemetery, as well, where he must ingest the ashes of her heart. To the delight of legend trippers, Mercy, the vampire, was not finally laid to rest on that cold March day in 1892: she continues to haunt the Chestnut Hill Cemetery and environs. The blue light reported by Everett Peck is still seen, hovering over Mercy's grave. In some recent variants, Mercy can be summoned by peering at her tombstone through the hole in a nearby gravestone and repeating three times, "Mercy Brown are you a vampire?"[276] In another variant, Mercy appears to the gravely ill, guiding them toward a peaceful death. (Internet narratives follow the same arc. In a Google search, "Mercy Brown vampire" yields millions of results, many of which contain recent iterations of contemporary Mercy legends.)

***

The tale of Mercy is not the only enduring vampire legend in Rhode Island. Another young woman from rural Rhode Island competes with Mercy.

A tombstone inscription that reads, "I am waiting and watching for you," does not seem out of place in the context of nineteenth-century epitaphs. A young woman dying of pneumonia at the age of nineteen, while tragic, would not stir a legend. But when these elements were combined into a legend trip sometime in the late 1960s, a new and vigorous local legend was born. At least, that is the commonly accepted hypothesis by those who have investigated the legend of Nellie Vaughn.[277]

Over a period of twenty years, I conducted fieldwork, including interviews, site visits, and research into published and unpublished materials to track down the legend of Nellie Vaughn. As I concluded in *Food for the Dead*, there are three interrelated legends: (1) Nellie Vaughn is a vampire; (2) Nellie is not now, nor has she ever has been, a vampire; (3) Nellie is a ghost on a mission to let people know she was never a vampire. The third legend combines the first two legends by assuming both the existence of the first legend (Nellie is a vampire) as well as the truth of the second legend

(Nellie is not, and never was, a vampire). The third legend explains why Nellie's ghost haunts the Plain Meeting House cemetery.

Nellie's status as a vampire began when a teacher at Coventry High School, in western Rhode Island, related the story of Mercy Brown, omitting names and exact locations. The students were told that the vampire was buried in a cemetery next to a church. High school students being high school students, they went on a quest to find the vampire's grave. Traveling down the main north-south thoroughfare in western Rhode Island, Route 102, they came upon the Plain Meeting House Church and adjacent cemetery (which is a few miles from the highway). They noticed Nellie Vaughn's gravestone, bearing the inscription, "I am waiting and watching for you." She was the same age at death as Mercy Brown (nineteen), and died about the same time (1889, as opposed to 1892). These details combined with the suggestive inscription to transform Nellie into a vampire.

The ensuing years of unrelenting legend tripping prompted residents of the town of West Greenwich (where Nellie is interred) to launch a counter narrative, refuting Nellie's status as a vampire. I considered this counter narrative to be, simply, "the truth." But as the months passed and I continued to be frustrated in my efforts to find, or even identify, the Coventry teacher responsible for initiating the teenagers' search for the vampire, I began to interpret the counter narrative as a legend itself. Certainly, it incorporates the defining features of a legend. Set in the real world of historical time, with authentic places and credible characters, the narrative is told as true. While the realistic setting with ordinary people in everyday situations reinforces its credibility, without concrete verification, its truth is debatable. The narrative's form is loose and shifting (tied, as it is, to conversation) that turns upon the stable core of mistaken identity. The story circulates primarily by word of mouth, communicated face-to-face, from one person to the next, although occasionally it had been published in a local newspaper or broadcast on local TV. The legend's author has long been forgotten, so it has become a community's story, providing the insider's view of the issue, countering the outsider's story that continues to attract legend trippers. Labeling this narrative a legend is not an indictment of its truthfulness. I am still convinced that this explanatory tale is essentially factual.

The third legend is more complicated. In 1994, local author Charles Turek Robinson published eyewitness accounts by people reporting supernatural incidents around Nellie's gravesite in his book, *The New England Ghost Files*. Several people (including two "former town officials") claimed to have seen a young woman dressed in Victorian attire in the cemetery. When approached, the specter disappeared. Some reported hearing a disembodied woman's voice repeating "I am perfectly pleasant" near Nellie's grave. Robinson recounts the experiences of a Coventry woman, Marlene, who visited the cemetery several times in 1993. On her first visit, Marlene attempted to make a rubbing of Nellie's gravestone, but her paper kept showing moisture stains even though the stone was dry. Adding to the mystery, her charcoal stick disappeared. She gave up on the rubbing and began taking photographs of gravestones, including Nellie's. When developed, the photos were unremarkable—except those of Nellie's gravestone: in each, the letters on her stone appeared backwards. Marlene later returned to the cemetery with a friend. In front of Nellie's grave, her friend could feel, but not see, someone poking her arm. Marlene's husband, during a subsequent visit, was the first to report hearing the female voice at Nellie's grave repeating, faintly, "I am perfectly pleasant." He also felt something "invisibly scratch him across the left side of his face." And, according to Marlene, "he had several red marks to prove it."

Although Marlene's husband never returned to the cemetery (can you blame him?), she returned alone several days later, where she encountered an attractive young woman. They walked, chatting, among the gravestones. The woman mentioned that she was a member of a historical society. Stopping at Nellie's stone, Marlene asked the woman her opinion of the vampire legend. The woman thought that the legend was foolish because Nellie had not been a vampire, at which point the woman's behavior changed abruptly and the woman began to repeat, "Nelly [sic] is not a vampire." Alarmed, Marlene returned to her car. When she turned and looked back at the cemetery, she was "astonished to see that the strange young woman was no longer there." Marlene theorized that Nellie was attempting to exonerate herself. Because of her gravestone's inscription, and perhaps also because her body *had* been exhumed from her family's small plot and then re-interred in the central church cemetery not long after her death, rumors began circulating that she was a vampire. Marlene believes that Nellie cannot find peace until her reputation as a vampire has been put to rest. For this reason, her apparition often appears agitated and begins repeating, "I am perfectly pleasant."[278]

However, as we see in the next two chapters, historians writing histories of their local New England towns found vampire incidents to be anything but "perfectly pleasant."

Part V   Historians

Newspapers include many more vampire stories than New England local histories. Town historians probably encountered these incidents during research, but deliberately decided to exclude them. The histories by Franklin Benjamin Hough (1822-1885) confirm this conclusion. Hough was born and died in Lewis County, New York. The son of a physician, he earned his M.D. degree from Western Reserve College in 1848 and began his medical practice in St. Lawrence County, New York. In 1852 he quit the medical profession to devote his time to writing. His first works were histories of New York's most northern counties: St. Lawrence, Franklin, Jefferson, and Lewis. He was adept at finding vampire incidents but averse to sharing them. In his *History of St. Lawrence and Franklin Counties, New York* (1853), Hough was candid about why he excluded vampires: "It had been our design to enumerate some of the evidence of superstition, as evinced in various enterprises of money seeking, by digging, draining the beds of streams, &c., &c., and searching for vampires, of which the annals of St. Lawrence county [sic] afford at least three instances. Our space forbids the details, revolting to humanity, and regard for the living, leads us to pass unnoticed these heathenish mutilations of the dead."[279] Hough teased again the following year in his history of Jefferson County (which abuts St. Lawrence County), using the superstition of burying executed criminals outside of consecrated ground to dangle the vampire lure: "The lamentable prevalence of superstition thus evinced, has its equal only in the popular belief in vampires, which, on more than one occasion, has disgraced the annals of this and neighboring counties."[280]

Hough had plenty of company among local chroniclers who avoided sullying the preferred images of their towns—and protected the reputations of their inhabitants (living and dead)—by not mentioning unpleasant elements, such as vampire exhumation. Most of these town histories were written during the latter half of the nineteenth century, when the "old New England" appeared to be slipping away, if not already gone. It lived in memories of the very old and in tangible artifacts that had been preserved. "Alarmed by the rise of urbanization, industrialization, and immigration in the region's seaports and river valleys," Kent Ryden and Simon Bronner wrote, "members of old New England families" constructed an ideal New England:

Fearing that they were losing their cultural influence and conceptual ownership of New England, they reinvented the region to reflect a narrow yet cherished version of the colonial past, inscribing it in the present by refurbishing old village centers (and defining as "colonial" a white New England village ideal), founding historical societies, preserving the homes of old Yankee farmers, and pursuing a colonial revival in architecture and decorative arts. The constructed historical region is still the pastoral version of New England—post-Puritan, premodern, rural, and Anglo-Saxon—that many people carry in their heads today.[281]

A Rhode Island newspaper article, "The Vampire Tradition," published about a year before the Brown family exhumations in 1892, suggested that the vampire practice deliberately had been kept from official historical interpretation:

> Every person well informed about Rhode Island history has deep down in his mind some knowledge of the vampire tradition. There is little known about it that can be learned from records of any description, and so what is known of it comes from other sources, from stories told by old people, from allusions contained in a number of writings upon Rhode Island lore and from too contracted newspaper sketches of actual cases of survival of the tradition. One may search through the regular histories of the State without finding a reference to it, and yet there is no question of the incompleteness of such works because of their want of attention to the subject. The character of a people is to be read to some extent in the shadows as well as the lights of its history, and the individualism, so to say, and the essential intellectual and spiritual fibre of the people of Rhode Island are indicated by the unaccountable, mystifying and weird vampire tradition, as well as by the records of a scientifically historical past.

The anonymous author of this article indicated that, although he disapproved of bowdlerizing the historical record (essentially advocating for "historicism" as opposed to "presentism"), he understood its rationale: "In all such instances the least noise made about the performance was the better, for nobody cared to spread the view that such an evil power had appeared in his family. As the burials were in family or remote country plots, the formalities necessary to secure disinterment were nil."[282]

Magical cures for consumption that did *not* entail exhumations, corpse mutilations, and cannibalism fared better in local histories, for they could be viewed as quaint instances of supernatural (even miraculous) experiences, devoid of horrific details and heathenish implications. In his *History of Sanford, Maine* (1901), Edwin Emery included a family narrative that traces the origin of "Jotham Weed" to Jotham Moulton (1771-1857), who was nearing the end of life with consumption. "One night he dreamed that he went over to Mrs. Batchelder's house, across the river, and that she accompanied him out into the pasture to find a certain weed, which he had dreamed of seeing, and which would cure him. Seizing upon the dream as an omen of health and life, the invalid went the next day to see Mrs. Batchelder, told his dream and together they sought the weed in the pasture. They found it, and gathered some. He carried it home, steeped it, and drank the extract,—and recovered his health. The weed was afterward known as 'Jotham weed.'"[283] Moulton afterward became a respected physician.

It's tempting to view the vampire tradition as "New England's dirty little secret." But the number of incidents argues that it was not "little." Many exhumation stories appeared almost simultaneously in dozens of different newspapers, suggesting that, despite being ignored by local historians, this "secret" was buried in plain sight. "Dirty?" Well, the vampire tradition was deemed an odious superstition in virtually every context, an interpretive frame I hope to counterbalance.

Gravestone of Rachel Harris Burton, Factory Point Cemetery, Manchester Center, Vermont. *Courtesy of Cyril Place.*

11 Heathenish Mutilations of the Dead

In "the sacrifice to the Demon vampire" that occurred in Manchester, Vermont, in 1793, the passage from event to historical narrative was rather direct, if not continuous and easy to follow. An eyewitness communicated (whether orally or in writing is unknown) the story of the exhumation and heart burning to John S. Pettibone (1786-1872), who then incorporated the narrative into the manuscript of his early history of Manchester. Internal evidence indicates that Pettibone wrote his history sometime between 1857 and May 24,1872 (the date of his death). After he died, Pettibone's family gave the manuscript to Judge Loveland Munson (1843-1921), Manchester's town clerk at that time. Upon Munson's death in 1921, Pettibone's manuscript was retained by his widow, Mary C. Munson. She noted that Judge Munson's "last piece of work was to annotate a copy of the 'Early History' in pencil."[284] In 1929, she donated the handwritten manuscript to the Vermont Historical Society. Pettibone's "Early History of Manchester" was published the following year in the *Proceedings of the Vermont Historical Society*. Although Pettibone's history passed through several hands, was annotated by Judge Munson, and "some errors of form" were "corrected"—presumably by the editors of the *Proceedings*—the eyewitness account of the exhumation that Pettibone included probably remained essentially as he recorded it:

> Esquire Powel's second wife was the widow of Joseph Harris and sister to Isaac Whelpley. Captain Isaac Burton married her daughter, Rachel Harris. She was, to use the words of one who was well acquainted with her, "a fine, healthy, beautiful girl." Not long after they were married she went into a decline and after a year or so she died of consumption. Capt. Burton after a year or more married Hulda Powel, daughter of Esquire Powel by his first wife. Hulda was a very healthy, good-looking girl, not as handsome as his first wife. She became ill soon after they were married and when she was in the last stages of consumption, a strange infatuation took possession of the minds of the connections and friends of the family. They were induced to believe that if the vitals of the first wife could be consumed by being burned in a charcoal fire it would effect a cure of the sick second wife. Such was the strange delusion that they disinterred the first wife who had been buried about three years. They took out the liver, heart, and lungs, what remained of them, and burned them to ashes on the blacksmith's forge of Jacob Mead. Timothy Mead officiated at the altar in the sacrifice to the Demon Vampire who it was believed was still sucking the blood of the then living wife of Captain Burton. It was the month of February and good sleighing. Such was the excitement that from five hundred to one thousand people were present. This account was furnished me by an eye witness of the transaction.[285]

Most authors of local histories recorded the minutiae of daily life. Pettibone constructed his narrative from detailed descriptions of the built landscape and the genealogies of those who peopled it; the actors and their scenery come to life in colorful and curious anecdotes, such as the narrative of Rachel Burton's exhumation.

John Samuel Pettibone was born in Manchester, May 18, 1786. He was the youngest of Samuel and Rhoda Bridgman Pettibone's nine children, and the only son who lived to maturity. He married Laura Graves, the daughter of Dr. Josiah Graves, who was the first physician to practice in Rupert, Vermont, where he settled in 1788.[286] The 1853 catalogue of graduates and biographical register of Middlebury College, from which John Pettibone graduated in 1811, listed his accomplishments: "He has always been a farmer, and resided in Manchester. He has been a member of the General Assembly, seven years; Judge of Probate for the District of Manchester, seven years; member of the Governor's Council, four years."[287] Pettibone also was a veteran of the War of 1812. At the time of the exhumation, he was a child of seven years, which may explain why he was not an eyewitness. Given the number of people reported as witnessing the heart-burning—which would have amounted to perhaps half or more of the entire population of Manchester at that time (the 1790 federal census lists the town's population as 1,276)—Pettibone's informant could have been virtually anyone in living in Manchester.

In a letter that accompanied the donated manuscript, Mary C. Munson wrote that she had a typed transcript made and sent copies to Miss H. Canfield and Mr. Harris Whipple, "who are more familiar with the early history of the town than any others I know." They assisted her in identifying places and names that she was unable to decipher in Pettibone's handwritten manuscript. In revealing her Burton family lineage, Mary Munson also alluded to the exhumation narrative's relatively limited circulation in oral tradition: "The Capt. Isaac Burton, whose wives were the subject of the witchcraft incident, was a brother of my great-grandmother and I never heard any tradition of the incident from any one of the family. The only trace of an independent source came through the son of old Judge Fowler who told Miss Canfield he had heard his father tell the story as one he heard from the older residents of the town in his day."[288] Mary Munson's great-grandmother, Bethia Burton, was the sister of Isaac Burton. She and her husband, Rufus Munson, had four children, including Cyrus Munson.[289] Loveland was the only child of Cyrus and Lucy Loveland Munson. Loveland Munson married his second cousin, Mary Burton Campbell, the Mary C. Munson who donated Pettibone's manuscript to the Vermont Historical Society. Since the narrative was not part of the Burton family's oral tradition (at least as far as Mary Munson was aware), and there is no evidence that it was active in the town's oral history, the exhumation narrative as it appeared in Pettibone's history likely was that which was related by his anonymous eyewitness.

Sources outside of Pettibone's history provide some missing dates and corroboration that the people named in the exhumation narrative were residents of Manchester during that period. Rachel Harris and Isaac Burton were married March 8, 1789; Rachel died February 1, 1790, at the age of twenty-one. Burton's second wife, Hulda Powel, was born in Manchester, August 3, 1770; she and Isaac married January 4, 1791. Esquire Powel, Rachel's stepfather, was Martin Powel. As noted above, the Judge Munson who received Pettibone's manuscript was Judge Loveland Munson. The "old Judge Fowler who . . . heard his father tell the story as one he heard from the older residents of the town in his day" likely was Judge Harvey Klapp Fowler, who was born in 1818 and arrived in Manchester in 1837; his son, who became Miss Canfield's informant, probably was Joseph Wickham Fowler. Timothy Mead (1723-1802) was one of the first settlers in Manchester, arriving in 1764 from southeastern New York with several other proprietors who had been given land grants. Mead was one of the town's leaders, along with his friend and political confidante, Martin Powel (Rachel's stepfather), and his fellow proprietor, Isaac Whelpley (Rachel's uncle). Mead married into the Burton family and was, according to Pettibone, not particularly religious, but generally good-natured though "overbearing" and "high tempered."[290] Timothy Mead's son, also Timothy (1755-1828), "officiated at the altar in the sacrifice to the Demon Vampire," and another son, Jacob (1762-1822), was the blacksmith on whose forge Rachel's excised heart was burned. John Pettibone's sister, Electa (1781-1854), married Amos Mead (1778-1865), the son of Timothy Mead, who presided over the exhumation ritual. The celebrated families of early Manchester, including the Burtons, Pettibones, Munsons, Meads, and Canfields, could trace common ancestry, often through several lines. Any member of these families easily could have been Pettibone's eyewitness informant.

The phrase, "Demon Vampire . . . sucking the blood," so out of place in New England's consumption narratives, compels a conclusion that it was added to the eyewitness account collected by Pettibone. He could have inserted it into his manuscript any time prior to his death in 1872. Or the phrase could have been added by Judge Munson, who annotated the handwritten manuscript shortly before his own death in 1921. Not as likely (but still possible), the editors of the *Proceedings of the Vermont Historical Society* might have introduced the phrase when they "corrected some errors of form" prior to the manuscript's publication in 1930.

Pettibone wrote that "nearly all the leading men of the first decade were immigrants." Many of them came from Amenia, Dutchess County, located in the southeastern portion of New York state. While these early founders were principled men, there seemed to be a decline following the Revolutionary War, when, in Pettibone's opinion, Manchester became "an immoral place" where "drinking, gambling, and whoring were common."[291] Pettibone, a practicing Baptist (the record shows that his children were enrolled in Sunday school classes at the Manchester Center Baptist Church[292]), attributed this moral slide to the effects of the Revolutionary War, when men became conditioned to gather in taverns and carouse. Manchester, like many of the towns that developed in southern Vermont in the years immediately preceding the Revolution, had the feel of a loose frontier town, certainly more tolerant than the older, decorous communities along New England's seaboard. Even the town's two doctors during the period seemed caught up in Manchester's degeneration: "Dr. Gould, a graduate of Yale College, and Dr. Asel Washburn were the physicians of Manchester. Dr. Washburn was esteemed as an excellent physician but his usefulness was much lessened by intemperance, and Dr. Gould became a drunkard."[293] Again, we see that the prospects of those in dire need of efficacious medical treatment—including, of course, consumptives—utterly lacked anything encouraging. Unfortunately, but not surprising, Hulda, the second Mrs. Burton, died on September 6, 1793, about seven months after Rachel's exhumation.

✻✻✻

Historical research can be a humbling experience. Tracking the exhumation narrative contained in David L. Mansfield's history of the town of Dummerston, Vermont (1884) offers a lesson in the value of thoroughgoing persistence, or expressed negatively, the hazards of complacency. The significance of Mahatma Ghandi's words, "I have humility enough to confess my errors and to retrace my steps," will become clear as we work through Mansfield's narrative, beginning with Lieutenant Leonard Spaulding. Mansfield described him as "a prominent and capable man in business affairs and in continual service for the town until the year 1788, when he was unable on account of sickness, to act as one of the committee in finishing the building of pews in the meeting-house."²⁹⁴ On July 2, the town chose a replacement for Lieutenant Spaulding. He died of consumption fifteen days later, at the age of fifty-nine, which is where we pick up Mansfield's story:

> Although the children of Lt. Spaulding, especially the sons, became large, muscular persons, all but one or two, died under 40 years of age of consumption, and their sickness was brief.
>
> It is related by those who remember the circumstance; after six or seven of the family had died of consumption, another daughter was taken, it was supposed, with the same disease. It was thought she would die, and much was said in regard to so many of the family's dying of consumption when they all seemed to have the appearance of good health and long life. Among the superstitions of those days, we find it was said that a vine or root of some kind grew from coffin to coffin, of those of one family, who died of consumption, and were buried side by side; and when the growing vine had reached the coffin of the last one buried, another one of the family would die; the only way to destroy the influence or effect, was to break the vine; take up the body of the last one buried and burn the vitals, which would be an effectual remedy: Accordingly, the body of the last one buried was dug up and the vitals taken out and burned, and the daughter, it is affirmed, got well and lived many years. The act, doubtless, raised her mind from a state of despondency to hopefullness.²⁹⁵

You may recall that the name Spaulding was associated with a vine growing from coffin to coffin in a vampire incident discussed in chapter 3. In an anonymous letter published in the *Chicago Tribune* in 1885, submitted by Mrs. B. M. Prince, the writer described the exhumation of three generations of an unnamed family. The exhumer found, covering each of these corpses, "a little vine, which had year after year spread its meshes like a web over the entire corpse, and seemed to be still alive and growing! It was snowy white and cold as ice, a fit inhabitant of the dark and silent grave." He wrote, "Now, there is a tradition among the Germans that if a descendant [sic] pluck from the corpse of the last victim of consumption this little vine, it will eradicate the disease from the family." While the exhumer does not mention the need to burn the vitals of the vine-covered corpses, an addition to his letter, presumably written by Mrs. Prince (who immigrated from Germany with her family, as a young child), offers another exhumation narrative from Smithfield, Rhode Island, in which the entire corpse, from which the vine was growing, was burned. Genealogical research showed that the family of Mrs. Prince's husband and that of Leonard Spaulding trace a common lineage back to Edward Spaulding (1596-1670), who immigrated to Massachusetts, between 1630 and 1633, from Spalding, Linconshire, England.

The list of the children of Leonard and Margaret Spaulding shows that their five sons and three of their six daughters died before reaching the age of forty. According to Mansfield's narrative, the body of either Reuben or Josiah was exhumed to save one of the three remaining sisters. The ritual must have appeared efficacious to the community, for all three sisters—Sarah, Olive, and Anna—lived well into old age. The Spaulding children, in order of their deaths, are:

1. Mary, b. October 11, 1761, m. David Laughton, d. May 12, 1782, age 20
2. Esther b. April 1767, d. July 1783, age 16
3. Timothy (twin) b. May 13, 1765, d. June 13, 1785, age 20

4. Betsey, b. June 22, 1758, m. Henry Stevens, December 19. 1779, d. February 2, 1790, age 31

5. Leonard Jr., b. March 14, 1760, m. Priscilla Gleason, December 9, 1779, d. September 3, 1792, age 32

6. John, (twin) b. May 13, 1765, d. March 26, 1793, age 27

7. Reuben, b. November 19, 1756, m. [Miss] Gates, d. January 20, 1794, age 37

8. Josiah, b. March 30, 1771, m. Eunice Skinner, d. December 3, 1798, age 27

9. Sarah, b. July 19, 1763, m. Charles Wilder, October 27, 1782, d. September 17, 1841, age 73

10. Olive, b. October 17, 1773, m. 1st. David Wilson, February 17, 1798, 2d. Daniel Mixer, d. November 21, 1842, age 68

11. Anna, b. April 7, 1767, m. Samuel Laughton, December 14, 1786, died, January 13, 1849, age 81

The author of the Spaulding family's exhumation narrative, David Lufkin Mansfield (1837-1905), was born in Salisbury, New Hampshire, on September 17, 1837. In a brief autobiography (written in the third person) that appears at the end of his history, we learn that Mansfield "had few advantages for schooling until after fourteen years of age," after which "he had the advantages of a high school." The "misfortune of ill health, caused by rheumatism, changed his course of life at the age of 22 years." Instead of farming, as he had intended, he "engaged in school teaching." After relocating to Dummerston from Walpole, New Hampshire, in 1861, Mansfield taught in the town's schools and served as town superintendent.[296] According to his obituary, Mansfield, who died on April 13, 1905, of "complications" due to chronic rheumatism, "possessed a studious, painstaking mind, and had done a large amount of valuable work in his study of local history and genealogy."[297]

Mansfield's devotion to genealogy is evident on nearly every page of his history of Dummerston. While he provided no introductory remarks describing how he approached his task, we can infer that he viewed himself as a chronicler. The major portion of his history consists of biographies of early town residents, many obtained from their descendants. Interspersed among the genealogies are anecdotes, most of them either direct reminiscences or drawn from received family or community tradition. Here and there, Mansfield included narratives taken from written accounts. The extracts from Lt. Spaulding's journal, relating his experiences in 1758 at Crown Point and Fort Edwards during the French and Indian War,[298] are mesmerizing in their sparse, off-handed narration of stark brutality: "the 7d. Nothing straing only one of the Regelers was in Swimming and was Drowned and two men was found Ded a bout half a mile from the encampment judged to have ben keld by the moohokes who thay be Longed to it is not Known[.]"[299]

Mansfield's history, as most other local histories of New England towns published during the nineteenth century, includes large segments that read like excerpts from the Book of Genesis: so-and-so begat so-and-so, who begat so-and-so. History emerges through the agency of people, of course, but wading through pages and pages of genealogy can be off-putting, especially for a researcher eager to capture data that are circumscribed by two primary issues: the folklore associated with consumption rituals and their meanings. When I was researching the Spaulding family exhumation, the grave plant that grew from coffin to coffin appeared to be a rare, if not unique, motif in the New England tradition. In *Food for the Dead*, I traced its various folkloric incarnations, writing, "Implicit in this motif is the belief, distributed throughout the world, that at death, the soul or spirit of the corpse may enter into a plant, especially one that grows from its grave."[300] If the vine was, indeed, such a "life token," I continued, then "the fate of the external soul or spirit within the grave would be bound up with the well-being of the vine. If the spirit was conceived to be benign or beneficial, then one would want to nurture the plant. On the other hand, if the indwelling soul was seen as evil or harmful, it would be prudent to destroy it."[301] I also summarized what was known of Lt. Spaulding and his family, closing my discussion of this exhumation incident by noting how ironic it was that a man, "though wounded several times, survived terrible battles fighting Native American, French, British and Hessian soldiers," only to succumb, finally, "to a microscopic organism."[302]

Revisiting this case with a focus on the narrative's context necessarily turned my attention toward storytellers and the "many different and interrelated aspects" of storytelling events[303] that still awaited discovery in Mansfield's history. His narrative of the Spaulding family exhumation begins with a phrase that points to oral tradition as his source: "It is related by those who remember the circumstance . . . ." A logical inference is that he (1) heard the story directly from people who were alive at the time of the exhumation, who could have been relating an event that they had either (a) witnessed, or (b) heard from others (who, in turn, may or may not have been present at the exhumation); or (2) read a written or printed account of the event that was recorded by a witness or listener. In the terminology used by folklorist Gary Butler, the narrative, whether received orally or in writing or print, could be a personal-experience, family-experience, or community-experience narrative.[304] Mansfield provided some clues regarding his informants. In 1882, during the time he was compiling his history, the oldest person living in Dummerston was ninety-four year old Wranslow Holton, who was born January 11, 1788.[305] Mansfield followed his brief account of Holton's life with a section ("Dying Full of Years") listing the names, with the dates and ages of death, of town residents who had attained the age of eighty; his subsection ("Persons Living in Town—1880") contains the names of nearly two dozen storytelling candidates who were living at the time of the exhumation. This list includes two Laughtons, probable relatives of Samuel Laughton, the man who wed Anna Spaulding, one of Lt. Spaulding's daughters, who survived the family's consumption epidemic (she died in 1849 at the age of eighty-one). According to Mansfield, the Spaulding farm was about two miles north of the meeting house on what was known as "Spaulding's Hill" (there is still a Spaulding Hill Road in Dummerston). I have yet to narrow the list of possible bearers of this oral narrative by identifying the proximity of their residences to Spaulding's Hill.

Mansfield's history obviously was published in installments, as narrative threads pause, then reappear later in the work. It is also apparent from reading these serialized texts that Mansfield incorporated feedback that he received from people who had read earlier versions of his narratives. In effect, Mansfield's history functioned as an extended storytelling event in which Mansfield, the narrator, continued to alter his narrative based on feedback from people in the community, his audience. His extended narrative of the Spaulding family exemplifies this process and illustrates how persistent and determined research rewarded both Mansfield and me with a better text. About one-hundred and fifty pages after his initial description of Leonard Spaulding, Mansfield added some corrections: "Lieut. Leonard Spaulding whose record begins on page 24 was commissioned captain in the Revolutionary war, as we have learned since the printing of his record, and he was not a citizen of the town as stated on p. 26, when it was organized. His old account book dates back to 1766 not 1779."[306] Twenty-eight pages later, under the heading "Notes And Brief Historical Sketches," Mansfield added to the Spaulding narrative: "The incident about the Spaulding family, page 27, was done by the advice of a physician. The sick daughter was Anna Spaulding. She recovered and became the wife of Samuel Laughton."[307]

Again, we have a physician involved in an exhumation, but here his role went beyond curious onlooker. Two obvious questions are: Who were the physicians practicing at that time in Dummerston, and which of them might have prescribed the consumption ritual? What was the state of medical care in late eighteenth-century Vermont? Mansfield included an anecdote about Anna Spaulding Laughton's mother-in-law that throws light on the latter:

> The wife of Samuel Laughton, sen., was often called on to visit the sick, as there were few physicians in those days. She responded to all the calls for aid in sickness, and often went in winter time on snowshoes across the lots to visit families 2 or 3 miles away. On one occasion she visited a sick woman whose husband had not got reconciled to the destruction of the tea in Boston harbor. Having brought some tea with her for the sick woman, she watched her opportunity to steep it while the man was out at work, gave it to her patient and then drank some herself. Before she had finished her cup, the man came in, smelt the tea and stormed furiously, saying it cost too much human blood, to drink it, it was like drinking human blood and he would not have tea used in his house. Mrs. Laughton could not be frightened and deliberately finished drinking her tea in the presence of the enraged man.[308]

This narrative reinforces an observation that Jane Beck made in her article on traditional folk medicine in Vermont: "Prior to 1860, medical care in Vermont was chiefly in the hands of midwives who traveled the local neighborhood not only birthing children, but administering to the sick as well."[309] Mrs. Laughton's staunch attitude also accords with Patricia A. Watson's findings in "The 'Hidden Ones': Women and Healing in Colonial New England." Mrs. Laughton and her female cohorts had parity with their male counterparts, at least in the arena of medicine. "A remarkable medical network existed during this period in which women and men freely exchanged health-care information, regardless of social class or sex," Watson wrote. "Before apprentice- and university-trained physicians increased in numbers and began to consolidate their authority during the decades surrounding the American Revolution, women were not excluded from the medical area."[310] Beck noted that, "Midwives were considered 'experienced' in contrast to 'learned' physicians."[311]

Even the combination of learning and experience in a single practitioner might not have been a predictor of successful treatment. One of the physicians practicing in Dummerston at the time of the Spaulding exhumation was profiled in a Mansfield anecdote. Dr. Thomas Baker, whom Mansfield labeled "the candid doctor," arrived in Dummerston from Oxford, Massachusetts in 1783:

> It is said, on one occasion in his practice, he was called to see a sick person in the evening, but not being in a condition to deal out medicine, intelligently, at the time, as members of the family noticed, his prescription was not used. Very early, next morning, the Doctor came in haste and asked excitedly about the patient and the medicine. After learning that the

medicine had not been given as he ordered, he said, after looking at it, "you did well not to give it to the woman, for if you had it would have killed her dead as the devil." "The fact is," said he, "people wait till they are almost dead, then send for drunken Dr. Baker."[312]

It is possible that the surviving Spaulding family members sent for Dr. Baker at the point of Anna's sinking into consumption. If nothing else, Mansfield's anecdote suggests the unsettled state of medicine at that time.

Mansfield wrote that Dr. Samuel Stearns arrived in Dummerston "about 1796, and practiced in this town several years," which places him in Dummerston at the time that the corpse of the last Spaulding child to be buried was exhumed—which records show was twenty-seven year-old Josiah.[313] The "famous Doctor Samuel Stearns" (1741-1809)—the title of John C. L. Clark's 1936 biographical sketch[314]—was a well-traveled man, having lived in or visited Scotland, Ireland, London, and Paris, in addition to much of the Northeast. By all accounts, he was pompous and egotistical, with an inflated sense of his accomplishments. In his well-known almanacs and other publications, he formulated tide charts, created aids to nautical navigation, and made apparently accurate observations of the aurora borealis. But this man of many disciplines, who was fond of writing in rhyming verse, also believed in apparitions and was enamored of animal magnetism and other theories now considered quirky. By today's standards he seems an odd combination of science and pseudoscience—astronomer as well as astrologer—but, in his time, such combinations were not unusual. Even so, he was viewed by many as dancing on the edges of legitimacy, and was, probably unfairly, characterized as a quack. He published a treatise on herbal medicine, the *American Herbal*, in 1801, and devoted nearly three decades to compiling a survey, never published, of all known medical knowledge. Stearns could easily have encountered the exhumation cure, given his extensive traveling and knowledge of a variety of medical practices. But would his generally empirical approach to science and herbal orientation to medicine argue against a conclusion that he was the one who prescribed the consumption ritual? Or could the ritual's connection to the consumption vine make the opposing argument? The identity of the physician who advised the Spaulding family to exhume Josiah remains a mystery.

***

John Langdon Sibley was one of the diligent historiographers who used, in his words, "a great variety of sources" to craft the history of his hometown. "To some persons," he noted, "it will seem open to objections of too great minuteness of detail, and of occasional violation of good taste."[315] We are fortunate that Sibley perhaps violated good taste by including a vampire story in his history of Union, Maine:

> Consumption, too, has called off one after another from some families, till but very few members remain to mourn over the departed. In such cases, it is not unnatural for those who are fast wasting away, eagerly to adopt any suggestion for relief from the destroyer. Accordingly, in 1832 and 1833, a few persons put in practice the proverb, that the burning of the lungs of relatives who died of consumption would cure that disease in the living. One body was exhumed several months after death, and the vital parts were burned near the grave, which was in the Old Burying Ground. The theory seemed to be, that the disease, being a family disease, would thus be burned out or exterminated. But death still claimed the fair and the beautiful as his own.[316]

His use of the words "not unnatural" suggests that Sibley had empathy for afflicted families, granting that they were reasonable to seek hope amid despair. Sibley also writes that the ritual was performed more than once, perhaps in more than one family.

It seems likely that Sibley knew the family or families involved in the exhumations, given that his history was published less than twenty years later. He did not name the "few persons" who performed the ritual, but he provided some clues, perhaps unwittingly, in a large family-register appendix. Many of these genealogies list a cause of death; between the years 1808 and 1844, nineteen residents of Union are tagged as dying of consumption. Six of those deaths were in the family of Thomas Nye, between the years 1822 and 1834, including five of Thomas and Anna Dunbar Nye's nine children: Stillman, born January 18, 1797, died April 4, 1822; Cyrus Crocker, born December 23, 1799, died May 27, 1828; Eliza, born June 22, 1809, died June 7, 1830; Charles Austin, born May 26, 1807, died April 27, 1832; Caroline, born September 22, 1804, died March 6, 1834.[317]

Sibley described Thomas Nye as "a carpenter, who had worked on the State House, in Boston, Mass., when it was building." He was born in Barnstable, Massachusetts, January 20, 1773, married Anna Dunbar (born in Bridgewater) in Warren, in 1796. Sibley wrote that Nye "settled on the west side of George's River, at the corner of the road about half-way between Hill's Mills and Sunnybec Pond."[318] Thomas, himself, died of consumption in 1827. If this is the family referenced by Sibley, then Charles likely was exhumed soon after his death to save Caroline, who, unfortunately, died in 1834. We cannot be certain that the Nye family was one of those that Sibley suggested were eager "to adopt any suggestion for relief from the destroyer." But Thomas Nye assuredly had every reason to do so.

12   Her Heart Should Be Consumed for the Benefit of Her Sisters

As narratives move from one context to another, their content and structure often change. This process may not proceed in a single direction, and feedback loops might be activated even decades later, as an exhumation narrative from Chazy, New York demonstrates. A version of this story appeared in Duane Hamilton Hurd's *History of Clinton and Franklin Counties, New York* (1880):

> In 1818 an event occurred in Chazy showing how blind is superstition. As old people will remember the notion was quite prevalent in those days that from the lungs of a person dying with consumption, there sprouted a growth which proceeding through the earth communicated the consumption to the blood relatives of the deceased, and that the only way to save the lives of surviving relatives who were predisposed to consumption was to burn the body. In the year named, Shepard Woodward died with a regular old-fashioned consumption. His sister, the wife of the Rev. Mr. Boynton, was quite feeble and threatened with the same disease.
>
> After much debate and mature deliberation, the consultation of the elderly bodies of large experience and observation, it was decided to exhume the body of Mr. Woodward and commit it to the flames, so a few days after the burial, Messrs. Chandler Graves, Aaron Adams and Seth Graves took up the remains in the night with lanterns dimly burning, and placed them on a pile near the burial-ground, where they were consumed by fire. Among those who observed their proceedings were Mariette and Maria Carver, who was attracted by their lanterns in the burying-ground, and went out to see what was being done, but were required to return. Maria is still living, the wife of Henry Gregory. But we do not learn that the "cremation" prolonged the life of Mr. Woodward's sister, who soon after fell a victim to the same disease.[319]

Although Hurd provides no source for his text, it is a verbatim rendering of one that appeared in the *St. Albans* [VT] *Daily Messenger* on May 11, 1875, which credited the *Plattsburgh* [NY] *Sentinel* as the source. Even the title—"'Cremation' to Prevent Consumption"—remained unchanged. These stories were based on an earlier newspaper account, published in at least sixteen newspapers within a span of days in 1819. Most texts adhere closely to the following variant, which appeared in the *Plattsburgh Republican* (February 27, 1819): "*Superstition.* —On the 23rd inst. at Chazy, in this county, the body of Shepherd Woodward, who was buried about 18 months since, was taken up, and burned on a pile previously erected for that purpose. The person whose body has thus been consumed, died of a consumption, and this measure was executed in pursuance of a superstitious belief that burning the body of the deceased would prevent the further progress of consumption in that family."

It seems logical that the stories published just after the incident would be more detailed than those published nearly six decades later. Why this is not the case opens fascinating insights into the narrating process. In 1875, the *Plattsburgh Republican* ran a series of articles that included the exhumation narrative. To verify the story's authenticity, the newspaper contacted Maria Carver Gregory, who, as a girl of about seven years of age (in the company of her nine-year-old sister), had witnessed the exhumation. Maria verified the account, then she supplied additional details. Hurd's narrative, therefore, consists of a newspaper account augmented, years later, by the personal narrative of a witness. If we had only Hurd's story, we would be unaware of its heterogeneous ancestry. The narrative appears in the section of Hurd's history that chronicles the town of Chazy. The section is organized by titles of anecdotes or names of individuals, with no transitions connecting these subsections. Given Hurd's sparse approach, with few contextual markers, it is unclear why he included the exhumation narrative. Yet, it comports with his general approach to local history compilation, which was to incorporate a potpourri of oral, written, and printed narratives based on an assortment of documents, genealogies, memoirs, and recollections. The exhumation story seems to function as a pinch of spice in a historical bouillabaisse.

Duane Hamilton Hurd (1850-1934) was ubiquitous and prolific, compiling dozens of local histories. J. W. Lewis, who published many of these histories, boasted that the company had "long made a specialty of this class of work."[320] It appears that Hurd subcontracted the labor, commonly having several people researching and writing the various sections of "his" histories. He wrote the prefaces for these compilations, often beginning with the same formulaic opening that is found in his history of Clinton and Franklin Counties: "The province of the historian is to gather the threads of the past, ere they elude forever his grasp, and weave them into a harmonious web, to which the art preservative may give immortality. Therefore he who would rescue from fast-gathering oblivion the deeds of a community, and send them on to futurity in an imperishable record, should deliver 'a plain, unvarnished tale.'"[321] When historical writing is a cottage industry, one questions the extent to which sources were examined and evaluated.

Despite Hurd's documentary nonchalance, his Chazy narrative contains enough information to allow fashioning a plausible account of the event. The unfolding drama commenced with Woodward's death, documented in his obituary in the *Plattsburgh Republican* (June 12, 1817): "Death notice. Last Saturday, of pulmonary tuberculosis, William Shepherd Woodward, merchant of this village, age 22." Woodward's sister, Maria, already was showing signs of consumption. She had married Joel Boynton, pastor of the Presbyterian Church in Chazy Village, who served from 1807 to 1832. The story stated that, "after much debate and mature deliberation, the consultation of the elderly bodies of large experience and observation, it was decided to exhume the body of Mr. Woodward and commit it to the flames."[322] Those participating in the exhumation were an august body, including Seth Graves and his brother, Chandler, and Amasa Adams (not Aaron, as printed in the published versions). Hurd wrote that Adams settled in the east part of town, north of Chazy Landing, about 1808 or 1809, and engaged in farming. He served the town in several capacities, including justice of the peace. Townspeople remembered him as, "a great hand for public occasions," being "especially regular in attendance at all public executions."[323] Seth Graves (1760-1838) was born in Durham, Connecticut. He and his wife, Elizabeth, were among the first settlers of Chazy. They eventually owned large tracts of land and built a gristmill, sawmill, and hotel. Years later, people recalled that the first loaf of bread from wheat flour in the town was made by Elizabeth.[324] Seth was a soldier in the Revolutionary War. Given his surname, one might surmise that

Graves would, indeed, wish to attend the exhumation. Certainly, it attracted the attention of two young sisters, Mariette (1809-1886) and Maria (1811-1886) Carver, daughters of Dr. Nathan Carver, who arrived in Chazy from Hartford, Connecticut in 1801 and later became a judge.

Hurd's narrative includes several motifs common to exhumation narratives:

- The "notion was quite prevalent in those days."
- The decision to exhume emerged from a "consultation of the elderly bodies of large experience and observation."
- The clergy was involved (in this case, the life of the Reverend Boynton's wife was at stake).
- The ominous grave plant reappears, but its expression in this case seems to suggest more than just a "sympathetic connection" between the dead and the living, implying that the plant was a contagious vector, spreading the disease ("from the lungs of a person dying with consumption, there sprouted a growth which proceeding through the earth communicated the consumption to the blood relatives of the deceased").
- The hereditary basis for consumption, then widely accepted, also finds expression: "the only way to save the lives of surviving relatives who were predisposed to consumption was to burn the body."
- The ritual failed to save Maria Boynton (she died two years after her brother, William Shepherd Woodward).

***

In his *History of Concord, Massachusetts* (1904), Alfred Sereno Hudson (1839-1907) described the typical medical procedures in the emerging towns of colonial New England, where "the doctor acted as druggist, and obtained his herbs from his own garden or from the neighboring fields and forest. . . . He obtained his leeches from the pond. His pills, powders and other compounds he prepared with mortar and pestle." "Arrayed on shelves" in what passed as his study, were "various jars, vials, and crude instruments for cupping, surgery, and extracting teeth; for he was dentist as well as doctor."[325] "Some early practitioners," Hudson wrote, were "supposed to be skilled in surgery . . . and sometimes served as barbers as well as bone setters."[326] He pointed out that Cotton Mather recommended "the efficacy of a dead hand for scattering wens" and touted "the healing virtue of sowbugs."[327] Alluding to Paracelsian corpse medicine, Hudson noted that these early remedies were "in accord with the practice of physicians in England at that day; for it is stated that there was forced upon Charles the Second when upon his deathbed a volatile salt extracted from human skulls."[328] Richard Sugg has corrected the misperception of coercion, writing that "Charles II made his own corpse medicine."[329] Hudson ended his discussion of medical practices with a familiar consumption cure: "Almost, if not quite, within the memory of the present generation, in a town adjacent to Concord, pills made from the ashes obtained from burning a human heart have repeatedly been administered as a cure for consumption."[330] The adverb "repeatedly" suggests an ongoing tradition, rather than sporadic incidents.

Which "town adjacent to Concord" would Hudson not name? Pursuing the answer led to pathways, winding and crossed. Strong clues appear in a published diatribe against "popular superstitions" that, its author asserted, had "greatly injured the cause of medicine."[331] The title of Samuel Bulfinch Emmons' 1853 book bluntly summarizes his view that folk beliefs imperil human culture: *Philosophy of Popular Superstitions and the Effects of Credulity and Imagination upon the Moral, Social, and Intellectual Condition of the Human Race*. In Emmons' consumption cures, Paracelsian corpse medicine again lies beneath the surface: "Some time since, in the State of Maine, the body of a female was taken from the grave, her heart taken out, dried, and pulverized, and given to another member of the family, as a specific against the consumption. And the same thing has more recently been done in the town of Waltham, Massachusetts. The heart was reduced to a powder, and made into pills, but they did not cure the patient; while the person who took up the remains from the grave, and removed the heart, came very near losing his life, from the putrefactive state of the corpse at the time."[332] The deleterious effects of contagion seem apparent in the last example, although in 1853, Emmons certainly was unaware of its microbial agency.

Concord is adjacent to Waltham, where Emmons wrote that powdered hearts were made into pills to cure consumption—a good match and, therefore, a good hypothesis concerning the location of Hudson's text. Emmons narrated a specific event in which the patient died, and the exhumer came close to death because of contaminating contact with the corpse. Hudson's text is a generalized statement of a recurring practice ("pills . . . have repeatedly been administered"), unlike Emmons' narration of an actual event. The indefinite time frame provided by Hudson ("almost . . . within the memory of the present generation") situates the practice in the early nineteenth century.

Emmons rewarded readers with a scanty story of a similar event in Maine. Subtracting Emmons' relative date of "sometime since" from the date of publication yields a supposition that the exhumation in Maine occurred a little before 1853. But evidence indicates that Emmons had merely copied, with no attribution, a much earlier text of the incident. Bernard Whitman's 1829 *Lecture on Popular Superstitions* included a nearly verbatim version of Emmons' text: "During the last season, in Maine, the body of a female was taken from the grave, her heart taken out and pulverised, and given to another member of the family as a specific for consumptive complaints."[333] The date of the Maine exhumation was pushed back to circa 1827.

A brief but more detailed story, published in several newspapers (crediting the *Bangor* [ME] *Register*) in 1828, included the same pulverized-pill motif. The following version was printed in the Worcester [MA] *National Aegis* (April 16, 1828):

> Three of a consumptive family, residing in a town adjacent to this, have been cut off by this dreadful malady, within a very short period of time. During the winter of 1827, whether by some whimsical dream, by some preternatural revelation, or whether it was one of the odd fancies of some quackish old woman, I know not; but the notion came into their heads, that the heart of one who had died of consumption, was a 'sure-cure' for the disease. Accordingly the last deceased who had slept in her grave for about one year, was actually disinterred, the heart extracted, and a tincture made for the cure of another sister.

The date corresponds to Whitman's text, which could have been drawn from any of several newspapers that published the article that year. Reducing the heart to a powder and incorporating it into a "tincture" or "specific" (or, in Hudson's text, a "pill") emphasizes the medicinal, as opposed to the "spiritual" or supernatural, aspects of the remedy. The attribution of the prescription to, possibly, "some quackish old woman" suggests both folk and "quack" (or commercial) elements regarding how knowledge of these remedies spread. Conforming to type, the last person to die was the vampire/scapegoat.

Many newspapers that ran this article attached it to a similar narrative, also from Maine. While the first was situated in a town adjacent to Bangor, the second described a woman seeking a compliant doctor in that city. Following is the narrative as it was published in the Worcester [MA] *National Aegis* (April 16, 1828):

> A woman about sixty years of age, walked some miles to a physician of this town, expressing a deep concern for a friend of hers, who, she said, was "just gone with consumption," and entreating the doctor to go and assist to procure the heart of another person, who had died of the complaint, for the relief of her poor friend. But the hard hearted doctor treated the matter lightly, and said, "he did not believe in the resurrection business," which caused the lady to go away with a heavy heart, (of her own) much grieved.

The physician "treated the matter lightly," but the woman's entreaty shows, again, how closely the medical establishment was involved in these incidents. The woman must have believed, or at least hoped, that a "real" doctor would assist in carrying out the procedure. The "resurrection business" mentioned by the doctor alludes to the commercial enterprise of illegally exhuming corpses for use by medical students.

The absence of source attribution is a bedrock issue in situating these narratives. In his *Philosophy of Popular Superstitions*, Emmons listed eighteen "principal works from which valuable and important matter has been selected for these pages." I investigated all of them. Not surprising, the first on the list was Whitman's *Lecture on Popular Superstitions*. The third was the *Christian Freeman and Family Visitor*, a Universalist weekly founded by Sylvanus Cobb (1798-1866), who was born in Norway, Maine. While an itinerant Universalist preacher in Maine, Cobb was "among the early occasional preachers of the new faith in Buckfield."[334] In 1821, he became the Universalist minister serving the towns of Waterville and Winthrop, Maine (1821-1828), then moved on to Malden, Massachusetts (1828-1838). Cobb established his religious and reformatory periodical (which he also edited) in 1839, while serving as the Universalist minister in Waltham, Massachusetts. Sylvanus Cobb's residence in both Waltham and Maine—including Buckfield, the hometown of Ezra Morton Prince (chapter 3) and Seba Smith (chapter 4), both of whom were connected to consumption rituals—leads to the conclusion that Cobb was the link between the narratives of Hudson and Whitman, and Waltham and Maine.

***

Henry Stedman Nourse (1831-1903), in his history of the town of Harvard, Massachusetts (1894), discussed, "superstitions which now seem strange" but "then held the majority in firm bondage." He set the scene for his consumption narrative by sketching in the supernatural folklore that held sway just a hundred years earlier:

> The people among whom credulity created such fear of warlocks and witches as the nod servants of a personal devil, naturally saw supernatural apparitions; experienced miraculous cures; consulted fortune tellers and secretly wore charms; entered upon no important enterprise or journey on Friday, and never stepped into a house left foot foremost; dug their wells under the direction of some gifted individual in whose hand the witch-rod would

"work;" administered all the ordinary affairs of farm and household with reference to the phases of the moon.

Nourse then began his consumption story, apparently collected from oral tradition:

> When that fell destroyer, consumption, broke into a family circle and began to bear away its victims in slow but sure succession, humiliating the most self-confident physicians with a sense of their impotence, there often came to light a strange delusion—the vulgar belief that if the heart of one who had died with that disease were burned, and the members of the household inhaled the fumes from it, they would escape the doom hanging over them. There is a well-attested tradition that about a century ago, in a consumptive-stricken family of Harvard already bereft of eight or more of its youth, a dying girl extracted from friends a solemn promise that her heart should be consumed for the benefit of her sisters, and her last wish was duly carried out. One of these sisters at least survived to acknowledge to her inquisitive granddaughter who heard this tradition, that the story was essentially true.[335]

Nourse maintained the convention of deliberate obfuscation, an artifice common in New England's vampire stories. I suspect that he knew the identity of the "consumptive-stricken" family, since, in the other oral narratives included in his section on superstition he identified both the narrators and the main characters. Labeling the story "a well-attested tradition" implies that it had become part of the community's oral tradition.

Nourse's biography helps answer questions about his narrative, including why he would bother to include it in his history, especially if he was intent on protecting the family's identity. Nourse was born April 9, 1831, in the village of New Boston (now known as South Lancaster), Massachusetts. After graduating from Harvard in 1853, he taught ancient languages at Phillips Exeter Academy for two years, then studied and began the practice of civil engineering. He enlisted as a private at the beginning of the Civil War, eventually attaining the rank of major. Nourse fought in forty battles, including Vicksburg and Shiloh, where he was wounded. After the war, he was elected to the Massachusetts House of Representatives, and served on several boards and commissions, including that of the State Library of Massachusetts. Nourse was beseeched in a letter from Harvard native, Warren Hapgood, to write the town's history, as Hapgood believed he was too old to complete it. Hapgood appealed to Nourse's qualifications, including authorship of several works on the history of his hometown, Lancaster, which adjoins (and once included) the town of Harvard.[336] Nourse agreed to undertake the project, but noted that many residents who could have contributed their memories have "long turned to dust." "There is little left for the historian's use," Nourse lamented, "save the curt and prosaic public records; but fortunately those of parish and town have been well preserved."[337]

Nourse must have found *some* sources that gave "romantic interest to local story," because the small section on superstitions contains, not only the exhumation narrative, but two others that obviously had a life in oral tradition, one about a witch and the other about a ghost. The first concerns Goody Pollard, who was injured while in the form of an animal (a spider in this case) and manifested the injury when she returned to human form, thus confirming a neighbor's conviction that she was truly a witch.[338] The second tells of a missing Revolutionary War drummer, named Hill, whose murdered remains were located by means of a mysterious drumming.[339] Nourse wrote that the latter tale was "given" to him (orally or in print is unknown) by Reverend John B. Willard. In a memoir of Nourse, Samuel Shaw wrote, "Mr. Nourse's literary work was distinguished not only by painstaking care and research, but by an agreeable humor and a keen appreciation of those incidents which gave a personal and human interest to dry details of local history."[340] Another biographer elaborated on Nourse's attention to detail and veracity, calling him "an antiquary of the best type, with the nature and attributes of the historian" who "separated the true from the false."[341]

Nourse's own genealogy may have inculcated an appreciation for family legends. "He was descended, in the eighth generation, from Francis Nurse [the spelling of the name must have changed over time to Nourse] of Salem Village, whose wife, Rebecca (Towne), was the unfortunate victim of the witchcraft delusion who was hung at Salem, July 19, 1692."[342] On the maternal line, he was descended from the *Mayflower* pilgrim, John Alden.

Nourse was discretely circumspect when he retold the story of the dying girl's gruesome request. By omitting the family's name, he could tell a fascinating story while protecting the reputation of the family's descendants. Yet Nourse embedded information, separated from the exhumation story by a few hundred pages in his two-volume history, that a determined researcher might discover and link to the vampire story. His discussion of epitaphs in Harvard's old burying ground pointed toward a likely family:

"Rows of graves where three or more of the same family died within a few years of each other, or even within a single season, like those of the Bowles and the Cole families, tell of the ravages of epidemics—the scourge of consumption for which Harvard was once held in ill repute, or the fatal endemic dysentery of 1746 and 1756. In the Bowles family there were nine deaths within seven years."[343]

The Cole family deaths cluster during the dysentery epidemic. But the Bowles family matches perfectly, with nine children of Baptist Deacon William Bowles and his wife, Sarah, dying during the 1790s:

Sarah, June 3, 1790 (no age given)

Mary, November 25, 1792, age 27

Elizabeth, January 9, 1793, age 20

Stephen, February 3, 1793, age 4

John, September 28, 1795, age 32

Nathaniel [Samuel?], January 24, 1797, age 21

Anna, February 20, 1797, age 17

Josiah, March 29, 1799, age 22

William, October 19, 1799, age 26

Both parents also died within the decade: Deacon Bowles on July 11, 1796 (between John and Nathaniel), at the age of 59, and Sarah on November 11, 1798 (between Anna and Josiah), at the age of 57.

The one obituary of the Deacon's offspring I was able to locate was that of William, published in the Boston *Constitutional Telegraph* (October 16, 1799): "At Harvard, the 14 inft. of a confumption, Mr. WILLIAM BOWLES, a Young Man of an amiable character. He bore his ficknefs with patience and resignation, and died in peace. He is the laſt but one of a family of fifteen children; the reſt being all dead, moſtly of conſumtions, within theſe ſix or ſeven Years." Williams's obituary appeared in the newspaper three days *before* the death date inscribed on his gravestone, reinforcing a general guideline of genealogists: dates on gravestones, while literally carved in stone, should not be accepted unconditionally. It is not difficult to imagine, in this case, that the stone carver interpreted the handwritten '4' as a '9', so the date on William's stone is October 19 instead of October 14. Like the printed record in Nourse's history of Harvard, the deaths of the Bowles family members, including that of William, published in the *Vital Records of Harvard* (1917), are taken from the grave records of Harvard's Old Cemetery. The obituary states that William was one of fifteen children, only one of whom was still living at the close of the eighteenth century. The gravestones of Deacon Bowles' family in the Harvard cemetery do not exactly match Nourse's published transcriptions; there are stones for Deacon Bowles and his wife, Sarah, and eight children. The *Vital Records* include a daughter, Sarah, who died June 3, 1790, but I did not find her grave in the Old Cemetery, which may explain why she is not listed in Nourse's transcriptions. It is likely that Sarah completes the nine children mentioned by Nourse. So, vital records and cemetery databases have accounted for the "missing" children.

Placing the pieces into place yields the following: Seventeen-year-old Anna, the seventh child to die, asked her friends to remove her heart following her death. But two of her brothers died: Josiah, newly married, and William. Anna's sister, Eleanor, survived to marry Abel Willard in 1797. Their son, William Bowles Willard, had a son, named Abel Willard after his grandfather. Abel's daughter, Florence Almira Willard, is likely the "inquisitive granddaughter" who heard the sad tale of the Bowles family. Although Florence was Eleanor's great-granddaughter, the other details correlate so well that I must conclude that this scenario best fits the narrative that Nourse provided in his history of the town of Harvard—published in 1894, which would put the "century ago" enumerated in the text squarely in the middle of the Bowles family's consumption epidemic.

The narrative from Harvard was a case of therapeutic *excision* rather than therapeutic exhumation; it was *preemptive action*, taken prior to interment. Although not common, such preemptive actions have analogs. More than fifty years after the Harvard event, a preemptive excision occurred in Robinson Hollow, Massachusetts (which we will consider in chapter 18). In 1875, several variants of a story describing a similar request, by a self-proclaimed vampire, residing in Paris, were published in American newspapers, including the *Palo Alto* [CA] *Pilot* (March 18, 1875). A Serbian nobleman, Nicholas Boralajova, had relocated to Paris, believing that he was destined to become a vampire since he was the eldest son in an afflicted family. He requested that, upon his death, "his heart should be taken from his body to prevent him from leaving the grave—it being supposed that a vampire cannot get along without a heart." The article closed with, "It is to be feared the showing would indicate the prevalence of more superstitious delusions even here in the United States than might be supposed without reflection. The fact is this special weakness is one which it takes long to eradicate." No evidence suggests a direct relationship between these two narratives, separated by an ocean and three-quarters of a century. Perhaps hidden somewhere in the record, or lost in the vast realm of unrecorded events, there is an indirect connection. The anonymous author's concern that there could be such "superstitious delusions even here in the

United States" has proved to be, ironically, better founded than he feared. His conclusion that these superstitions take "long to eradicate" proved correct.

***

Frederic Denison included an exhumation story in his history of Westerly, a town that occupies the southwestern corner of Rhode Island, in Washington (South) County, abutting the Connecticut town of Stonington:

> Among the delusions and superstitions that, at different times, have tarnished the medical profession, one has strongly lingered, among the ignorant, even to the present generation. It consists in the whim that in some mysterious way the dead, or the diseases of the dead, may feed upon the living, coupled with the idea that diseases have their seat in the vitals of the body. Hence the bodies of persons, dying of a dreaded disease, have been opened, and the heart, lungs, liver, and other parts have been burned as a means of protection to the living.
>
> The black man, Bristoe Congdon, and three of his children, died with the consumption. The body of one of the children was exhumed, and the vital parts were burned in obedience to the *dicta* of this shallow and disgusting superstition. Similar cases have occurred in more enlightened families.[344]

In the preface to *Westerly and Its Witnesses* (1878), Denison wrote that he had "chosen to mention . . . principal authorities in the text of the work"—rather than in notes, presumably—because he believed that approach would be "more convenient to the reader." Unfortunately, no authorities are associated with the exhumation text. Denison also wrote that he was "obliged in some cases to depend upon the memories of the aged." Because Denison omitted his research materials and the story lacks particulars, several important facts are unknown: the story's source, the form of its transmission, when it occurred, and who authorized the exhumation. A search of newspapers, including the *Narragansett Times*, the community's most important local newspaper prior to the publication of Denison's history, failed to provide the missing information.

Denison informed readers that Bristoe Congdon and his children were interred in "Indian Ground (9)," which "lies on the cross road from the post-road to Dorrville, about one hundred rods west of the road, on the Wells farm, about southwest from the residence of Libbeus Sisson, Esq., in a meadow." Denison noted that there was no wall or fence surrounding this cemetery, and that the "graves are few and unlettered." He also wrote that it was "reported to contain the bodies of Indians" as well as "the remains of blacks."[345] Since the stones were not inscribed, one wonders how he knew that the Congdons were buried there. The entry for Bristoe Congdon in the Rhode Island Cemetery Database (1647-1900) provides no dates of birth or death and states that he is the only person known by name to be buried there. The database characterizes the cemetery (designated Westerly Historic Cemetery #511) as "lost." Two bits of evidence, combined with the history of intermarriage among residents of South County, leads to the conclusion that Bristoe could have been Native American, African American, or a mixture of the two: (1) he is buried in an "Indian Ground" that was believed to contain both African Americans and Native Americans and, (2) the choices for racial identity in the 1840 census, as we will see momentarily, were "free white persons," "free colored persons," and "slaves." A free African American, Native American, or mulatto would have been subsumed under "free colored person."

Information about Bristoe Congdon (and his family), though scant, is tantalizing. In the 1820 federal census, a "Brister Congdon" is living in Westerly, with five "free colored persons" in the household, including one male aged 14 through 25, one male 26 through 44 (probably Brister), two females below the age of 14, and one female aged 14 through 25 (perhaps his wife). He appears again in the 1830 federal census, with seven "free colored persons," including one male under 10 (a new son?), one male 36 through 55 (Brister), two females under 10, two females 10 through 24, and one female 24 through 36 (his wife). The two new householders probably are a son and daughter born after the 1820 census. The 1840 federal census shows a "Bristow (Brister) Congdon" residing in Westerly, with seven "free colored persons" in the household, including one male aged 10 to 23, one male aged 36 through 54 (presumably Brister), three females below the age of three (three new daughters or other kin?), one female aged 10 through 23, and one female aged 36 through 54 (Brister's wife). Two people in the household were employed in agriculture, probably Brister and, perhaps, his son, who would have been in his late teens. Assuming that Bristoe, Brister and Bristow Congdon are the same person, he was born sometime between 1786 and 1794 and his presumed wife was born between 1795 and 1804. In the 1850 federal census, there is a Sarina Congdon, black female, age 54 years, born about 1796, who was listed as a widow (she had her own "dwelling house" and was alone in it). Also on the same page in the 1850 census for Westerly (therefore, probably a neighbor of Sarina), the household of white farmer Oliver Davis and his wife, Nancy, includes George Sweet, a fifty-four-year-old white laborer, and Sarah, a fourteen-year-old Black female. Sarina and Sarah are the only two Blacks listed on the census page.

A reasonable scenario emerges from the data: Bristoe Congdon died between 1840 and 1850, and three of his children also died prior to 1850. His widow, Sarina, allowed their daughter, Sarah, to live with the Sweets, perhaps in exchange for housekeeping and other domestic duties. (A Miss Sarah S. Congdon is listed in the Westerly City Directory, living on Main Street, from 1881 through 1911.) Since the traditional exhumation ritual directs that it is the male head of the family who makes the decision regarding if, and whom, to exhume—often in consultation with other people linked to the family—that role may have been enacted by Bristoe's son, who would have been about thirty years of age in 1850. An educated guess is that the exhumation was performed about 1858, twenty years before Denison's history was published.

The life of Frederic Denison (1819-1901), author of the Westerly history, was not the drab tedium one might expect of a nineteenth-century Baptist minister in New England. He not only preached, he also soldiered, wrote history, poetry and songs, taught people formerly enslaved to read and write (and transcribed their songs), kept detailed journals, and chronicled the institutions in his life. Sometimes he stirred things up with unorthodox projects. Yet, for all his eclecticism, Denison was fundamentally a man of his time and position, conventional and conservative. In 1867, according to an article in the *Norwich* [CT] *Aurora* (February 19, 1868), he was appointed secretary of the Rhode Island Temperance Convention but the next year he created "quite a commotion" by exchanging pulpits in Westerly with an Episcopal minister, an action that was "condemned and prohibited" by the Episcopal bishop in Providence.[346]

Denison was born in Stonington, Connecticut, on September 28, 1819, and received his early education at Bacon Academy in Colchester, Connecticut. He then worked as a carpenter to pay for his tuition while attending the Connecticut Literary Institute in Suffield. Denison began his career as a Baptist preacher at the First Baptist Church in Westerly, Rhode Island, after graduating from Brown University in 1847. The next year, he married Amey Randall Manton. In 1850, he received an A.M. degree from Brown and fathered a daughter, Frederica. In 1859, he became the pastor of a church in Central Falls, Rhode Island, where he remained for two years. Denison, who was strongly anti-slavery, enlisted when the Civil War began, serving as Chaplain to the First Rhode Island Cavalry, and later, the Rhode Island Heavy Artillery. Prior to his enlistment, Denison had already authored two books, and he continued writing throughout the war, describing his wartime experiences in South Carolina, Georgia, and Florida in lengthy letters to Providence newspapers. After the war, he served as pastor to Baptist churches in New Haven, Connecticut, and Westerly, Woonsocket and Providence, Rhode Island. He wrote histories of the two Army units he served, and published "sermons, poetry, hymns, newspaper articles, memorial addresses, tracts on Baptist history, and several works of antiquarian local interest." He was also the official poet of the Rhode Island Society of the Sons of the American Revolution.

When he is in storytelling mode, Denison's writing is engaging. When he holds forth on religion, manifest destiny, or the white man's burden (not employing that phrase, of course)—topics that infuse his writing——his mode shifts from narrative to oratory, and one can imagine his passionate voice rising from the pulpit. Both modes of discourse are evident in *Westerly and Its Witnesses*. After depreciating the indigenous peoples of North America, Denison took up the story of "The First Whites" with this opening: "Since the red men failed to fulfil [sic] the commission given to mankind to subdue and cultivate the earth, and make it a theatre of moral culture, Providence determined to supplant them, and give the vineyard to another people who should bring forth fruits thereof. Considering the greatness of the change, and the established laws of human nature, the expulsion and replanting have been rapidly progressing and are nearly accomplished."[347] Clearly, in his view, the Indians failed because they were not Christians.

Despite his attitude toward New England's first inhabitants, Denison deplored slavery and argued against racial prejudice. In his history, he wrote, "But long after slavery had disappeared from amongst us, there strangely lingered a strong prejudice against the people who had suffered, —a prejudice against their color, or condition, or capacity. A black man, even a mulatto, is still regarded as belonging to a class not entitled to all that is bestowed upon a white person."[348] So, when Denison wrote in his exhumation narrative that "similar cases" of "this shallow and disgusting superstition" had "occurred in more enlightened families," he seemed to be saying that we might excuse Congdon for engaging in this "delusion" because he did not have access to the civilizing institutions that were available to white people. When we contextualize Denison's text, it seems obvious that the object of his scorn was superstition—the dark side of humanity, he supposed—that has always lived in the shadows. As both a chronicler of history, and a churchman, his task was to bring superstitions to light so that they could be eliminated. He wrote, "The history of any township would be devoid both of a portion of its vital facts and of its instructive lessons, if no mention were made of the fancies and follies, the superstitions and delusions of the people. The errors of the past are a part of our warnings for the future."[349] A few pages later, Denison deployed his oratorical mode to punch the lesson home, admonishing chroniclers who cleanse these delusions from their narratives:

> It is only to be regretted that annalists, in recording the life of towns and communities, should so often judge it proper to ignore the superstitious side of society; a faithful portraiture is ever the best. That we may wisely and safely steer our barks on life's tempestuous voyage, it is well to know the shoals and rocks on which preceding voyagers have struck. And the people of the present day may innocently smile at the weakness of former generations only when they prove themselves sufficiently intelligent to avoid the pretentious arts and schemes of quacks, theorists, fanatics, and impostors. Hypocrisies and errors never bear the test of continuous history; only truth shines with a perpetual light.[350]

Civil War service left Denison with indelible memories. One, especially, may have colored his description of the of the Congdon family exhumation. In March 1862, Rhode Island Governor William Sprague, who had participated in the First Battle of Bull Run the previous July, received permission from the War Department to return to the battlefield with a detachment of the First Rhode Island Cavalry to properly bury Rhode Island's dead soldiers. One of Denison's duties as chaplain was to tend to the physical and spiritual needs of the soldiers—living, wounded, and dead. "While the work of exhumation was going on," Denison stated, "the Governor and certain of his staff . . . learned from the colored people that some of the dead bodies had been exhumed and barbarously treated by the Confederate army."[351] They found the body of Colonel Slocum "uninjured and easily recognized. But on opening the grave of Major Ballou we found neither coffin nor body, and the grave itself bore witness of disturbance and violence." The burial detail followed their Black informant a short distance to a stand of trees, where, Denison wrote, they

> were horrified by discovering ashes, dead coals and brands, and in these the bones of a human body, save the head. As was proved, a Georgia regiment, that had suffered in the battle from the fire of Rhode Island troops, had exhumed the body of Major Ballou, supposing it to be that of Colonel Slocum, beheaded it, denuded it, and burned what remained. Words may not describe the indignation revealed in the face of the Governor, and of all who gazed upon that rifled grave and those bones protruding from the ashes and dead coals.[352]

Denison continued, "No part of the skull, not even a tooth, was discernible after searching the ashes. All the remaining bones of the body were found and fathered."[353] They placed what remained of Ballou's body into a coffin and returned the following day to complete their work. "We immediately opened the muddy pits containing the bodies of the brave dead, which were packed, like cord wood, into the pits, with faces downward—meant as a mark of indignity—and scarcely covered with earth."[354] Cremation, decapitation, and prone burial were regarded as indignities.

Denison's experience with vengeful exhumations at Bull Run might have prevented him from bringing empathy to the Congdon exhumations. On the surface, therapeutic and vengeful exhumations are identical: bodies are dug up, mutilated in various ways, burned, buried face down. These two kinds of exhumation are separated by intention, context, and meaning, which is expressed in anthropological terms as *endo* versus *exo*, where endo-exhumation is the exhumation of a corpse from one's own community (whether family, tribe, or other social group) and exo-exhumation is the exhumation of a corpse from an outsider group. Memories of encountering the bodies of comrades that had been exhumed, decapitated, burned, and buried face down may have resurfaced when, less than a decade later, Denison wrote about the exhumation and mutilation of the Congdon child's corpse. He denigrated the "shallow and disgusting superstition" undertaken by the Congdon family, but was "horrified," in equal measure, at the "barbarous treatment" of the Union corpses by the Rebels from Georgia. Both represented, for Frederic Denison, the uncivilized "errors of the past" that "are a part of our warnings for the future."

***

New England's large compendium of local histories mostly ignored consumption rituals, so an author who dared include one that prescribed cannibalism was extraordinary. Inattention to the existence of cannibalism in Europe and European America is long-standing and extends into recent times. Karen Gordon-Grube observed, in her article on traditional medical cannibalism, that "anthropologists have been so preoccupied with seeking evidence of institutionalized cannibalism among primitives that they have overlooked or ignored a well-documented cannibal tradition in their own camp. I refer to medicinal cannibalism, which was practiced more or less extensively throughout Europe in the 16th, 17th, and even 18th centuries. Furthermore, such cannibalism, involving human flesh, blood, heart, skull, bone marrow, and other body, parts was not limited to fringe groups of society, but was practiced in the most respectable circles"—including, we might add, among Britain's royalty.[355] But, in another article on medical cannibalism, Gordon-Grube, herself, seems unaware of the prevalence of New England's vampire practice : "It must be added, however, that although Paracelsian ideas in general were present in New England, medicinal preparations made of the human body do not seem to have been widely used."[356] The general unfamiliarity with New England's consumption tradition, which often included the medicinal use of human corpse material, may be the result of deliberate masking. Much of this book is, to borrow Sugg's characterization of his own work on corpse medicine, "a kind of 'Dirty History'—a recovery of habits and beliefs which have been effectively white washed from so many history books."[357] I expand on New England's corpse medicine traditions in the next part.

PART VI Physicians

When the first colonists arrived in New England, as historian Louis Mazzari noted, they brought health-care practices that "differed little" from those of "ancient Greece and Rome." He wrote, "Responsibility for health care largely fell within the family unit, most often to the mother. When families could not treat their own medical problems, they called in a neighbor or sought the advice of local persons reputed to have medical expertise: grocers, booksellers, midwives, nurses, bone setters, and ministers, as well as apothecaries, surgeons, and physicians."[358] The number of options available to those needing treatment is not a testament to efficacy. Even as the eighteenth century waned, the afflicted continued to grope in the dark. Yet, for families adrift in a sea of despondency, the consumption ritual was a lifeline of hope. Medical historian Richard A. Meckel observed that early colonial New Englanders "were not always fatalistic in the face of death and disease. As numerous surviving diaries show, they prayed, pleaded, and occasionally bargained with God, dosed themselves and their children with folk and family medications, and sometimes called in a local medical practitioner," who was "rarely a trained physician."[359]

The two strands of medicinal cannibalism in New England have been delicately disregarded. One, a fading tradition that had competed with other establishment medical theories in Europe, has been attributed to the Swiss-German scientist and astrologer, Theophrastus von Hohenheim (1493-1541), better known as Paracelsus. The Paracelsian tradition existed in a commercial context, where pieces of human corpses—augmented with herbs, minerals, and other chemicals—were bought and sold as medicinal commodities in the open market. The other strand of cannibalism was a family-based folk tradition, handed down in communities by oral tradition and practice, existing outside the bounds of official culture, where parts of deceased relatives were sometimes ingested to cure ailing family members. About fifteen percent of the New England consumption rituals include a cannibalistic prescription: most instruct the consumptive to ingest the ashes of an incinerated organ; a few require inhaling the smoke of the burning body or its parts.

In the previous chapter, we saw that Alfred Sereno Hudson seemed unaware of the folk medical tradition of cannibalism, conflating it with the Paracelsian tradition.[360] By the time Hudson incorrectly asserted that Charles the Second (1630-1685) was an uncooperative user of corpse medicine, the cannibal cover-up had been long established. In truth, as Richard Sugg observed, "Charles himself, far from being 'helpless', was in fact one of the country's keenest users of spirit of skull."[361] Hudson viewed the consumption cure simply as an outdated pharmaceutical prescription. The making of pills to be administered that Hudson reported emphasizes the commercial rather than the altruistic aspects of healing. The two texts from Maine discussed in the previous chapter veer toward the Paracelsian; in one case a tincture was made of the extracted heart and in the other a woman was supposed to "procure" a heart to use as medicine.

Of the two competing, establishment schools of Western medicine, the Paracelsian and the Galenist, it was the latter that transformed into the dominant biomedical paradigm and, because winners write history, controlled the historical narrative. Certainly, a significant factor in marginalizing the Paracelsian theory was cannibalism. "For a long time historians have . . . sought to protect medicine as a whole, and perhaps the past itself, from the potent slur of cannibalism," Sugg wrote in his article on corpse medicine. "Some twentieth-century interpreters of the corpse medicine that was used in Britain during the seventeenth century have characterized it with such phrases as 'irrelevant and nasty' and 'loathsome and insignificant'," he wrote. It was impossible for them "to admit that the seemingly bizarre remedies of the past were real to those who prescribed or used them."[362] As late as 1979, anthropologist William Arens (*The Man-Eating Myth*) asserted that evidence establishing the existence of anthropophagy (a technical term for cannibalism) was almost entirely anecdotal. If he was arguing that the use of cannibalism was not factual, he overstated his case. We need look no further than both Old and New England for verification of its existence. Another anthropologist, Karen Gorden-Grube, wrote in "Evidence of Medicinal Cannibalism in Puritan New England," that the Paracelsian medical theory "was fundamentally opposed to that of the official, Galenist school of medicine."[363] As we have seen, "Galenist physicians sought to heal by restoring a supposed imbalance of the four humors, based on the precept that 'contraries cure.' . . . Paracelsians believed medicines to heal by the occult 'influences' or 'spirits' they contained."[364]

The presence of the Paracelsian paradigm in New England—exemplified by the "dispensatory" of Edward Taylor (ca. 1642-1729), an early Puritan minister, poet, and medical practitioner, residing in Westfield, Massachusetts—may have paved the way for the acceptance of the vampiric version of cannibalism. Taylor's pharmacopoeia incorporated a variety of human body parts, including skin, fat, bones, marrow, skull, brains, and heart. Regarding the latter, Taylor wrote (emphasis his) "*Man's Heart*, dried and *took* cures epilepsy."[365] Also included in Taylor's book of medical prescriptions is the following: "*The blood drunk warm and new* is held good in the falling sickness. It moves sweat, Stops all bleeding drunk new or in ashes."[366] The two consumption cures noted above ("pills made from the ashes obtained from burning a human heart have repeatedly been administered as a cure for consumption" and "heart extracted, and a tincture made for the cure") easily could have come from Taylor. Gordon-Grube did not shrink from citing Taylor's cannibalism: "Taylor advocates cannibalism, pure and simple, and that the purpose is medicinal does not change the fact. Nor can it be claimed that the contemporary mind did not recognize the ingestion of mummy and human blood as cannibalism."[367] The presence of cannibalism in the consumption ritual was not impersonal, for it incorporated social, particularly family, relationships.

Both of New England's cannibal traditions rest on the belief in a spiritual dimension of the physical world. In the vampiric version, a nonmaterial presence in the vital organs of a corpse causes its kinfolk to die. The narrator of an incident from Chesterfield, Massachusetts, seemed on the brink of offering a theory when she wrote: "The story said that, when one member of a family died with consumption, his or her 'vitals'. . . became animated after burial and came back to earth in invisible form to prey upon the 'vitals' of others in the family until they, in turn, wasted away with the same disease."[368] Writing about medical superstitions, in 1871, a physician related "an old superstition" which supposed that "after the death of one individual of the family, the vitality of some remaining member was gradually consumed, by a mysterious sympathetic process, in retarding the decomposition of the lungs of the deceased."[369] The best explanation that the narrator of a consumption ritual from West Stafford, Connecticut, could suggest was that "the vital organs of the dead still retain a certain flicker of vitality, and by some strange process absorb the vital forces of the living."370

While the Paracelsian theory never caught on in New England, traces of it appeared in unlikely American contexts. On May 12, 1857, the *Charlotte* (NC) *Democrat* described the "discovery that several children, who died recently, had been disinterred and their bodies removed." The exhumations were traced to a quack doctor (that is, an untrained practitioner who profited from deliberately misrepresenting his medical qualifications), A. L. Nugent, who was using corpses "for the purpose of extricating medicinal properties from their flesh and bones." The newspaper suggested that "his theory appears to have been that a medicine could be made by boiling the liver of a human being that would cure liver complaint; and so with regard to other diseases." The "boogey man" aspect of robbing the graves of young children to use as corpse medicine had an ancient toehold in Europe. In his *Teutonic Mythology* (1883), Jacob Grimm wrote that "in churchyards they [witches] dig up the *bodies of young children*, and cut the *fingers* off; with the fat of these children they are supposed to make their salve. This seems to be their chief reason for entrapping children."371 The *Märchen* (that is, fairy tales) about witches luring young children into their homes to boil them in cauldrons does, indeed, reflect the authentic Paracelsian tradition of Europe, as do the large number of cannibal motifs found in these tales.372 The European witchcraft belief system deems the remains of unbaptized children as especially efficacious for healing or performing other miraculous or diabolical feats. Since unbaptized children were not yet Christians, they were "in greater danger from witches than the baptized."373

Aside from being a startling late example of Paracelsian medicine in America, the Nugent case raises the question of why quack medicine and practitioners were still viable in the mid-nineteenth century, when the biomedical paradigm was consolidating its hold on disease treatment. The short answer is that quacks and other competing practitioners continued to be a thorn in the side of establishment physicians because, when the latter were unable to deliver the desired results (quite often), patients looked elsewhere. Some turned to quacks while others relied on community-based folk healers. Both traditions, now perhaps viewed as "alternative," had long been in the mix of available treatments. In Elizabethan England, Wallace Notestein wrote, "the realm was alive with men who were pretenders to knowledge of mysteries." From the sixteenth through the early eighteenth centuries, the separation between "reputable physicians" and quacks or "good witches" was fluid. Notestein asserted that even well-respected physicians sometimes resorted to witchcraft to explain difficult cases. So, "it was altogether easy to believe that good witches who antagonized the works of black witches were more dependable than the family physician, who could but suggest the cause of sickness. The regular practitioner must often have created business for his brother of the cunning arts."374

Quacks have been on the scene for a very long time, plying their trade alongside practitioners of both folk and establishment medicine. Since the membranes separating fraudulent, folk, and official medical systems were porous, agents and procedures percolated through in all directions. Yet, as early as the seventeenth century, quacks were not only prevalent, but despised, preying on those who found little relief from other medical channels. In 1605, an anonymous defender of "true physicians" railed against "counterfiet [sic] mountebanks" in the following alliterative, archaic, yet resonant, prose:

Runagate Jews, the cutthroats and robbers of Christians, Slow bellied monks who had made escape from their cloisters, Simoniacal and perjured shavelings, busy St John-lack-Latins, thrasonical and unlettered chemists, shifting and outcast pettifoggers, light-headed and trivial druggers and apothecaries, sun-shunning nightbirds and corner creepers, dull-pated and base mechanics, stage-players, jugglers, pedlars, prittle-prattling barbers, flithy graziers, curious bathkeeepers, common shifters and cogging cavaliers, bragging soldiers, lazy clowns, one-eyed and lamed fencers, toothless and tattling old wives, chattering charwomen and nursekeepers, 'scape-Tyburns, dog-leeches and such-like baggage. In the next rank, to second this goodly troupe, follow poisoners, enchanters, wizards, fortune-tellers, magicians, witches and hags.[375]

Did he omit anyone? People were simultaneously repelled and attracted when scam artists came to town. Their strange ways hinted at wisdom beyond the ken of locals. Did they know secrets that could unlock the mysteries of the prevalent incurable illnesses? An anecdote published in the *Independent Democrat* (June 7, 1849), a New Hampshire newspaper, parodies the allure of foreign knowledge: "'The orifice of the typantum [sic] appears to be enlarged, and the globules of the abdominal functions are much deferred in this patient,' remarked the Quack Doctor, as he held a sick man's arm. 'Lor, Doctor, where was you born?' inquired the old lady. 'In Germany, Madam; in Germany. Finished my education in France.' 'I thought so. *I knew so*,' replied the old woman, 'nobody was ever born in this country that knows as much as you do!'"

Like pollinating bees, traveling quack doctors, many foreign-born, spread treatments from community to community. A story from the *Bangor* (ME) *Daily Whig and Courier* (November 29, 1843) illustrates why desperate people in hopeless situations resort to quacks:

> We learn that in a recent case of sickness where the patient must evidently die, the attending physician after exhausting all his skill, informed the patient and his friends that he could do nothing more and that he was rapidly approaching his end. The friends in great distress gathered around and were desirous of trying some remedy not yet attempted. An aged and superstitious woman, a foreigner, entered the chamber of the sick man, and as if commissioned by Heaven, informed the company that the man could be cured! All eyes were turned upon her shriveled countenance, when she directed them to skin a cat alive and place the warm skin upon the bowels of the patient while the cat should be allowed to run, and at the same time to cut open a live chicken and instantly place the halves to the feet of the patient. The direction was instantly complied with and the patient—died.

The medical establishment of all eras had issues with quacks. Not only were they competitors, but they also sullied the reputation of medicine as a noble enterprise. A commentary in the *Journal of the American Medical Association* (1896) depicted consumption quacks as vampires. Its opening sentence is a self-serving exaggeration:

> Most cases of consumption can be recovered from within three months. These opportunities are lost under the use of quack medicines. The vampire, in its original conception, was a dead man restored to life but without soul and with only life enough to suck the blood of its victim when asleep. In zoology the vampire is a bat which sucks the blood of persons and beasts when asleep. The consumption quack are the vampires which feed upon the slow tortures of this disease.[376]

Henry Nourse, in his history of Harvard, Massachusetts, offered a more realistic assessment of the medical establishment's ability to treat consumption: "When that fell destroyer, consumption, broke into a family circle and began to bear away its victims in slow but sure succession," even "the most self-confident physicians" were humiliated "with a sense of their impotence."[377]

Changes in the Western medical system during the nineteenth century exceeded those of all previous centuries combined. This unprecedented awakening was the result of careful observation and experimentation. The empirical mode of addressing disease and health was not as unequivocal two hundred years ago as it is now. The chasm between not understanding and finally knowing the cause of consumption was bridged by a century of eclectic searching. The way forward is far easier for me to describe here than it was for the seekers to traverse. They toiled in a fog of mystery that invited compelling and intriguing speculation and, though they sometimes found common ground, usually they were at odds.

Reenactment of vampire exhumation. *Courtesy of Cyril Place*.

13   The Charm Worked No Good and the Patient Died

The ancient Galenist tradition of medicine did not go unchallenged in the nineteenth century. Like the Paracelsian theory, the expanding homeopathic system was based on a premise that "like cures like." Analogic magic, the fundamental logic of homeopathy, can be traced back to antiquity. As folklorist Wayland D. Hand explained in his discussion of folk medical magic and symbolism in the West, homeopathic magic assumes that "external similarity rests on what would seem to be an apparent internal connection and a basic inner unity and dependence. Under this premise, cures are undertaken on the theory that similar things are cured by similar means, as set forth in the Latin phrase *similia similibus curantur*."[378] The German physician Samuel Hahnemann (1755-1843) founded the homeopathic school of medicine after experimenting with cinchona, a flowering plant whose bark contains quinine. He came to believe that "if a patient had an illness, it could be cured by giving a medicine which, if given to a healthy person, would produce similar symptoms of that same illness but to a slighter degree. Thus, if a patient was suffering from severe nausea, he was given a medicine, which in a healthy person would provoke mild nausea. By a process he called 'proving', Hahnemann claimed to be able to compile a selection of appropriate remedies."[379] Orthodox, Galenic practitioners objected especially to the notion that the extremely diluted amount of medicine that homeopathic doctors prescribed contained therapeutic power as a "dematerialized spiritual force."

    Galenist physicians regarded the homeopaths as quacks, an attitude that understandably put homeopaths on the defensive. In reaction, some homeopaths raised the cudgel against those they viewed as the real quacks, including advocates for perceived lingering medical superstitions. This is where we find Austin Wells Holden, in 1871, giving the Address to the Semi-Annual Meeting of the Homeopathic Society of the State of New York. As a convert from conventional medicine to homeopathy, Holden was prepared to zealously defend his chosen medical field.

    Holden was born in White Creek, Washington County, New York, on May 16, 1819. After graduation from St. Lawrence Academy, in Potsdam, he relocated with his parents to Glens Falls, in Warren County, at the southern tip of Lake George. He began the study of law but, lacking the economic resources to continue, worked in his father's cabinet shop, and studied on the side. At about the age of twenty-two, he entered Albany Medical College, graduating with distinction in 1848. He opened a medical practice in Warrensburg, where he married Elizabeth Buell. She was the daughter of the Honorable Horatio Buell, a former judge of Warren County, and a niece of Sarah Josepha (Buell) Hale, who was for many years editor of *Godey's Lady's Book* (which, as we will see, was not averse to publishing New England vampire stories). In 1852, Holden returned to Glens Falls, where, after five years of conventional medical practice, he became a homeopathic practitioner. Like other educated men of his generation, Holden was eclectic and community oriented. In addition to working as a lawyer, cabinetmaker, and physician, he also wrote poetry and local history (publishing *A History of Queensbury, N. Y.* in 1874), served in the Civil War, and was a member of his town's board of education.[380]

In the ornate oratory expected of people in authority at that time, Holden opened his address to the Homeopathic Society by acknowledging the medical community's schisms, as well as the dismal record that all factions exhibited in their attempts to heal. He targeted the medical community in general, but obliquely indicted the establishment Galenist tradition for its failures. He was felicitous in denigrating "superstitions," but, treading between regular physicians and outlier practitioners, Holden pointedly suggested that the former had not been appreciably more successful in their treatments than the latter. As a homeopathic doctor harried by mainstream physicians, Holden was on the front lines of mid-century medical rifts. He likened the dogmatism of medicine to that of religion: "From the earliest times, medicine, like religion, has been divided into sects and schools as hostile and inimical to each other, as opinionated and dogmatical in the declaration of their doctrines as any of the religious denominations of the same age. Thus we read of the Galenists and Paracelsians, the empirical and the rational schools."[381] Holden then relinquished any subtlety regarding the medical community's efficacy: "Pereira, it is alleged, said, on his death bed, that he was a wise man who should discard all medicines, and leave the business of curing to nature alone." He repeated a commonplace quip attributed to Oliver Wendell Holmes, who "startled the medical public by asserting, that leaving out opium, certain specifics, wine and anaesthetics, 'if the whole materia medica, as now used, could be sunk to the bottom of the sea, it would be all the better for mankind and all the worse for the fishes.'"[382]

The target of Holden's consumption narrative was the common enemy of both Galenists and homeopaths—the quack:

> It is less than twelve years ago, that in one [of] the suburbs of a neighboring village, at the instance of a self-styled doctor, a corpse was exhumed, and, though the body was far gone with decomposition, it was opened, the heart taken out and burned to ashes, to affect, as was alleged, the cure of a sister of the deceased, who had the consumption. The heart of the dead man, as the doctor stated, and the family believed, preyed upon the vitals of the living sister, and until this was utterly destroyed, she would continue to waste and pine away. It is, perhaps, superfluous to add, that the charm worked no good, and that the patient died.[383]

Despite a predictable lack of detail, we can conclude that this incident occurred sometime around 1859, probably in or near Holden's hometown of Glens Falls, New York. The newspaper story of Salmacious Adams' widow (chapter 6) is strikingly similar: both narratives occur in the same locale at the same time, prescribe the same ritual procedures, and include a physician (or at least a "self-styled doctor") at the scene. Two differences are the sex of the dying sibling (brother in the news article and sister in Holden's narrative) and whether the prescribed ritual was ever consummated (aborted in the news article, but completed, and failing, in Holden's narrative). If Holden relied on memory to recount the event described in the 1858 news article, we should expect discrepancies. In the newspaper text, the widow of the dead man assumes the significant role of belief specialist, giving her center stage; in Holden's narrative, her role shifts from helpful widow to dying sibling. Finally, a ritual not completed—and, therefore not failing—would dilute Holden's argument against superstition. If his narrative was a different event, and *not* the Adams incident, then we can add another case to the New England vampire inventory.

Holden's interpretation of the vampire practice was in step with the era's medical establishment: "While we smile at the narration of these witch cures, and wonder at the credulity which fed and fostered the imposition, a moment's reflection will teach us that even the most outre, thrilling and bizarre incongruities and monstrosities of doctrine and belief of those days may find their parallels among the superstitious practices of this enlightened age."[384] Yet, his refutation of superstition did not put him in the establishment Galenist camp. Near the close of his oratory, Holden remarked, "a rough, harsh layer of superstition . . . is continually thrusting its unsightly angles and asperities to sight."[385] Beneath the surface of Holden's address runs a current of rebuke of the medical establishment's hubris, a tone that he established near the beginning of his address, asserting that "the history of medicine is a chronicle of mistakes and blunders; a record of experiments and failures, to which science has only capriciously lent its light."[386] From this perspective, one might judge the consumption ritual as another stage in the evolution of medical practice; and, we might ponder what future critics will make of our current medical situation.

***

Coincidentally, the same year that Holden's address was published in a homeopathic journal from New York, the *American Journal of Homeopathic Materia and Record of Medical Science* (September 1, 1871) included two additional exhumation narratives. Both texts appeared as unannotated commentary in the "Editorial Department" section under the heading of "Medical Superstitions," which began with: "In looking over some of the old volumes of medicine, one is amazed and even sickened at the revolting prescriptions there met with for the cure of disease. Nearly every disgusting object in nature is recommended as an infallible remedy for the most serious forms of disease."[387] There followed a short series of excerpts from *Dr. Willis' Receipts for the Cure of All Distempers* (London 1701). Bringing the matter closer to home, the editor then wrote: "But we need not look to the distant past for illustrations of gross superstitions in these matters. At the present time, and in our own midst, we may find an abundant evidence of superstitions scarcely less wonderful than those already referred to."[388] Citing several well-known folk cures, labeled "innocent whims," the editor turned to the "other notions equally revolting as any of the superstitions of the past," including the following:

> A few years ago, the papers contained detailed accounts of the exhuming of the body of a woman in the central part of this state, who had died of consumption, that her liver might be removed and burned, and the ashes used as a cure for the same disease.
>
> In another case, where several members of a family had died from this disease, the survivors became impressed with an old superstition that after the death of one individual of the family, the vitality of some remaining member was gradually consumed, by a mysterious sympathetic process, in retarding the decomposition of the lungs of the deceased. Laboring under this delusion, the body of the last departed was disinterred, and the lungs removed and burnt, that the disease in those still living, might thus be arrested.[389]

Once again, lack of narrative detail inhibits further investigation.

***

A. W. Holden saw the continuing use of "superstition" as "human nature" in an ongoing search for "wonder-works and miracles." How such an aspect of human nature might function was addressed a little more than three decades earlier by an establishment physician, Dr. John Clough, in an article written for the *Boston Medical and Surgical Journal* (1840):

> I cannot omit to mention a circumstance which occurred in this town (New Ipswich), not thirty years since, and similar occurrences probably occurred in many other towns in New England. This was disinterring a human body, which belonged to a family all strongly predisposed to consumption, for the purpose of removing the heart, which was burned, the ashes of which were considered a sovereign remedy to those of the family who were still living, and might be afflicted with the same disease. This only illustrates the fact that those elements of character, which held such a magic sway over the minds of men in ancient times, have not ceased altogether to influence the community in our comparatively enlightened day.[390]

Clough was connecting his exhumation narrative to, as the title of his article informs us, "The Influence of the Mind on Physical Organization." More specifically, he was discussing the effects of "magic, incantation, amulets, and holy relics." These, he argued, "had their influence in an age of extreme ignorance and superstition—when the darkness which shrouded the human mind was so thick that it could be even felt."[391] He believed that the power of such deceptions emanated from their ability to enrapture the human mind. Although Clough seemed unaware of the term "psychosomatic" (a word that first appeared in the language about six years before his article[392]), he undoubtedly understood that the mind could affect the physical being in ways that science could not explain (perhaps also analogous to the modern placebo effect): "The influence of imagination, in aiding the happy effects of medicine in curing disease, should not, however, receive our unqualified censure."[393] The lesson that Clough seemed to draw from his discovery is a rather weak advisory that physicians ought to be aware of the possible effects of their mannerisms on patients, whose distraught minds may magnify any cue, whether positive or negative.

John Clough (1809-1879) was born in Gilmanton, New Hampshire on January 23, 1809. He received a medical degree at Dartmouth College in 1839 and began general practice in New Ipswich, New Hampshire. In 1845, having abandoned medicine and taken up dentistry, he relocated to Boston. After twenty years of a successful dental practice in Boston, he moved his practice several miles northwest to Woburn, where he died on November 25, 1879. When he wrote the article that included his exhumation narrative, in 1840, Clough was a medical doctor living in New Ipswich. Internal evidence in the narrative yields 1810 as the approximate date of the exhumation.[394] Since Clough was then an infant, he must have learned about the incident from a source other than personal experience.

A provocative connection might shed light on Clough's knowledge of the event: his wife, Ellen Eliza Champney (1815-1888), was born in New Ipswich. Her family had been there for quite some time, so he could have learned of the exhumation from her or her family. Her family ties to the town's medical community suggest a credible channel of oral tradition. John Preston (1738-1803) was one of New Ipswich's earliest physicians. His sons, John and Timothy, followed him into medicine in New Ipswich. After graduating from Dartmouth with a medical degree, the younger John (1770-1828) returned to New Ipswich to practice with his father, where he married Elizabeth Champney (1779-1869). Elizabeth was Ellen's aunt, the daughter of the brother of Benjamin Champney (1764-1827), Ellen's father. So, the younger physician, John Preston, was Ellen's uncle by marriage. Did she hear the story of the exhumation from him? The most prominent physician in town surely would know about a local family's consumption epidemic; it would be unfathomable if they had not first consulted the established medical community. In a town with a population of just 1,393 in 1810,[395] Dr. Preston also would be aware of any alternate cures that were attempted by that family. As far as I know, the record does not indicate that Dr. Preston, who also served the town in various capacities, including town clerk from 1802 to 1819, commented upon a therapeutic exhumation in the town. Bear in mind that oral tradition relies on the spoken word for its transmission, a medium that eventually obscures its origins, as repetition over time tends to reduce a narrative to its dramatic core. Specific details, such as names, dates, and places, are easily shed during this process.

There were at least two consumption-plagued families living in New Ipswich during that period, the Breeds and the Bartletts. In their history of New Ipswich (1852), Kidder and Gould wrote that Allen Breed (1744-1806) became a wealthy farmer, but, unfortunately, "the seeds of consumption seem to have been inherited in this family from their birth; they all fell victims at an early age—a long row of stones in the south burying-ground tells the sad story."[396] The stones in the Smithville Cemetery (as it is now called) do, indeed, relate a sorrowful tale, as eight members of the family died within a nine-year span. Allen Breed and his wife, Lydia Mansfield, gave birth to six children; four of them died while in their thirties. Elisha on his thirtieth birthday, March 6, 1802. His son, also Elisha, died at the age of three in 1805, and his wife, Rebecca, died a year later. Elisha's father, Allen, died in the same year. Three of Allen and Lydia's remaining children, and one of their spouses, subsequently passed away: Lydia, in 1807, at the age of thirty-nine; John, later that year, at the age of thirty-eight, and his wife, Abiah Lampson, in 1808, at the age of thirty; Enoch died in 1811, at the age of thirty-three. In their history of New Ipswich (1914), Charles Henry Chandler and Sarah Fiske Lee wrote that "the early disappearance of this family [Breed] from the town" was "probably largely due to the tendency of its members to consumption."[397] The Breed family has every reason to perform the consumption ritual. If they did, it is likely that Enoch, the last to die of this family epidemic, was the patient requiring a cure. Perhaps it was John, the last sibling to die before Enoch, who was exhumed. If the ritual was performed in 1810 (as Clough suggested in his text), then John would have been dead for about three years.

The other local family known to be afflicted with consumption, the Bartletts, fared as poorly as the Breeds. The Bartlett's gloomy story is evident in the Old Burial Ground (now, Hill Cemetery) of New Ipswich. Of the nine children of Samuel and Elizabeth Appleton Bartlett, six had died by 1809: Elizabeth (1773-1790), John (1779-1802), Sarah (1776-1803), Samuel (1763-1805), Isaac (1761-1806), and Noah (1768-1809). Two children of Noah and his wife, Mary Hills, died at the age of two. Their parents died within the same time frame: Mary in 1806 and Noah in 1809. According to Chandler and Lee, a son of Noah and Mary, John, "had proposed to enter the ministry, and had studied at Andover Theological Seminary; but before completion of his course of study he, the last survivor of a family which perished from consumption, was attacked by the same disease, and while on his way south in hope of relief, he died in New Jersey."[398]

Whether or not the Breed or Bartlett families were involved in the exhumation in New Ipswich described in John Clough's narrative, their stories exemplify the terrible consequences that consumption visited on many families. Clough does not relate the outcome of the exhumations, but if these families *had* resorted to the ritual, it failed to loosen the vampire's grasp.

***

In 1831, Josiah Holbrook (1788-1854) and Daniel Webster (1782-1852) founded the American Lyceum Association to promote adult education (a concept later rebranded as "continuing education" and "lifetime learning"), mainly through public lectures and publications. Webster, of course, was a New England statesman and orator of legendary, if not always appealing, status. Three years later, with Jerome V. C. Smith, M.D., as editor (or "conductor" as he was labelled on the title page), Holbrook launched a journal whose title reflects the eclectic scope of learning that educated elites hoped would captivate the masses. The first volume of *Scientific Tract and Family Lyceum Designed for Instruction and Entertainment and Adapted for Schools, Lyceums and Families* included articles on a range of topics, from astronomy and animal magnetism to waterpower and wolverines, incorporating along the way, barking dogs, fossils, luminous appearance of the sea, phrenology, rat-catching, and the vulture.

Also included in the journal's 1834 debut was an article entitled, "Popular Whims and Superstitions, Relating More Especially to the Practice of Medicine in the Nineteenth Century." Its first sentence underscores the journal's goal "to enlighten the public mind, and to emancipate it from the thralldom of superstition and ignorance," implicitly accepting that "the shackles of superstition always fall before the light of science." The article's author remains masked, credited only as "W."—perhaps Webster, himself. "Some of the popular superstitions of the present day, respecting the practice of physic," the article begins, "may show the errors with which physicians have to contend." The essay consists of nineteen numbered entries, "curious relics for succeeding generations," each consisting of a separate medical superstition, including transference of warts, divining rods, and the influences of the moon.[399]

In W's article, the following "Whim 18th" was headed, "Consumption":

A superstition somewhat common among the most illiterate classes of the community is, that if the portions of the lungs which remain after a friend has died with the consumption, be taken out and burnt, the remainder of the family of the diseased will never afterwards be liable to the complaint. I have known the bodies of three or four people, who had died of consumption, disinterred, some of whom had been buried six or seven years, and their viscera burnt to ashes. What is remarkable, a physician attended as master of ceremonies.[400]

The author's phrase, "I have known," strongly suggests that he had firsthand knowledge of the events, even if he was not actually at the scene. And, when he terms the practice "somewhat common," he appears to preemptively deflect any suggestion that it was an extraordinary occurrence. Again, a physician was present— "as master of ceremonies" in this case. What is the role of an MC at a consumption ritual? Perhaps it was just to oversee the disinterment of the corpses and excision of the lungs, much like that of Dr. Harold Metcalf in the Mercy Brown case from Exeter, Rhode Island. The absence of locations or surnames attached to these exhumations leads to the familiar dead end. Worth noting is "Whim 6th": "To cure epilepsy, or falling-sickness fits, procure the upper part of the skull of a dead man, pulverize and take every day until the patient is cured." Ingesting portions of a human cadaver (mummia or mummy) as a routine curative prescription is, as we have seen, pure Paracelsian medicine.

***

Dr. Henry I. Bowditch (1808-1892) was convinced that consumption was caused by living in damp areas. In 1862, he published the results of ten years' research correlating soil moisture with the incidence of consumption in Massachusetts.[401] Despite the inclusion of numerous maps, based on contributions by physicians from more than half of Massachusetts' 325 townships, showing that consumption deaths appeared to cluster in damp, low-lying areas— "consumption-breeding districts," as Bowditch termed them[402]—his pamphlet did not address *how* soil moisture caused consumption. "How such effects are produced," Bowditch admitted in his preface, "I do not and cannot pretend to explain."[403] Nor was he able to convince more than a small contingent from New England and Britain that this "association" with consumption was a "cause" of consumption.

Bowditch was fighting an uphill battle against contagion. Two years later, in his book, *Is Consumption Ever Contagious, or Communicated by One Person to Another in Any Manner?* (1864) Bowditch summarized the history of medical thought regarding his question. From the time of Aristotle until 1775, Bowditch wrote, there was "very general consent to the proposition of the contagious nature of consumption."[404] There followed a brief period, until 1800, of "evidently great indecision." Since that time, Bowditch asserted, "with some very few exceptions, Medical Opinion and Public Opinion have been gradually becoming settled in opposition to contagion."[405] His judgment that contagion was no longer an accepted explanation for consumption missed the mark. The first issue taken up at the inaugural International Medical Congress in Paris, in August 1867, was consumption. According to correspondence from the Paris congress, summarized in the *San Antonio Express* (October 1, 1867), "The discussion turned mainly on these two points: Is tubercular consumption contagious, and may it not be prevented by inoculation?" The correspondent's summary directly contradicts Bowditch's assertion. "Enough new facts have been arrived at," the correspondent wrote, "to warrant a certain number of the experimenters to declare that tubercular phthisis is contagious."

Bowditch had written to the secretary of the International Medical Congress two months prior to the meeting, requesting that the congress consider his soil-moisture theory when discussing "the causation of phthisis." In his letter, which included a copy of his paper on consumption in New England, Bowditch outlined his theory and summarized the data that he argued "proved" that "this law of soil-moisture is a cause of consumption certainly in New England, and possibly in other parts of the globe."[406] The Medical Congress was indifferent to Bowditch's theory. "The results of my effort was simply an abortion," he complained. "The Secretary and the Congress virtually ignored the whole subject."[407] Bowditch's chagrin was stirred by an increasingly unitary paradigm regarding consumption's etiology; Continental Europe's physicians and scientists led the way, testing hypotheses generated from the contagion theory through controlled experiments. In 1869, for example, Jean Antoine Villemin injected tuberculosis matter from human cadavers into laboratory rabbits. When the rabbits became infected with the disease, the theory that consumption was contagious was on its way to validation. Though disappointed, Bowditch was willing to entertain competing theories; still, he would not be dissuaded from his belief that one of consumption's "chief causes" was soil-moisture.

In "Consumption in New England," Bowditch augmented his Massachusetts sources with "confirmatory facts, statistics and opinions from Rhode Island, Maine and New Hampshire." One of his correspondents was Dr. John Larrabee Allen, of Saco, Maine. The final sentence in Bowditch's summary of Allen's contribution is striking, yet delivered nonchalantly:

> Dr. Allen, of Saco, Maine, than whom there lives no more intelligent witness—a practitioner of long standing—assures me that, in his own practice, for fifteen years past, he has noticed that, on two ridges of land, whose only difference consists of this characteristic of moisture of the soil, almost every family has been decimated on the wet part, while almost all upon the dry portion have escaped. This statement is a most striking exemplification of the views already given. It seems likewise to show that it is not *elevation* or *exposure*, or a certain amount of cultivation or woodland, nor any peculiar trade, &c. that is the cause of the malady; for, in all these respects, the two localities seem alike. One ridge is quite dry, the other is literally filled with springs. Nowhere can a spade be driven a few feet into the ground, without meeting water. In fact, in former times, the superstitious frequently had their friends, who had died of consumption, disinterred, and Dr. Allen invariably found the coffins filled with water, however shallow may have been the graves.[408]

This allusion does not relate a specific exhumation, nor does it say what was done following disinterment. Yet it does disclose that therapeutic exhumations occurred frequently "in former times"—which, if we take "fifteen years past" literally, would have begun about 1847. Dr. Allen apparently was routinely present at the exhumations (probably not to corroborate Bowditch's soil-moisture thesis). Perhaps the consumptive families and their friends requested his assistance in excising vital organs and identifying unusual signs of life among the deceased via post-mortem examinations. Maybe he was there simply out of professional curiosity.

John Larrabee Allen was born in Cornish, Maine, on April 26, 1814. After attending the local schools, he studied medicine with Dr. Stephen C. Brewster, of Buxton. Allen graduated from the Maine Medical School in Brunswick, in 1836, then settled in Springvale, where, according to a Maine historian, "he enjoyed a large and successful practice until the fall of 1852, when he removed to Saco. There he became one of the best known physicians in this section."[409] During the Civil War, he served at the Fairfax Seminary Hospital, near Alexandria, Virginia. For nearly twenty years after the war, he was the United States examining surgeon for pensions. Allen died in Saco on September 4, 1897.[410] Apart from Bowditch's summary (which, undoubtedly, he received from Allen by way of personal correspondence), I have found no other references to therapeutic exhumations in Saco.

A tragic incident, reported as "Fatal Encounter Near Union City," in the San Francisco *Daily Globe* (May 3,1858), gives expression, without design, to the impact that consumption had on some Saco families. The article relates a fight that had recently occurred between James F. Whitton and "a man named Horner." Whitton, armed with a knife and a six shooter, faced off with Horner, who had a Bowie knife. Whitton fired two shots, the second of which entered Horner's thigh. At the same time, Horner lunged at Whitton, striking him in the abdomen with his knife. Whitton died soon after. The narrative concludes:

> James F. Whitton was from Saco, Maine, where he leaves a father and a mother. He was the eldest of several children, the rest of whom have died with consumption within the past few years. It has been but a few weeks since he received intelligence of the death of his last sister, accompanied by an earnest appeal from his mother to come home. He was making preparations to obey the summons, and would have been on his road to the already desolate home of his parents in a few weeks more, only for his untimely death. He leaves

considerable property. He was 29 years of age, and had been a resident of this Bay for the six years preceding his death.

The corpse of James F. Whitten (1827-1858)—not Whitton, as rendered in the newspaper article—did return to Saco; he is interred in the family plot in Laurel Hill Cemetery, along the banks of the Saco River. Also interred there are his siblings: the sister mentioned in the article, Melinda Ellen Whitten Hawkes (1829-1858), who died February 5; Frances Jane Whitten (1831-1845), Samuel Emery Whitten (1834-1856), and Leonard Whitten (1836-1857). It would be ironic, indeed, if James had gone to California hoping to avoid the consumption epidemic that plagued his family only to succumb to Horner's Bowie knife.

14   The Village Witch Told Them ... Cut Her Heart out and Bury It

Dr. Henry Bowditch's desire to explain consumption, as well as his devotion to other medical and social concerns, led to his adamant lobbying of the Massachusetts legislature to establish a state Board of Health. An eventual, certainly unintended, consequence of this effort led to the recording of yet another vampire narrative—which was, not surprisingly, all but swept into the dustbin of history.

In his 1862 treatise on consumption in Massachusetts, Bowditch wrote, "Certainly, a Board of Health, composed of the ablest of the profession, and of others interested in vital statistics, is quite as important as Boards of Agriculture or of Education."[411] Bowditch argued for a Board of Health from a position of distinguished authority. Henry Ingersoll Bowditch (1808-1892) was born in Salem, Massachusetts, on August 9, 1808. His father was Nathaniel Bowditch, the renowned, self-educated mathematician, astronomer, and businessman whose book, *The New American Practical Navigator* (1802), is credited with bringing maritime navigation into the modern era. The younger Bowditch graduated from Harvard College in 1828 and received his medical degree there four years later. To complete his medical studies, he went to Paris, where he allied himself with the "new clinicians," a group of empirical epidemiologists who were challenging the old order of Galenic "heroic medicine" that included purges and bloodletting. His mentor, Pierre Charles Alexander Louis, stressed "the importance of science, statistics, and observation of large numbers of patients with similar symptoms." Louis's "numerical way," as Bowditch called it, "enabled him to combine his interest in science as a field to help patients and medicine."[412]

In a letter to his father, Bowditch explained Paris's liberating influence: "I want to see everything more free than it is now . . . that is what is done here and what ought to be done in America."[413] Back home, Bowditch became an ardent abolitionist and friend of Frederick Douglass. He was active in Boston's close-knit antislavery community, serving as an agent of the Underground Railroad and shielding runaways from slaveholders. His espousal of the natural rights of all people also led him to become an early advocate for women's rights. For most of his professional life, Bowditch was a practicing physician at the Massachusetts General Hospital. He was also a professor of clinical medicine at Harvard (1859-1867), president of the American Medical Association (1877), and a fellow of the American Academy of Arts and Sciences. Bowditch finally prevailed in his push for government support of public health, becoming the first chairman of the Massachusetts State Board of Health, the first such organization in the country, serving from its inception in 1869 to 1879.[414] Bowditch may have been behind the curve when he advocated for his soil-moisture theory of consumption's cause, but he was a pioneering reformer who fought for the rights of all people—including the poor, minorities, and immigrants—to live in a healthy environment, with clean air, water, and soil, and to have equal access to health care.

In his new, official role, Bowditch had the authority and resources to apply his social philosophy. He compiled data that he believed would reveal the perplexing link between the human condition and consumption; if that could be established then a remedy could soon be at hand. The

Fourth Annual Report of the State Board of Health of Massachusetts, published in January 1873, contained an extraordinary discourse about consumption. Labeled "Medical Public Opinion," and occupying more space than any other single health issue in the report, were the tabulated responses to a questionnaire that Bowditch had mailed, in July 1871, to prominent physicians in active practice in New England (predominantly, Massachusetts). Of the twenty questions that he asked, Bowditch wrote, "Some . . . are evidently connected with what are usually deemed antecedents of consumption in Massachusetts, while others may seem to have little bearing upon them, and may be deemed futile or irrelevant."[415] The questions addressed the following issues: heredity, overstudy or overwork, occupation, bodily injury, mental trouble, marriage and child-bearing, sexual indulgence, and wet or exposed location of dwelling. Many of the "preventatives" that the 210 responding physicians cited are today's common-sense approaches to wellness: exercise, fresh air, adequate and healthy nutrition, good hygiene, proper clothing, keeping occupied in body and mind, avoiding excesses, and getting sufficient rest and sleep. Dr. Spalding's response is a good summary of this approach: "No specifics, but such general treatment as will promote sound physical health."[416] Several physicians had their own pet peeves, ranging from specific tonics or patent medicines they deemed efficacious, to the avoidance of dancing. Dr. Ballou, for example, wrote that "corsets and thin shoes, or whatever deteriorates the blood or destroys nervous sensibility, are causes of consumption."[417] Since Dr. Ballou did not elaborate, we might presume that tight corsets restricted blood flow in the lungs, while thin shoes allowed cold, damp soil (Bowditch's consumption bugaboo) to impede the sensibilities of the feet.

Bowditch's comments regarding his eighteenth question—Is consumption ever caused or promoted by contagion or infection? —suggest that a significant number of physicians in New England had begun questioning whether the long-assumed hereditary basis for consumption was, instead, the effects of contagion. One wonders if Bowditch had the 1867 Paris Medical Congress in mind when he wrote, "Under the light of modern investigations as to the inoculability [sic] of tuberculosis, the question of communication of the disease from one person to another, becomes a vital question." One of his corresponding physicians believed that "the dried sputa have particles in them of the real '*contagion*' of this disease, and that they must be floating about in every atmosphere in which a consumptive is living." In cases where "the disease runs through many members of a family," this New England physician gave credence to contagion rather than "a hereditary tendency."[418]

By this time, Bowditch, himself, seemed on the cusp regarding heredity versus contagion. In one of his statements, he advised treating the disease as if it was transmissible, by having patients physically isolated. Yet he did not explicitly endorse the contagion theory, apparently unwilling to completely abandon the hereditary theory favored by English and American practitioners.[419] In contrast to the more impersonal theory of contagion embraced by Continental Europeans, which emphasized conducting controlled experiments, Americans clung to notions increasingly held to be outmoded. The belief that consumption resulted from life habits acting upon hereditary tendencies led American physicians to stress the importance of anecdotal evidence gleaned from years of treating consumptives. One fortunate outcome of their reliance on such evidence (at least for this book) was the chronicling of consumption narratives. The hypothesis that consumption was hereditary corresponded with the folk belief that vampires preyed primarily upon their own families.

A convergence of inherited taint and lifestyle choices accorded with an American worldview that stressed personal and societal reform. Brought to life in the Second Great Awakening (ca. 1790-1850) following the American Revolution, it was especially resonant in New England, hub of the temperance, antislavery, and women's rights movements. Medical theory echoed theology: though one had inherited a ghastly tendency, one might escape a dreadful end through proper living, just as redemption from original sin is possible for one who embraces inner faith and lives a moral life. (It is tempting to interpret Roderick Elliston's inherited "bosom serpent"—in Nathaniel Hawthorne's short story, "The Bosom Serpent"—as hereditary consumption that gnawed away at Elliston until he changed his lifestyle by turning away from self-absorption and embracing society.)

In the *Fourth Annual Report* of the health board, Bowditch saved the soil-moisture issue for the last of his twenty questions: "Is consumption ever caused or promoted by a wet location?" After showing the tabulated results, where 80 percent of responding physicians (168 of 210) answered "yes," Bowditch wrote: "Upon this question the profession is more nearly unanimous, than upon most of those that preceded it. This is probably owing to the fact that investigations carried on in Massachusetts, many years since, by myself, and subsequently in England . . . have fully proved that residence on a damp soil tends to the production and promotion of consumption, in New and Old England."[420] The source he cited that "fully proved" his theory was his own *Consumption in New England* (1862). In this monograph, Bowditch contrasted the differences between dry and wet locations and posed a question that is eerily evocative of supernatural nocturnal assaults: "Does it therefore seem incredible to you that such differences, even when not so appreciable during the waking hours, may materially influence our health during sleep, when our natures seem most susceptible to evil influences?"[421] Instead of being more susceptible to vampire assaults, however, the absence of the "caloric, luminous, and . . . actinic rays of the sun" allows "evil influences"— consumption, in this instance—to work unabated.

A self-described "obscure country practitioner," Dr. John Colby Manson, known as "the local doctor" in Pittsfield, Maine, concluded a protracted and lengthy correspondence with Dr. Bowditch, humbly opting for contagion: "Now, my dear sir, I repeat that I am firmly of the opinion, after about twenty years' observation in this *consumptive district*, that this disease is *surely communicable*, and it would be almost impossible to convince me to the contrary. . . . I believe it is so produced by inhalations of exhalations from the lungs and body—but this opinion is somewhat timidly expressed by an obscure country practitioner, striving for reliable information in any matter conducing to the physical betterment of suffering humanity."[422]

If it seems remarkable that twenty-five percent of Bowditch's questions concerned the relationship between alcohol use and consumption, one must recall that, by the mid-nineteenth century, the Temperance movement had become a mass movement. One of the questions was framed this way: "Is consumption prevented by the drunkenness of an individual? Is a drunkard less liable than others to consumption?" Odd as it may seem, twenty-seven percent of the respondents answered in the affirmative. It was Dr. Abell's reply that included an illustrative vampire narrative:

> I should be sorry to be understood as recommending drunkenness as a cure. But I have known several instances where nearly all the family, from five to nine children have successively died of phthisis. Finally, one of the boys, from sheer desperation, turned to excessive drinking of alcoholic stimulants. These boys are now past middle life, and enjoying

good health when last heard from. In two families, not less than five or six victims in each were carried off by consumption. In each there was always one sick, and a short time before death another would be prostrated. In one family they resorted to that horrible relic of superstition, the burning of the heart, etc., of the *dead*, and the ashes were swallowed by the survivors, in the hope that the fatal demon would be exorcised from the family, but it did not avail. But another son fell a victim; and then the alcoholic treatment was tried, not as an expected remedy, but as a means of forgetfulness of impending doom, and no deaths in the family have to my knowledge since occurred."[423]

Who was Dr. Abell, and how do we interpret his narrative? Telling clues appeared in the 1874 Massachusetts State Board of Health annual report (p. 287), where Abell's comments regarding "cerebral-spinal meningitis" appear under the subheading of "Chicopee"—indicating that he was a physician residing in Chicopee (Hampden County), Massachusetts, at that time. A search yielded Erasmus Darwin Abell (1817-1899), who began practicing medicine in Chicopee, in 1852. He was born in Lempster, New Hampshire, January 26, 1817, the son of Dr. Truman and Sarah (Lane) Abell. His father was a physician and a noted astronomer and botanist, who, for fifty years, authored Abell's *New England Farmers' Almanac*. The younger Abell was educated in public and private schools (assisted by tutors for Greek and Latin), attended lectures at the Vermont Medical College, Woodstock, then graduated from Dartmouth Medical College in 1838. He also studied medicine with his father and his maternal uncle, Dr. Robert Lane. From 1839 to 1852, Abell practiced medicine in Sutton, Lempster, and Ringe, New Hampshire; he then relocated to Chicopee, where he remained in practice until 1876. Abell was involved in community affairs wherever he resided, serving, variously, as surgeon of New Hampshire's militia, postmaster, superintendent of schools, and health officer.

Abell introduced his anecdotes about consumptive families and alcohol with the phrase, "I have known several instances." Yet, it is not clear if he treated the consumptive families, if he knew of them through other physicians, or if the stories of their plights simply were circulating in the community; Chicopee's population in 1870 was 9,607. The best conclusion is that the family of consumptives who resorted to the "horrible relic of superstition" resided in or near Chicopee, and that the consumption ritual occurred after 1852, the year that Abell began practicing in Chicopee, and before 1872, the year the report was published. Chicopee lies within the area of the Connecticut River Valley that was fertile ground for the vampire tradition.

In the *Fourth Annual Report of the State Board of Health of Massachusetts* (1873), in addition to addressing the possible influence of alcohol on consumption, Abell also discussed the effect of different climates on lung functions. He wrote that when he and his wife visited Minnesota (indicating that they did so on a regular basis) their colds and hay fever were less severe.[424] Abell's eventual relocation to Minnesota essentially proves that the Abell who appeared in the 1873 Health Report was, indeed, Erasmus Darwin Abell. An article in the *Springfield Republican* (April 11, 1868) cements that connection: "Dr. Abell is about to remove to Minnesota for a permanent residence on account of his continued ill health in Chicopee." Abell did, indeed, relocate to Farmington, Minnesota, in 1876, where he died on October 2, 1899. Both Erasmus Darwin Abell and Henry David Thoreau were born in 1817. And both traveled to Minnesota to alleviate their respiratory ailments. It seems to have worked for Abell but, unfortunately, not for Thoreau (see chapter 6).

***

"In my opinion the following story from New Hampton is even better," wrote Dr. Harold D. Levine, after sharing an eyewitness account of a man who surreptitiously inverted the coffin of a deceased relative, hoping to prevent further consumption deaths in the family (discussed in chapter 3).[425] Levine had included two consumption narratives in his article, "Folk medicine in New Hampshire," which appeared in the *New England Journal of Medicine* (1941). The article was derived from a paper he had read before the Merrimack (New Hampshire) County and Center District Medical Society on July 10, 1940. Levine's second, "better" story, while short, was agreeably detailed:

> The two daughters of a prominent farmer, Marshall Bowen, both contracted consumption. They tried everything known at the time, and finally went to the village witch, who told them that one girl was past saving, but that when she died, they should cut her heart out and bury it by Dickerman Brook. This would prevent the death of the other daughter. Tradition is vague as to whether this dictum was followed, but the other girl died of the same disease.[426]

Marshall Bowen was born about 1804. He married Lydia P. Beatty (born about 1803) in April 1834. A matrimony notice, published in the *Boston Traveler* (May 9, 1834), included a curious phrase, "In Holderness, N. H. after a courtship of one evening, Mr. Marshall Bowen to Miss Lydia Beatty." (What transpired on that single evening of courtship?) Lydia was from Holderness, a small town in Grafton County, which is adjacent to Belknap County, the location of New Hampton. Vital records show that the Bowens had four children: Charles L. (born 1835), John F. (born 1836), Julliette (born 1838), and Mary A. (born 1843). Lydia died at age sixty on May 6, 1862, in New Hampton, with "inflammation on stomach" listed at the cause of death. Charles married Abbie A. Emmons in September 1858. Charles enlisted as a private during the Civil War, the day before Christmas in 1861, serving in Company M, New Hampshire 1st Cavalry Regiment. A death notice the *Philadelphia Inquirer* (December 1, 1863) states he died on November 22, 1863, of typhoid fever while in the Union Army, at Annapolis, Maryland. Marshall Bowen died of "inflation of the bowels" at age sixty-six in April 1870.[427] Julliette and Mary were among the students attending the New Hampton Literary and Biblical Institution, between 1856 and 1861, which, at that time, was under the auspices of the Freewill Baptist Church. Siblings John (apparently known as Frank, probably his middle name), Julliette, and Mary were in the same household in the 1870 federal census. Julliette married George Shaw, in 1875, and lived until 1918. The fates of John F. and Mary A. are unknown. Given this limited amount of information, an educated guess is that the events described in the narrative occurred near Marshall Bowen's death in 1870.

Who was the "village witch" that recommended removing the heart from the daughter's corpse? The authors of *A Small Gore of Land: A History of New Hampton, New Hampshire* (1977) identified "Granny Hicks" as "New Hampton's only bona fide witch, as far as we know."[428] Granny Hicks certainly found a place in the town's folklore, as her story has all the attributes of a local legend that circulated long enough to become variable. She was said to have lived alone in a rustic shack that she had built for herself in an isolated part of town near the Pemigewasset River. Her credentials as a folk witch seem impeccable: of mysterious origins, [429]she violates the social norms of her village by living a lonely, isolated existence, yet children and animals are attracted to her; she is a shapeshifter, who can cast evil spells, punish those who incur her ill will, and make accurate prophesies. Researchers have concluded that Granny Hick's actual name was Esther Prescott Hyde.

Born as Esther Rollins, in Epping, New Hampshire. Following the death of her first husband, John Prescott, she married John Hyde (or Hide) from Lee, New Hampshire. Esther died in New Hampton on April 14, 1817, at the age of 64; her body is buried in the New Hampton village cemetery in the Prescott family lot.[430]

Since Granny Hicks died in 1817, and the Bowen family exhumation ritual occurred more than fifty years later, Granny Hicks could not be the witch in question. But "bona fide" or not, there must have been a "witch" in New Hampton when the heart excision was prescribed. Perhaps she was simply New Hampton's version of the folk crone, a familiar figure in New England communities during the early years, who took it upon herself to learn and apply the cures and other formulae that had been passed down in the region for longer than anyone could remember. If the consumption ritual was, indeed, prescribed by "the village witch," then we can reasonably infer that it was a folk cure known to at least one person in the community, who then communicated this knowledge to the desperate family of Marshall Bowen.

In the introductory remarks to his article on folk medicine, which includes the Marshall Bowen narrative, Dr. Harold Levine wrote, "I have undertaken to describe how the people of New Hampshire, particularly in the central portion, have treated and, to some extent, continue to treat disease at home. I have obtained most of the material directly from them and many of the quoted passages are verbatim accounts of what they have said."[431] Since none of the eight published references he cited included the Bowen narrative, one is left to speculate that Levine collected it from oral tradition. If so, from whom, and how did Levine obtain it? One likely source is Cornelius Weygandt (1871-1957). While the Bowen narrative is not included within Weygandt's *New Hampshire Neighbors: Country Folks and Things in the White Hills* (1937), Levine cites Weygandt for "personal communication." Weygandt had a lengthy, intimate relationship with New Hampshire. Although he was born in Germantown, Pennsylvania, and spent his career teaching English at the University of Pennsylvania, in childhood he summered in New Hampshire at his parents' cabin in the White Mountains. In addition to *New Hampshire Neighbors*, his books about New Hampshire include *The White Hills* (1934) and *The Heart of New Hampshire* (1944). However, I was not able to locate Bowen's narrative, nor that of Granny Hicks, in any of these works. One of Weygandt's students was MacEdward Leach, a folklorist and moving force behind the establishment of folklore studies at the University of Pennsylvania. The late dean of American Folklore, Richard M. Dorson, characterized Weygandt as "a twentieth-century writer and collector of New Hampshire traditions"—praise that underscores the significance of Weygandt's contributions to the field of folklore.[432] Weygandt's interest in folk traditions easily could have led him to the Bowen narrative.

Harold D. Levine (1908-1993) was a 1925 graduate of Brookline (Massachusetts) High School, a 1929 graduate of Harvard College, and a 1932 cum laude graduate of Harvard Medical School, the year he was elected to Harvard's Boylston Medical Society. Following graduation, an internship in pathology, and an assistant residency in medicine, Levine had a general practice and worked as a surgeon until 1941. He was a leading heart specialist and associate clinical professor at Harvard Medical School and Senior Associate of medicine at Peter Bent Brigham Hospital in Roxbury, MA, as noted in the *Boston Herald* (February 26, 1964). He also served as president of the Massachusetts Heart Association. Levine's connection to New Hampshire is not certain. He read his paper on folk medicine (prior to its publication) before the Merrimack (New Hampshire) County and Federal District Medical Society. The *New England Journal of Medicine* article lists him as "Librarian of staff, Franklin Hospital" (which was located in Bristol, New Hampshire); an Associated Press article about his published essay (which appeared as "Ancient Remedies Discussed in New England Journal of Medicine" in the Baton Rouge *State-Times* on March 22, 1941) also gave his home as Bristol, New Hampshire.[433] As the hospital's librarian, Levine had access to numerous materials that he could have culled for his research on the state's folk medical practices. While his obituary in the *Boston Globe* (April 10, 1993) described him as a resident of Brookline, perhaps he spent summers in New Hampshire, as did Weygandt, and then retired there.

It is fair to note that Levine was not entirely dismissive of the folk remedies he collected. "One may be amused at the faith that these people placed in the healing virtues of their herbs," he wrote, "but it is nonetheless true that modern research has substantiated many of the old ideas."[434] Since he did not share his feelings about the Bowen exhumation, we cannot know if he found the consumption ritual an amusing idea from the past or, in concert with most physicians who encountered it, a disgusting superstition. But I would guess that he would not believe research would validate this old idea.

Some aspects of the Marshall Bowen narrative separate it from ordinary vampire rituals (if any of those can be characterized as "ordinary"). The preemptive removal of the girl's heart after death but prior to interment is implicit in the witch's instructions. The burial of the vital organs apart from the corpse also is unusual; another instance is from Belchertown, Massachusetts, which, you may recall, included placing the lungs and liver in a separate box, then burying it in the same grave, about a foot above the coffin (chapter 2). Burying a heart that is possessed by an evil power/demon near a brook has the force of tradition behind it. George Lyman Kittredge, in *Witchcraft in Old and New England*, wrote, "There is a manifest connection with the ancient and all-but-universal belief that water (particularly running water) dissolves a spell or interposes an obstacle to the passage of uncanny beings."[435] I would not be surprised if the family was instructed to bury the troublesome heart on the *opposite side* of Dickerman Brook.

The Bowen narrative reported that "tradition is vague as to whether" the ritual was performed—a situation that strikes me as unusual. To think that "tradition"—presumably, *oral* tradition—would preserve the story without including (at least in a more sharply defined way) whether the girl's heart was cut out seems implausible. Ponder that scenario: someone ("tradition") knew about the witch advising Bowen to excise his daughter's heart but was uncertain if he acted upon her advice. Two depressingly common motifs in this consumption story include the ritual as a last resort ("they tried everything known at the time") and the unfortunate outcome ("the other girl died of the same disease").

***

Desperate people who take desperate measures are not necessarily fools. Since the failure rate for consumption cures was essentially the same among folk, quack, and establishment systems—equivalent to doing nothing—people had little reason to reject any possible cure. Belief is not an essential ingredient for acting; in the face of almost certain doom, action itself may suffice. The dual nature of the folk healer, or the quack—embodying both hope and doom—was not lost on people. During New England's arduous formative years, Yankee humor played on life's equivocal duality. Grim humor reflects grim realities whose consequences, even when inescapable, need not be tyrannical. Folklorist C. Grant Loomis included the following anecdote in his article, "Some Lore of Yankee Genius" (1947): "A pedlar, with his cart, overtaking another of his clan on the road was thus addressed: 'Halloa, friend, what do you carry?' 'Drugs and medicines,' was the reply. 'Good,' returned the other. 'You may go ahead. I carry grave stones.'"[436]

Part VII Evolutionists

Nineteenth-century news writers, historians, and physicians were nearly unanimous in their condemnation of New England's consumption ritual. Contributors to new journals of the emerging disciplines of folklore and anthropology were providing a "scientific" rationale for marginalizing beliefs and practices of those deemed less than civilized. The cultural evolutionists' perspective was evident in the second volume of the *Journal of American Folklore* (1889), which contained Jeremiah Curtin's narrative of an exhumation in Vermont and John McNab Currier's first of three articles on New England folklore (his second included a vampire superstition from New Hampshire). That volume's untitled introduction noted that, although "the importance of the study of popular traditions" was "recognized by men of science," it was "not yet understood by the general public."[437] In an eerie if awkward foreshadowing, the introduction's anonymous author(s) (presumably general editor W. W. Newell or editors Franz Boas, T. Frederick Crane, and J. Owen Dorsey) hinted that vampire practices fit squarely within the argument for considering even "rude and shapeless" traditions, that is, "ore scarcely impressed by the die [a device used to impart a desired shape, form, or finish to a material]": "Man is a child of the soil; the figures which labor where he stands, which lie where he will be buried, these spirits which rise and walk in his fancy."[438] The introduction's closing argument asserted that the only way to determine if similarities between New World and Old World folklore are due to "the common procedure of human imagination" or the result of diffusion is to amass "abundant means for comparison" through diligent collection: "the report of one collector must be supplemented by the report of others."[439] By collecting folklore internationally, scholars could make the requisite cross-cultural comparisons and also identify "survivals" from lower stages of evolution, such as vampire practices, often for the stated purpose of eradicating them. The comparativist perspective used by folklorists and anthropologists was thus crucial to the theory of unilateral cultural evolution, which posited that human culture evolves through progressive stages, from savagery to barbarism to civilization. The comparativist approach is more interested in "patterns of thought" revealed in the "items of folklore" than in situational contexts.[440]

Chapter 15 focuses on the vampire narratives of Jeremiah Curtin (1835-1906) and John McNab Currier (1832-1919). Chapter 16 examines George R. Stetson's "The Animistic Vampire in New England," which appeared in the *American Anthropologist* in 1896. His article not only put New England on the vampire map, it also launched Rhode Island's status as the "Transylvania of America." The impact of Stetson's vampire article was immediate and continued to reverberate in the popular press for years. In chapter 17, we scrutinize seven newspaper articles whose interpretations of Stetson's vampires ranged from serious rebuke to absurd parody.

Ransom family gravestones, Ransom-Kendall Cemetery, South Woodstock, Vermont. *Courtesy of Cyril Place.*

15   A Peculiar Kind of Vampirism

The title of Jeremiah Curtin's brief communication in the *Journal of American Folklore*— "European Folk-Lore in the United States"—reveals his assumption that the vampire tradition in New England had European origins:

> Even in New England curious and interesting material may be found among old people descended from the English colonial settlers. About five years ago an old lady told me that fifty-five years before our conversation the heart of a man was burned on Woodstock Green, Vermont. The man had died of consumption six months before and his body buried in the ground. A brother of the deceased fell ill soon after and in a short time it appeared that he, too, had consumption; when this became known the family decided at once to disinter the body of the dead man and examine his heart. They did so, and found the heart undecayed, and containing liquid blood. Then they reinterred the body, took the heart to the middle of Woodstock Green, where they kindled a fire under an iron pot, in which they placed the heart and burned it to ashes.
>
> The old lady who told me this was living in Woodstock at the time, and said she saw the disinterment and the burning with her own eyes.[441]

Curtin characterized this event as "a peculiar kind of vampirism quite worthy of some Oriental country." While *Oriental* came to be synonymous with *Asian*, Curtin's use of the word is tinged with the nineteenth-century perspective of "Orientalism," which juxtaposed West versus East —a reductive dichotomy forged during Western Europe's imperial expansion into contiguous areas of southeastern Europe and beyond, to the Middle East, North Africa, and Asia. "The Orient was almost a European invention, and had been since antiquity a place of romance, exotic beings, haunting memories and landscapes, remarkable experiences," Edward W. Said wrote in *Orientalism* (2003), his seminal book on what he termed a "semi-mythical construct."[442] Yet, as the title of Curtin's article indicates, he saw the Woodstock vampire incident as rooted in the folklore of Europe. His observation that it was found among "people descended from the English colonial settlers" leaves one wondering if Curtin supposed that the early settlers brought the practice from England. In other works, Curtin compared similar motifs from disparate cultures to illustrate his conviction that "in mythology and folklore we find documentary history of the human mind." "Mythology and philology, taken together," Curtin elaborated, "form a science bearing the same relation to the history of the human mind that geology does to the history of the earth." Just as geological strata form an archive of our planet's history, our cultural history ("mythology, folklore, and languages of nations") records "stereotype impresses of the mental condition of these nations at successive period of their existence, beginning in time centuries beyond the first written history."[443]

Curtin's Woodstock informant indirectly revealed her community's belief that consumption could be caused by either ordinary or supernatural agency. The presence of liquid blood in a corpse's heart was interpreted as "fresh" blood, signaling an ongoing, unnatural, and malignant connection between living and deceased family members:

> The diagnostic procedure was to check the condition of the deceased's heart. The old lady informed me that the belief was quite common when she was a girl, about seventy-five years ago, that if a person died of consumption and one of the family, that is, a brother or sister, or the father or mother, was attacked soon after, people thought the attack came from the deceased. They opened the grave at once and examined the heart; if bloodless and decaying, the disease was supposed to be from some other cause, and the heart was restored to its body; but if the heart was fresh and contained liquid blood, it was feeding on the life of the sick person. In all such cases, they burned the heart to ashes in a pot, as on Woodstock Green.[444]

That Curtin collected a vampire narrative from an old woman in Vermont makes sense, considering his biography. The son of Irish immigrants, Curtin was born in Detroit on September 6, 1835, but grew up on a farm near Milwaukee. After graduating from Milwaukee University, he studied folklore and mythology at Harvard under the tutelage of Francis J. Child, compiler of the ten-volume *English and Scottish Popular Ballads* (1882-1898). Curtin's residence in Russia for six years (1864-1870), and his two folklore collecting trips to Ireland (1887, 1891-1893), resulted in the publication of several folk narrative collections, one of which he dedicated to Child. While at the Bureau of American Ethnology (now the Smithsonian Institution), Curtin did fieldwork among Native American groups in New York, Oklahoma, and California (1883–91), leading to his publication of *Creation Myths of Primitive America, in Relation to the Religions, History, and Mental Development of Mankind* (1898). Financially, however, his most significant work was a best-selling translation (from Polish) of Henryk Sienkiewicz's *Quo Vadis*. In 1908, Theodore Roosevelt deemed Curtin "one of the two or three foremost scholars" in the country, citing a wide-ranging career in the social sciences, humanities, and public service.[445] Curtin was, at various times, a diplomat, folklorist, ethnographer, linguist, author, lecturer, and translator.

Curtin married Alma Cardell (1847-1938), whose parents were also Irish immigrants. "For the next three decades," a biographer of the Curtins wrote, they "were working partners who spent virtually all their time together, traveling incessantly and having no home or fixed address. Alma Curtin served as Jeremiah's stenographer, editor, and perhaps, at times, writer; her work, however, was never acknowledged by him, and he took credit for many aspects of her work."[446] Curtin apparently had a knack for fieldwork and valued, especially, the lore that old women could bestow, as one of his acquaintances recalled: "Differences of race and age . . . gave way before him, and wherever he went he had access to intimacy. Old women were among his friends. He said that he always found ancient crones, mammies, and withered squaws the best story-tellers."[447]

Alma was her husband's link to Vermont. The couple spent time in both Warren, Vermont—her birthplace and the home of her father, James Cardell—and Bristol, Vermont, where, after Cardell's death, they spent their time in the home of Alma's sister.[448] One of Curtin's biographers noted that, "As a rule the Curtins traveled to Vermont in the winter season. . . .. In 1884 the Curtins spent three weeks' holiday in the state."[449] That holiday coincided with Curtin's interview of the old woman who recalled the Woodstock exhumation and heart-burning (if the time frame provided in Curtin's text is accurate). After Curtin's death, in 1906, Alma wrote several books under his name, including *Memoirs of Jeremiah Curtin* (published in 1940, two years after her death). Unfortunately, the *Memoirs* include no description of the vacation of 1884; so, both the identity of the old woman, as well as the context of the apparent encounter between her and Jeremiah (or, perhaps more aptly, Jeremiah *and* Alma), are cloaked. Perhaps the best chance for discovering the old woman's identity resides in clues contained in another, more unusual exhumation narrative that she shared:

> The same old lady said that her uncle, a physician of good standing and repute, was present, with other physicians, at the opening of a grave in the town of Malone, New York. The reason of the opening was as follows: A "bone auger" had been observed making its way through one of the grave mounds in the church-yard, increasing in height day by day. At length it was determined to dig down and trace this auger to its origin. They dug through the earth to the coffin below, the auger had bored its way through the coffin lid. The lid was removed, and the people found that the "bone auger" was growing out of the heart of a man buried some time before.[450]

More unanswered questions emerge. What was a bone auger? Was it made of bone? Was this auger (that is, a boring tool) an augur (that is, a prognosticator or omen)? Was it like the grave plants growing from corpses or coffins that, as we have seen, foretold (or worse, caused) yet more mysterious deaths in a family? Were other events in the town linked to this phenomenon? Who "determined to dig down and trace this auger to its origin," and why? How did people react after they learned of its origin? Since I have been unable to document the existence of graveyard bone augers in folk tradition, I can offer no cognates that might assist in uncloaking this mystery.

The physicians' attendance at the disinterment was one of the least unusual elements of this narrative. Henry S. Waterhouse, "a brilliant surgeon" who arrived in Malone before 1807, may have been one of those physicians. "But there was a mystery about him never solved," a retrospective article in a local newspaper, the *Malone Palladium* (May 7, 1884), pointed out. During the War of 1812, he allowed his garden on Webster Street to become a burial place for dead soldiers; not all of them were, or remained, buried. "He was charged with dissecting not only dead soldiers, but also with robbing the graves of near neighbors. When a loved one was buried, there was always apprehension," the article stated. And there was "a Nemesis" that "seemed to follow him," as "one member after another of his family died. There lie in that old cemetery to-day—unvisited by mortal relative so far as I know—his first wife and six children." Waterhouse remarried, had another son, and eventually relocated to Key West, Florida. His second wife died shortly after arriving in her new home. "Suspicion was excited, but no one could tell whether justified or not. Not many months later the New York papers announced that . . . Henry S. Waterhouse and his son were sailing in a yacht, on a Sunday afternoon, near Key West; there came a squall, the boat was upset, and both were drowned."

Frederick Seaver, in his chatty *Historical Sketches of Franklin County* (1918), included the anecdotes about Dr. Waterhouse, as well as several others from the town of Malone about murders, prophetic dreams, the Underground Railroad, fortunetellers, and witchcraft. Although he did not mention bone augers or vampires, in the equivalent of a nineteenth-century meme for "absurd medical superstition," Seaver included a story about a local farmer who, in 1824, had fallen onto a pitchfork, which went through his breast and protruded from his back. Sixty years later, he was in good health; he stated that he healed his wounds by medicating the tines of the fork and wrapping them in red flannel.451 This instance of 'weapon-salve" (also called "weapon-ointment") is one of a number of folk cures that were incorporated into the Paracelsian system of sympathetic healing, based on tapping into "the invisible effluvia and influences with which the world vibrated." Thus, "it made sense to apply the ointment to the weapon rather than the wound because then the vital spirits in the blood congealed on the weapon would be drawn along in the air to rejoin the body."452 By deploying this story, Seaver tells us, indirectly, that some people in Malone might have been more than ready to trace the source of a graveyard bone auger. And the anecdotes about Dr. Waterhouse suggest there was at least one physician who had the experience and eagerness to participate.

In his article, Curtin directed attention to a rich vein of imported tradition "derived from nearly every country in Europe," just waiting to be mined. He noted that he had "found folk-lore in a dozen States from the Atlantic to the Pacific" and encouraged scholars to join him in recording "curious beliefs and stories whenever we come upon them."453 Through his folklore fieldwork, Curtin was seeking to discover recurring motifs of "primitive thought," that is, "stereotype impresses of the mental condition" of the human mind. Although he did not address this point explicitly, Curtin noted similarities among some of the nocturnal supernatural assault narratives that he published. In the Irish tale, "The Blood-Drawing Ghost," a young woman named Kate must retrieve a blackthorn stick that her foreordained husband had placed on the tomb of a recently dead old man. Arriving after nightfall, Kate is instructed by the dead man to remove him from his tomb and carry him on her back to a house, where he draws blood from three sleeping young men. He mixes the blood with food and eats it, commanding Kate to do the same. She only pretends to eat the blood-soaked meal, hiding it instead. The dead man then commands Kate to return him to his tomb before the cock crows. On the way, she asks him if there is a cure for the three young men whose blood was drawn. He says she must give them three bits of the food containing their blood, which Kate does, releasing them from their unnatural death. Predictably, pots of gold buried on the young men's property come to Kate after she marries the eldest.454

In a headnote to this tale, Curtin discussed its similarity to Slavic tales about "dead men who dwell in their tombs as houses." These malignant tomb dwellers rise at night and devour people, after which they must return to the grave by cockcrow. To kill the monster, one must stake it to the earth through its heart or burn it to ashes.

> The burning, as described in Russian tales, is performed by a great crowd of people armed with bushes, long brooms, shovels, and rakes. These gather round the fire to drive back everything that comes from the body. When the body is on the fire a short time it bursts, and a whole legion of devilry rush forth in the form of worms, snakes, bats, beetles, flies, birds; these try with all their might to get away. Each carries the fate of the ghoul with it. If only one of them escapes, the dead man will be eating people the next night as actively as ever, but if the crowd drive everything into the fire again he will be destroyed utterly.455

The creators of Hollywood's early monster films could have been channeling Curtin's description of the rustics, armed with rude farm implements, beating back a hoard of creatures emerging from the burning corpse.

"A peculiar kind of vampirism quite worthy of some Oriental country," is how Curtin summarized the old woman's Woodstock narrative. But in his description of the Slavic tradition (which he probably regarded as "oriental") Curtin does not mention vampires, nor does the word appear in his published Slavic tales. When William Ralston published a variant of the above tale as "The Soldier and the Vampire" (*Russian Folk-Tales*, 1873), however, he included a lengthy discussion of vampirism.[456] Yet a blood-sucking creature appears in another tale included in Curtin's collection. In "The Footless and Blind Champions," the protagonist, an adopted daughter, says of Baba-Yaga: "Whenever ye go away to hunt, an ancient old woman comes, evil-faced, long-haired, gray; she makes me search in her head, and then she sucks my blood."[457]

Curtin's brief account of the Woodstock vampire, in company with the blood-drawing Irish ghost, the blood-sucking Russian Baba-Yaga, the Slavic tomb-dwellers, and the Russian ghoul, exemplify the quicksilver qualities of folk monsters. Within this jumble of creatures are shared motifs that Curtin might have employed to advance his argument that the ancient and universal traits of the human mind are revealed in folklore. For example, a common result of attacks by these denizens of folktales and legends, bent on causing human misery, is the gradual wasting of their victims; they become "haggard" (a term with the same semantic root as the noun and verb form of "hag"). In the elegant, zero-sum universe of folk philosophy, if someone is dwindling, another must be flourishing. Anthropologist George M. Foster coined the phrase "the image of limited good" to characterize this implicit traditional worldview.[458] Thus, when a person begins to waste away for no apparent reason, one searches for a phantom opposite, thriving unnaturally. A traditional hideout for these monsters is the grave, inside the vital organs of a corpse, where "fresh" blood is damning evidence.

Even though Curtin did not specify the place of vampires in a pantheon of creatures of lower mythology, his postulation of the universal human mind that he believed was incorporated into people's folklore anticipated the later views of psychologists, such as Ernest Jones, as well as some contemporary folklorists, particularly David J. Hufford. Being attacked by a vampire or incubus, or ridden by a witch or hag, or even experiencing a nightmare or sleep paralysis, all seem to share some common traits. Jones employed a psychoanalytic perspective in his influential study, *On the Nightmare* (1931). Hufford, a leading scholar of the supernatural assault tradition, argues for taking a step back from this and other sorts of culturally loaded interpretations. Because of the "theoretical baggage" carried by every term for this experience, Hufford chooses to use the term *Mara*, which refers "simply to the experience of finding oneself awake and paralyzed in the presence of a frightening being. Nothing more—neither interpretation nor cause—is implied." Hufford asserts that since Mara experiences occur even in cultures where traditions such as vampire assaults or witch-riding are absent, then "it seems fair to infer that the Mara experience itself has given rise to a variety of similar beliefs in different cultures." Hufford also found "other experiential categories with a similar cross-cultural distribution and interpretation" that showed "great similarity from one tradition to another."[459] His enumeration of these categories, in addition to supernatural assaults, included "mystical experience, miraculous healing, consoling visits by the deceased to the grieving, near-death experiences, and haunted houses."[460]

European vampire folklore appears to be a good fit with Hufford's definition of the Mara experience. But in New England, the vampire tradition did not retain the defining Mara features. Is it reasonable to apply the label "vampire attack" to an extra-corporeal, indirect assault on a person in cases where that person does not report nocturnal manifestations of being awake, yet paralyzed, and in "the presence of a frightening being"? Most people who have an opinion about what a vampire is probably see the motif of *nocturnal assault by a frightful creature* as the salient feature of vampirism. The New England vampire tradition rests squarely on the symptoms of consumption, some of which may mimic Mara manifestations. Yet, as we shall see, there are other, perhaps less sensational, explanations associated with this tradition.

A little more than a year after its publication in the *Journal of American Folklore*, Curtin's vampire narrative from Woodstock appeared in several newspapers. In the *Boston Evening Transcript* (September 6, 1890), Curtin's text was published verbatim, down to the final lines that included his name and place of residence (Washington, D. C.). In place of the original article's title, "European Folk-Lore in the United States," however, the *Evening Transcript* ran the headline, "Vermont Vampirism," with the subhead, "A Curious and Interesting Trace of European Folk-Lore in the United States."

The next iteration of Curtin's collected tale appeared about one month later in the *Vermont Standard* (October 9, 1890) with startling additions. To flesh out Curtin's spare narrative, the anonymous author provided much more in just two paragraphs:

> We may as well help the old lady's recollections in this matter and fill in with further details what she has left incomplete. To be particular in dates, the incident happened about the middle of June, 1830. The name of the family concerned was Corwin, and they were near relatives of the celebrated Thomas Corwin, sometime Senator in Congress from Ohio, well known for his wit and attractiveness as an orator. The body disinterred was buried in the Cushing cemetery. With regard to the cause of the illness that had seized the brother of the deceased, there was a general consensus of opinion among all the physicians at that time practicing in Woodstock. These embraced the honored names of Dr. Joseph A. Gallup, Dr. Burnwell, Dr. John D. Powers, Dr. David Palmer, Dr. Willard who recently died in New York, not to mention other members of the profession at that time residing in Woodstock and held in high repute at home and abroad. These all advised the disinterment as above described, all being clearly of the opinion that this was a case of assured vampirism. Only there was a slight controversy between Drs. Gallup and Powers, as to the exact time that the brother of the deceased was taken with consumption. Dr. Gallup asserted that the vampire began his work before the brother died. Dr. Powers was positively sure that it was directly after.
>
> The boiling of the pot on Woodstock Green, spoken of by the old lady, was attended by a large concourse of people. The ceremonies were conducted by the selectmen, attended by some of the prominent citizens of the village then residing on the common. It will suffice to name Honorable Norman Williams, General Lyman Mower, General Justus Durdick, B. F. Mower, Walter Palmer, Esq., Woodward R. Fitch, of old men of renown, sound minded fathers among the community, discreet careful men. The old lady has forgotten to state what was done with the pot and its ghastly collection of dust after the ceremonies were over. A hole ten feet square and fifteen feet deep was dug right in the center of the park where the fire had been built, the pot with the ashes was placed in the bottom, and then on top of that was laid a block of solid granite weighing seven tons, cut out of Knox ledge. The hole was then filled up with dirt, the blood of a bullock was sprinkled on the fresh earth, and the fathers then felt that vampirism was extinguished

forever in Woodstock. Eight or ten years after these events some curious minded persons made excavations in the Park, if by chance anything might be found of the pot. They dug down fifteen feet, but found nothing. Rock, pot, ashes and all had disappeared. They heard a roaring noise, however, as of some great conflagration, going on in the bowels of the earth, and a smell of sulphur began to fill the cavity, whereupon, in some alarm they hurried to the surface, filled up the hole again, and went their way. It is reported that considerable disturbance took place on the surface of the ground for several days, where the hole had been dug, some rumblings and shaking of the earth, and some smoke was emitted.

This narrative includes some well-known and widely distributed folk motifs. The traditional nature of these motifs is indisputable. The issue here, however, is their connection to the old woman's tale. The remains of the heart that were reinterred under a huge block of granite has been documented in York, England (E437.4 - ghost laid under stone) as well as in folktales from Switzerland and Iceland (E431.10 - corpse buried under stone so that sun will not shine on him again). Those familiar with the Old Testament will recognize the blood of a bullock sprinkled on the earth (found in Leviticus 16:14 and other OT books) as part of a purification ritual utilizing the blood of a sacrificed scapegoat. When the site was excavated some years later, all traces had disappeared,[461] but there was an odor of sulfur—a widespread motif in both Europe and the United States that indicates the presence of the Devil.[462] The accompanying rumbling disturbance that lasted for several days is a motif found among Native Americans.[463] This hodgepodge of motifs casts serious doubt on this text's authenticity, yet some of the new additions are historically credible. Of the individuals named in the *Vermont Standard* narrative, all, save one, can be documented as having lived in Woodstock in 1830. The single person whose existence cannot be verified in the historical record is (no surprise!) the narrative's central character, the vampire named Corwin.

Is the *Vermont Standard* text a fictionalized elaboration of the brief eye-witness account collected by Curtin? Decades ago, folklorist Richard M. Dorson cautioned that "critics must tread warily to distinguish folk from literary or personal inspiration," for "poet and novelist and playwright fit to their own imaginative purposes the folk materials they know."[464] Yet another explanation for the *Standard*'s evidently exaggerated narrative follows a course between history and fiction—the well-worn path of folk legend.

Folklorist Steve Glencarella theorizes that a legend, woven from strands of history, hearsay, and oral tradition, provided the principal ingredients for the *Standard*'s intricate story. The following six additions to the old lady's sparse narrative, he argues, were culled from several motifs in an evolving oral lore: (1) the name Corwin; (2) the location of Cushing cemetery as the site of the disinterment; (3) the role of local physicians in the exhumation and their subsequent arguments; (4) the presence of a large gathering that included prominent citizens; (5) details of the vampire ritual; and (6) the excavation of the site years later to supernatural effect. Against the backdrop of Anti-Masonic sentiment in Vermont, Glencarella describes convicted rapist Joseph Burnham's burial in the Cushing cemetery and subsequent exhumation (twice!), in October 1829, the second time before numerous spectators, to quell rumors that he was still alive. Through normative regional linguistic alteration, Glencarella contends, the name of a man who disappeared and then, perhaps, mysteriously reappeared, Russell Colvin, was transformed into Corwin. And he asserts, it was controversy over control of the local medical college, rather than when the vampire attacked its victim, that pitted the doctors named in the *Standard*'s text against one another. Glencarella establishes links to two other vampire rituals in Vermont (which we have already discussed). Frederick Ransom's exhumation in 1817, also in Woodstock (chapter 1), finds relevance through his brother, Royal Makepeace Ransom, a lawyer who took up the Anti-Masonic cause. He then became a central character in a satirical play, performed in Woodstock, that included images of legendary monsters, notably vampires, and fearful tales of abductions and murders, such as those of Burnham and Colvin. The 1793 exhumation of Rachel Harris Burton (chapter 11) is introduced through the person of John S. Pettibone, who recorded the exhumation in his history of Manchester, and who became one of three men assigned by the Vermont legislature to investigate Woodstock's political upheavals during the Anti-Masonic controversy. Glencarella argues that the process of telling oral stories, including those of hunting for buried treasure and robbing graves for medical colleges, incorporated other supernatural motifs, particularly those attached to the search for the iron pot about a decade after the ritual. Glencarella's explanation for the *Standard*'s narrative, based on extensive research in combination with his solid academic background in folklore, shows an admirable faculty for bringing coherence to a seeming jumble.[465] While possible, the number of intricate threads that would have had to come together in just the right configuration to create the proposed legend makes that outcome seem improbable. Even more damning, however, is that no version of this hypothetical legend has been recorded. Maybe equally plausible is that writers at the *Vermont Standard* were aware of these various strands of history and folklore and wove them into a narrative that had the patina of an actual legend.

***

The eclectic and inquisitive mind of Dr. John McNab Currier (1832-1919) permeates his biography. Throughout his career as a physician, he actively pursued other passions, as an antiquarian, historian, folklorist, natural scientist (especially mineralogy, paleontology, and archaeology), poet, publisher, and editor. Currier was a Renaissance Man, educated in an era when specialization had just begun pressuring scholars to know more about less.

At first glance, Currier's contribution to the vampire folklore of New England seems minimal: "If the lungs of a brother or sister who has died of consumption be burned, the ashes will cure the living members of the family affected with that disease. (Grafton County.)"[466] This meager text is far from barren. Sometime prior to 1891, the date it appeared in the *Journal of American Folklore*, someone in Grafton County, New Hampshire, was aware of a folk prescription for consumption that used the ashes from the burned lungs of a deceased sibling. Currier was a physician who undoubtedly wrote many prescriptions, so it is likely that the ashes were to be ingested by the ailing sibling.

Currier contributed a generous number of texts, with a bit of context, in three short installments on New England folklore, published from 1889 to 1893 in the *Journal of American Folklore*. Since the "hardships of pioneer life" precluded "any great interest . . . in educational matters," he concluded that "it was not strange that myths, beliefs in witchcraft, and reliance upon signs, should exist in a certain measure." Friends gathered to share neighborhood gossip and converse about farming, politics, and religion. "Now and then a bit of folklore received their attention, and that, too, without any reserve." One of those "memorable and pleasant occasions" from his "youthful days" created an indelible image. The family and friends, sitting around a "roaring and snapping fire" in the one large room, lighted only by the fire and the moon shining through "snow-laden evergreens," shared "several ghost and witch stories." Currier related a few of those, along with others told on different occasions—just as he heard them, he assured readers.[467]

Currier received a standard mid-nineteenth century classical education at Newbury Seminary (a "literary high school") and McIndoe's Falls Academy, both located in Vermont, just across the Connecticut River from his home in Bath, New Hampshire. After studying medicine with several physicians in the area, he graduated, in 1858, from the medical department of Dartmouth College. He then established a practice in Newport, Vermont, where, in 1869, he assisted in reorganizing the Orleans County Society of Natural Sciences. For five years, he published and edited the *Archives of Science and Transactions of the Orleans County Society of Natural Sciences*, a quarterly scientific journal. Currier continued this pattern as he relocated his medical practice to other Vermont counties, including Caledonia, Addison, and Rutland. Throughout his career, Currier actively nurtured the systematic study of medicine, natural science, and history.[468]

Currier's interest in local history and, especially, antiquities, directed his attention to the oral traditions of the several communities in which he lived. It is plain in his three *Journal of American Folklore* articles that, long before their publication (1889-1893), he had settled upon remembering, if not actually setting down in writing, "stories" and "brevities" that circulated among family, friends, and neighbors, many of which have been well-documented in folklore.

In 1860, two years after receiving his medical degree and opening his first practice, in Newport, Vermont, Curtin married Susan Havens Powers, the eldest daughter of Dr. John D. Powers and Jane B. Carleton, of Woodstock, Vermont. Perhaps improbably, this Dr. Powers was the one who, according to the undocumented assertion in the *Vermont Standard* (October 9, 1890), argued with Dr. Joseph Gallup over exactly when the vampire attacked its victim. While the *Standard's* assertion is undocumented, Dr. Powers was as real as the other physicians named in the narrative.

Researching Currier's role as a poet led to another physician linked to a vampire narrative. And that narrative led to yet another physician linked to a vampire narrative. And . . . well, the common element in each of the following apparently unrelated events is the vampire tradition. Enter this episodic chain at any point and you eventually travel full circle. Let us begin with a brief news article from the *New York Statesman* (December 1830). The body of a woman who died of consumption was exhumed and decapitated, but its relevance to vampires is not conspicuous:

> Robbing a Grave—Week before last, the wife of Mr. Penfield Churchill, of Hubbardton, died in that town, of consumption, and was buried on Friday or Saturday of the same week. Last week on Wednesday, it was discovered that, some marks which had been placed on the grave, had been removed, which induced the belief that the body had been stolen—on examination, the supposition proved to be correct; the grave contained nothing but an empty coffin—Suspicions were excited that the body was brought to this town; and on Monday last a warrant was issued, and a general search of the medical buildings in this village took place. The remains of a person, with the exception of the head, were discovered, which the husband recognized as those of his deceased wife, and which were taken to Hubbardton to be reinterred. Two students attached to the medical institution, were

immediately arrested on suspicion. Their examination we understand took place yesterday. We have not yet heard the result. *Castleton, (Vt.) Statesman.*

More than a century later, Frederick C. Waite supplied additional information, first in an article on grave robbing in New England (1943) and later in a book detailing the history of Castleton Medical College (1949). The following account is from the latter:

> One of the most dramatic episodes in the history of grave robbing in New England occurred in Castleton in late November 1830. A woman died at Hubbardton, seven miles distant, and was buried on a Saturday. The grave had been marked so that disturbance could be detected and on Sunday the grave was found empty.
>
> Three hundred men, led by the sheriff, marched from Hubbardton to Castleton and surrounded the medical college building at nine o'clock on Monday morning and demanded the body. The dean, on the plea that he would have to send a messenger to his home to get his keys, delayed the crowd until time had been given to conceal the body, after decapitating it to prevent identification if discovered. Then a committee was admitted to make a search of the building, which was unsuccessful until a member noticed a loose nail in a board and discovered the headless body when he removed the board in the floor.
>
> Meanwhile a student with a bundle under his overcoat had sauntered through the crowd and gone to a neighboring barn, where he deposited the bundle in the haymow. The searchers demanded the missing head and the dean said that it would be delivered if the sheriff would agree that there would be no arrests. This guarantee being given, the same student went to the haymow and returning with the bundle under his overcoat, handed it to the sheriff. The body was taken to Hubbardton and reburied. This episode was called the "Hubbardton Raid."
>
> Details of it vary in different published accounts. Two students were expelled as a result but one of them was admitted to another medical college at once and was graduated when he would have been graduated at Castleton. This shows that medical college authorities did not regard expulsion for participation in grave robbing as a serious offense, and that the expulsions at Castleton were only a gesture to the public.[469]

Waite did not identify the medical student who confronted the aggrieved folks from Hubbardton, but at least one other source named Charles Volney Dyer (1808-1878). Tom Campbell wrote of the raid in his history of the Chicago abolitionists, among whom Dyer played a leading role: "As the students panicked and tempers raged, Dyer stood and faced the mob with a 'do your worst' indifference that calmed everyone down." Dyer would later quip that the Hubbardton raid was a "grave offense."[470]

The link between Dyer and vampires was made by Moncure Daniel Conway in his 1879 book, *Demonology and Devil-Lore*: "Dr. Dyer, an eminent physician of Chicago, Illinois, told me (1875) that a case occurred in that city within his personal knowledge, where the body of a woman who had died of consumption was taken out of the grave and the lungs burned, under a belief that she was drawing after her into the grave some of her surviving relatives."[471] Dyer's biography shows that he had ample opportunity to learn about vampires. He was born, in 1808, in Clarendon, Vermont, a state with its share of vampire incidents. In his history of Castleton Medical College, Waite lists

Dyer as graduating in 1830.[472] Dr. Joseph A. Gallup (1769-1849), whom Waite described as the "leading" and "most notable" physician in eastern Vermont, was a founder of the Medical College, in 1818, and served several administrative and teaching roles until 1824.[473] Gallup, of course, was one of the prominent physicians said to be in attendance at the vampire exhumation in Woodstock, in about 1829. As we saw, according to the *Vermont Standard's* article, it was he and Dr. John D. Powers who quibbled over whether the vampire attacks began before or after the death of its victim.

Summarizing the chain of events to this point: Jeremiah Curtin's sparse vampire narrative in the *Journal of American Folklore* (1889) was expanded in the *Vermont Standard* (1890) to include, among other people and incidents, Doctors Gallup and Powers. In the same issue of the *Journal of American Folklore*, John M. Currier began a series of folklore contributions that included a vampire prescription. Currier had married Dr. Power's daughter. Dr. Gallup was a founder and faculty member of the Castleton Medical College, from which Charles Volney Dyer had graduated the year of a sensational grave-robbing scandal. After relocating to Chicago, Dyer related a vampire incident, which apparently took place in that city, to Moncure Daniel Conway. At the 1879 annual meeting of the Rutland County Medical and Surgical Society, Dr. James Sanford of Castleton gave a talk entitled "Reminiscences of Castleton Medical College," during which he referred to the Hubbardton Raid. After the talk was printed in the local newspaper, some of the college's former students argued that the date of the raid was incorrect (which is why the supposed fiftieth anniversary of the event was held only forty-nine years later). Since opinions on the actual date varied, it was proposed that Dr. Currier, a member of the society who "was interested in all antiquarian matters," be assigned the task of ascertaining the exact date of the raid. Currier located copies of the 1830 newspaper article, "thus setting the matter at rest."[474] At the medical society's monthly meeting in November, Currier read a poem he composed to commemorate the raid. The poem, written in the style of Longfellow's "Hiawatha" and running to some 500 lines, was subsequently published in a pamphlet whose title rivals its contents for length: *Song of Hubbardton Raid, Delivered on the 50th Anniversary of the Raid of the Citizens of Hubbardton, Vermont, on Castleton Medical College, Held at the Residence of J. Sanford, M. D., Castleton, Vt., November 29, 1879*. Currier undoubtedly was a better surgeon than poet.

The birth of Currier's father-in-law, Dr. John D. Powers, in 1769, and Currier's death in 1919, bookend the known timeframe of the New England vampire tradition. Within that relatively short period, scores of consumption rituals were performed. As this tradition wound down, another comparative evolutionist, George R. Stetson, undertook to summarize and interpret it. In the next chapter, we examine his consequential study, "The Animistic Vampire in New England."

16   In New England Consumption is a Spiritual Visitation

The seminal commentary on New England's vampire tradition is "The Animistic Vampire in New England," by George R. Stetson (1833-1923), which appeared in the *American Anthropologist* in 1896. This journal, under the auspices of the Anthropological Society of Washington, began publication in 1888, the same year as the *Journal of American Folklore*. Since 1902, when the American Anthropological Association was founded, the *American Anthropologist* has been the Association's principal journal.

Knowledge of George Rochford Stetson's educational and professional background is not readily available. He was born in Braintree, Massachusetts, on November 28, 1833. By the age of forty-one, he had accumulated enough wealth as a leather merchant in Boston that he was able to retire. He and his wife began traveling, eventually relocating to Washington, D. C. Freed from the constraints of the business world, Stetson pursued his interests in literary and scientific topics, especially anthropology. He was an active member of the Anthropological Society of Washington, serving as President between 1912 and 1914.[475] Stetson was also elected a member of the American Folklore Society in 1892.[476] His publications disclose an overriding concern with the intersection of race, intelligence, and education. He has been labeled a eugenicist, a view that was not extreme in the late nineteenth century. For many intellectuals, it was a natural and practical application of evolutionary theory: both the human species and its cultures could move forward through selective breeding, a view that dovetailed logically with the then-current anthropological theory of cultural evolution. Just as the eugenics movement argued that the human species could be improved if weak and unsuited genes were culled, so, too, cultural evolutionists asserted that human cultures would be better positioned to advance if illogical beliefs and practices from lower stages of cultural evolution were identified and eliminated, presumably through education. The meaning that Stetson drew from New England's vampire tradition is apparent against this background: he characterized the tradition as "an extraordinary instance of a barbaric superstition outcropping in and coexisting with a high general culture."[477] In Stetson's view, civilization had yet to fulfill its promise, and the vampire tradition was one more "illustration of the remarkable tenacity and continuity of a superstition through centuries of intellectual progress from a lower to a higher culture, and of the impotency of the latter to entirely eradicate from itself the traditional beliefs, customs, habits, observances, and impressions of the former."[478] Stetson died on March 4, 1923, in Washington, D.C.[479]

Stetson did not discuss how he gathered data for this article, but it is apparent that he visited Rhode Island to engage in fieldwork which, in this instance, entailed talking to people who were willing to share information regarding exhumations with which they had firsthand experience or had heard about in the community. Shortly before the article was published, Stetson exchanged at least three letters with Rhode Island publisher and historian, Sydney S. Rider (whose own vampire narrative we encounter in chapter 19). This correspondence indicates that Stetson was conducting fieldwork in Rhode Island during the summer of 1895. In his first letter to Rider (December 11, 1895) Stetson wrote, "I found in the neighborhood of Exeter Hill last summer a dozen or more families who had shown their faith in it [i.e., "the vampire superstition"] by exhuming the dead."[480] Stetson's second letter to Rider (December 13) makes it clear that he was tapping into local oral tradition: "I found in R. I. the last summer a dozen or more well authenticated cases of families that had followed the demands of the superstition and was told on excellent authority that the area of its adherents is not particularly limited."[481] If Stetson had consulted published materials for his Rhode Island vampire narratives, he did not cite them in either the text or sparse notes of his article.

The title of Stetson's article, "The Animistic Vampire in New England," implies that he conducted research throughout New England, yet his consumption narratives are restricted to Rhode Island. Unfortunately, Stetson provided information for only three of the "dozen or more cases" that he encountered, omitting surnames and specific locations. Fortunately, he included sufficient information to allow nearly certain identification of two of these events.

The first account is a good fit for the exhumations in the family of William R. Rose (discussed in chapter 8) of Saunderstown, a village in the town of North Kingstown, Washington (South) County. Stetson interviewed "an intelligent man, by trade a mason," who described his family's consumption ritual.[482] Stetson's informant surely was Benjamin C. Rose, the second son of William, born in 1851. The elder Rose, who had the bodies of a son and daughter exhumed in 1872, died in October 1894, about a year before the Stetson arrived for his fieldwork in Rhode Island. As we see below, Stetson quoted from Moncure Daniel Conway's *Demonology and Devil-Lore* (1879), which raises the question: Why did Stetson not cite Conway's mention of the Rose family exhumation? Perhaps for the same reason it took years for me to find the primary source for Conway's reference to the event. If Stetson had even noticed Conway's brief citation, he might not have connected the two events since Conway gave "Peacedale" as the site of the ritual, even though the family resided in Saunderstown. Probably more significantly, Conway provided a date of 1874 and the source as the *Providence Journal*.[483] As I finally discovered, the Rose event was reported in 1872, in the *Providence Herald*.

Stetson places the second narrative is the same village as that of the concealed Rose family, a place he characterized as "a small seashore village possessing a summer hotel and a few cottages of summer residents not far from Newport"—which describes Saunderstown. Stetson wrote: "The —— family is among its well-to-do and most intelligent inhabitants. One member of this family had some years since lost children by consumption, and by common report claimed to have saved those surviving by exhumation and cremation of the dead."[484] Although we do not know the identity of the family, it is apparent that the event had become fixed in local memory, perhaps even as a cure that worked.

The third incident, described in some detail (except for names and exact locations) by Stetson undoubtedly is that of Mercy Brown, which appeared in local newspapers in March 1892, a little more than three years prior to Stetson's first visit to the community (See chapters 8 & 9). While Stetson's time frame of "within two years" is a bit off, other corresponding details cement the connection, including: a doctor at the scene "who made the autopsy" (Harold Metcalf); the reported fact that the father (George Brown) initially objected to the exhumation, but finally consented in the face of pressure from his extended family and neighbors; and the subsequent death of another family member (Mercy's brother, Edwin).[485] Dr. Metcalf apparently confirmed Stetson's conclusion that vampire exhumations were not extraordinary in Exeter and surrounding communities: "Dr

—— declares the superstition to be prevalent in all the isolated districts of southern Rhode Island, and that many instances of its survival can be found in the large centers of population. In the village now being considered [Shrub Hill, Exeter?] known exhumations have been made in five families, in the village previously named [Saunderstown, South Kingstown] in three families, and in two adjoining villages in two families."[486] It is unfortunate that he chose not to pursue seven of those ten cases. Yet Stetson seemed almost prescient when he concluded: "It does not by any means absolutely follow that this barbarous superstition has a stronger hold in Rhode Island than in any other part of the country. Peculiar conditions have caused its manifestation and survival there, and similar ones are likely to produce it elsewhere."[487] The "peculiar condition" of consumption did, indeed, produce vampire rituals in other communities in New England and beyond.

Stetson's Rhode Island field trip was a trending break from "armchair anthropology," which relied mainly on previously collected and published data. By today's standards, Stetson's procedure for gathering information seems inexact. But, in his defense, an approach to methodical field collecting was just being formulated. His *interpretation* of the vampire practice, however, adhered to contemporary anthropology's notion of unilineal cultural evolution, promulgated by Edward B. Tylor (1832-1917) in *The Origins of Culture* (1871), the first volume of his *Primitive Culture* series.[488] Modeled on Charles Darwin's (1809-1882) theory of the evolution of biological species, this paradigm, as we saw in previous chapters, asserted that human culture evolved through three increasingly complex cultural stages, from the simple and illogical "savagery" through "barbarism" to "civilization," the pinnacle of rationality as epitomized by the Western educated elite of the late nineteenth century. For the commonly used word "superstition," Tylor substituted a term he supposed to be nonjudgmental: *survivals* were conceived to be the leftover fragments of previous stages of cultural evolution that were carried into a "higher level" of culture. Devoid of their original context, however, survivals lacked meaningful associations; they were merely irrational, vestigial relics, unconnected behaviors—beliefs, practices, texts, symbols, and so on—that were created by "primitive" people. Tylor's theory was elaborated by Sir James G. Frazer (1854-1941), who presumed a concomitant progression of thought, from magical to religious to scientific.

Stetson plainly viewed the vampire superstition as a survival, writing, "Other abundant evidence is at hand pointing to the conclusion that the vampire superstition still retains its hold in its original habitat—an illustration of the remarkable tenacity and continuity of a superstition through centuries of intellectual progress from a lower to a higher culture, and of the impotency of the latter to entirely eradicate from itself the traditional beliefs, customs, habits, observances, and impressions of the former."[489] Stetson presupposed the illogical naïveté of the people responsible for the creation of the vampire in the first sentence of his article: "The belief in the vampire and the whole family of demons has its origin in the animism, spiritism, or personification of the barbarian, who, unable to distinguish the objective from the subjective, ascribes good and evil influences and all natural phenomena to good and evil spirits."[490] To reinforce his supposition, Stetson quoted the opinion of Moncure Daniel Conway (from his *Demonology and Devil-Lore*, 1879) that the vampire belief was "perhaps, the most formidable survival of demonic superstition now existing in the world."[491]

The theory of unilineal cultural evolution provided America's educated elite—the scholars, physicians, historians, publishers, and others who managed the established narrative—a "scientific" rationale for marginalizing the consumption ritual and its practitioners. If Tylor expected his notion of survivals to be judgment free, he failed to reckon with the establishment's view of them as abhorrent pagan superstitions. Most of the defamers were at least nominally Christian, North America's established religion, founded on the belief that a virgin gave birth to the son of a deity who subsequently was killed but returned to life. To commemorate his existence and enter communion with him, adherents regularly consume portions of his body and blood. Would these beliefs be judged rational under the same "scientific" interrogation imposed upon "primitive survivals," such as the vampire practice? The rational and moral exceptionalism claimed, at least implicitly, by the civilized elite, put their own spiritual beliefs beyond such harsh scrutiny. They would not grant others—seen as less evolved primitives and peasants—the same capacity to regulate themselves according to the variety of cultural systems that exist in every human culture.[492]

Anthropologists now acknowledge that *all* human beings share an innate capacity to think rationally, interact both symbolically and practically, and organize for a variety of purposes—and that these behaviors are interconnected in various ways.[493] Farming in New England was historically almost always challenging and difficult, and thus is a good example. To be successful, families—the basic farm unit—participated in informal cooperative arrangements, called *changing works*, with neighbors to share labor and other resources. Division of labor was based on role relationships within families. Farmers implemented agricultural practices that incorporated the latest technological innovations, yet their work continued to be regulated according to the seasons, as it had been since before memory.[494] Many of these farmers paid attention to signs and omens passed down in oral tradition or published in almanacs, including phases of the moon, to govern tasks such as fertilizing, plowing, planting, and harvesting.[495] And some, as we have seen, performed variants of the vampire ritual to stop the spread of consumption in their families and communities.

Yet, Stetson deemed this ritual to be an "extraordinary instance of a barbaric superstition outcropping in and coexisting with a high general culture, . . . in the closing years of what we are pleased to call the enlightened nineteenth century."[496] Even though he viewed southern Rhode Island as an area in decline, Stetson's encounter with vampire exhumations in "civilized" New England disturbed his belief in the ever-upward evolution of culture. He settled on South County's detachment from mainstream establishment culture—*archaism of the fringe*, as it was termed—to explain this disconcerting survival: "Naturally, in such isolated conditions the superstitions of a much lower culture have maintained their place and are likely to keep it and perpetuate it, despite the church, the public school, and the weekly newspaper."[497]

What *was* this persistent vampire survival? In the specialized world of taxonomists—scientists who classify of life forms (or undead forms, in the case of vampires)—there are two varieties: steadfast and literal *splitters*, who use fine distinctions to create numerous categories; and generous *lumpers*, who apply liberal criteria to unite, into fewer categories, a more heterogeneous assemblage. There is little room for deviation in the splitter's expressway to defining a vampire. Folklorist Alan Dundes, in *The Vampire: A Casebook* (1998), upheld a centuries-old depiction of the vampire as a reanimated corpse rising from the grave to suck the blood of the living, thereby maintaining "some semblance of a life." While "widespread . . . throughout Eastern Europe," he argued, it is not distributed throughout the world.[498] His line of argument followed a precedent established in the mid-eighteenth century by Dom Augustine Calmet: "Vampires . . . are men who have been dead . . . and issue forth from their graves and come to disturb the living, whose blood they suck and drain."[499] Jacob Grimm continued along this well-traveled road in the nineteenth century with "dead men come back, who suck blood."[500]

When one is looking for universal instincts and underpinnings of the human mind, as Stetson and other evolutionists were, it is difficult to be a splitter. Stetson was, almost by default, a lumper, granting universality to the vampire as a specific type of malignant revenant. While he accepted "the general belief that the vampire is a spirit which leaves its dead body in the grave to visit and torment the living,"[501] he argued for contextual variability: "The character, purpose, and manner of the vampire manifestations depend, like its designation, upon environment and the plane of culture."[502] As a result, Stetson argued, we have a variety of subspecies:

> Under the names of vampire, were-wolf, man-wolf, night- mare, night-demon—in the Illyrian tongue *oupires*, or leeches; in modern Greek *broucolaques*, and in our common tongue ghosts, each country having its own peculiar designation—the superstitious of the ancient and modern world, of Chaldea and Babylonia, Persia, Egypt, and Syria, of Illyria, Poland, Turkey, Servia, Germany, England, Central Africa, New England, and the islands of the Malay and Polynesian archipelagoes, designate the spirits which leave the tomb, generally in the night, to torment the living.[503]

Depending on the belief system, vampires were either corporeal or spiritual. The ancient "Rabbins" [Rabbis or other spiritual leaders] had a difference of opinion on this point, Stetson wrote, as some believed vampires "were entirely spiritual" while others asserted that they "were corporeal, capable of generation and subject to death."[504] Stetson connected the early Greek word *diabolus* ("a calumniator, or impure spirit") to "modern Greeks," who "are persuaded that the bodies of the excommunicated do not putrefy in their tombs, but appear in the night as in the day, and that to encounter them is dangerous."[505]

Citing the work of Calmet, Stetson pressed the diversity of vampire beliefs, touching on the evil eye superstition and German shroudeaters, for which he also cited Michael Ranft's work, *De Masticatione Mortuorum in Tumulis* (Leipzig, 1728) without mentioning the author's name. With no transition, Stetson abruptly moved to Polynesia and the "Malay peninsula," home of the "water demon, who sucks blood from men's toes and thumbs."[506] He then circled back to the more familiar European-based vampires, summarizing Edward B. Tylor's "two theories" of the superstition. In the first, the vampire is the soul of a living man (often a sorcerer), while in the second, it is the soul of a dead man. In the former, the soul of a sleeping man travels visibly as a piece of straw or fluffy down that can enter homes through keyholes to attack its victims. There are two versions of the living vampire theory. In one, the vampires (or "*Mauri*") assault men at night by sitting upon their breasts (presumably while they are asleep) and sucking their blood. In the other, while children are physically assaulted, adults encounter the vampires simply as "nightmares."[507] Tylor's second theory corresponds to the now "classic" definition of a vampire, in which "the soul of a dead man" leaves its "buried body and sucks the blood of living men," who become "thin, languid, bloodless," and fall into a "rapid decline," then die.[508]

Stetson appeared to be narrowing his vampire focus by citing Tylor, but he quickly returned to full-fledged lumping, by citing the "belief in the *Obi* of Jamaica and the Vaudoux or Vodun of the west African coast."[509] He seemed to be setting the stage for his eventual speculation regarding how the practice arrived among apparently decorous New Englanders. Scapegoats loom in his declaration: "It is a common belief in primitive races of low culture that disease is caused by the revengeful spirits of man or other animals—notably among some tribes of North American Indians as well as of African negroes [sic]."[510] Near the end of his article, Stetson reaffirmed his uncertainty regarding origins: "Of the origin of this superstition in Rhode Island or in other parts of the United States we are ignorant; it is in all probability an exotic like ourselves, originating in the mythographic period of the Aryan and Semitic peoples, although legends and superstitions of a somewhat similar character may be found among the American Indians."[511]

Having established the ancient roots and worldwide distribution of the vampire belief, Stetson turned his full attention to New England. There, he asserted, "the vampire superstition is unknown by its proper name."[512] His argument echoed that made four years earlier in the *Providence Journal* (March 21, 1892) when its editors castigated a correspondent for omitting "all mention of the vampire" in his story, published in their newspaper, of the Brown family exhumations in 1892. The correspondent simply "failed to get to the bottom of the superstition," the editors argued. They bolstered their argument by pointing out "the files of the Journal when reference is made in them to the practice of the tradition in Rhode Island, without exception speak of the search of the graves in such cases as attempts to discover the vampire." The editors, like Stetson, implicitly assumed that, because insiders did not use the "proper name," they were unaware of the ritual's true meaning. Stetson reasoned that New Englanders did not label a troublesome corpse a "vampire" because they viewed consumption's lethal assault as spiritual rather than corporeal: "It is there believed that consumption is not a physical but a spiritual disease, obsession, or visitation; that as long as the body of a dead consumptive relative has blood in its heart it is proof that an occult influence steals from it for death and is at work draining the blood of the living into the heart of the dead and causing his rapid decline."[513]

Stetson's conclusion that the vampire is a universal demon rests on the general concept that the dead live on after death.[514] The next logical conclusion is that the dead *cause* death.[515] The belief that the dead prey on the living has been documented from ancient civilizations to the present. In her description of Romanian vampirism, Agnes Murgoci (1926) wrote, "It would seem that the most primitive phase of the vampire belief was that all departed spirits wished evil to those left, and that special means had to be taken in all cases to prevent their return."[516]

Stetson noted that death-bringing demons assume different forms. Vampire, werewolf, succubus, ghost, witch, and hag often receive similar treatment in folklore regarding their etiology, method of assault, identification, and destruction, as well as protective devices used to deflect their assaults.[517] Even the belief usually associated with vampires—that they are revitalized corpses—is nearly worldwide in distribution and dates from ancient times.[518] Murgoci found that, in Romania, vampires sometimes are linked to witches and wizards.[519] In Italy, the *strega*, or witch, can also act as a vampire, as "she sucks the blood of sleeping people through the little finger, thus inducing an inscrutable and therefore incurable marasmus" (a gradual loss of flesh and strength for no apparent cause).[520]

The widespread connection of shapeshifting to vampires introduces even more uncertainty into the task of separating them from other harmful demons. Not only might a vampire appear in its human form, but it may also, like the witch of folk tradition, take the form of various animals, including wolves, dogs, oxen, sheep, and insects, particularly butterflies. Bats, the preferred alternate form of literary and pop culture vampires, are less common in folklore. Bruce A. McClelland (2006) glumly acknowledged the vampire's elusiveness, in both folklore and taxonomy: "The vampire . . . is a literal shapeshifter, changing his name, his features, his activities and so forth over both time and space. Even the names of the vampire are legion, and it is often difficult to know whether one is dealing with the same or a different entity, when speaking of this or that vampire-like demon."[521] Most of the disparate conceptions of a vampire probably can be subsumed in the two-part definition offered by J. A. MacCulloch nearly a century ago: "A vampire may be defined as (1) the spirit of a dead person, or (2) his corpse, re-animated by his own spirit or by a demon, returning to sap the life of the living, by depriving them of blood or of some essential organ, in order to augment its own vitality."[522]

Stetson did not cite the *Encyclopaedia Britannica* as one of his sources, but its "vampire" entry functioned as a bridge between academic and popular vampire narratives. It was cited explicitly in the *Providence Journal*'s account of the Brown family exhumations and, later (as we see in the next chapter) in the *Boston Globe*'s extended treatment. Directly and indirectly, the following text has had a sustained effect on the vampire concept:

> VAMPIRE, a term, apparently of Servian [Serbian] origin (wampir), originally applied in eastern Europe to bloodsucking ghosts, but in modern usage transferred to one or more species of blood-sucking bats inhabiting South America. [At the end of the next paragraph, a discussion of vampire bats is given more than three times as many words as the leading *vampire* entry.]
>
> In the first-mentioned meaning a vampire is usually supposed to be the soul of a dead man which quits the buried body by night to suck the blood of living persons. Hence, when the vampire's grave is opened, his corpse is found to be fresh and rosy from the blood which he has thus absorbed. To put a stop to his ravages, a stake is driven through the corpse, or the head cut off, or the heart torn out and the body burned, or boiling water and vinegar are poured on the grave. The persons who turn vampires are generally wizards, witches, suicides, and persons who have come to a violent end or have been cursed by their parents or by the church. But any one may become a vampire if an animal (especially a cat) leaps over his corpse or a bird flies over it. Sometimes the vampire is thought to be the soul of a living man which leaves his body in sleep, to go in the form of a straw or fluff of down and suck the blood of other sleepers. The belief in vampires chiefly prevails in Slavonic lands, as in Russia (especially White Russia and the Ukraine), Poland, and Servia [Serbia], and among the Czechs of Bohemia and the other Slavonic races of Austria. It became specially prevalent in Hungary between the years 1730 and 1735, whence all Europe was filled with reports of the exploits of vampires. Several treatises were written on the subject, among which may be mentioned Ranft's De Masticatione Mortuorum in Tumulis (1734) and Calmet's Dissertation on the Vampires of Hungary, translated into English in 1750. It is probable that this superstition gained much ground from the reports of those who had examined the bodies of persons who had been buried alive though believed to be dead, and was based on the twisted position of the corpse, the marks of blood on the shroud and on the face and hands, -- results of the frenzied struggle in the coffin before life became extinct.

The belief in vampirism has also taken root among the Albanians and modern Greeks, but here it may be due to Slavonic influence.[523]

17 Explain It? I Can't. It Is an Old, Old Belief

The publication of George Stetson's "Animistic Vampire in New England," in January of 1896, had an immediate (in addition to a long-lasting) impact. Within three months, six separate newspaper articles flowed from Stetson's wake; their tone arced from castigation to caricature. A seventh article appeared in late 1902. Apparently based on fieldwork, it was well-written (anonymously) and surprisingly detailed, with new information.

"Vampires In New England," Rene Bache, *Boston Transcript* (January 18, 1896)

On January 18, 1896, Rene Bache's syndicated vampire article, kindled by Stetson's "Animistic Vampire in New England," appeared in the *Boston Evening Transcript*; it was subsequently reprinted in several other newspapers.[524] In his summary of Stetson's article, Bache, whose byline read "Special Correspondent for the Transcript," added a few world-wide examples of his own, including a discussion of being buried alive—a fate, which, as we saw in the *Encyclopaedia Britannica* article, was supposed to account for the great attention given to the vampire tradition. Alluding to the vampire scare in Hungary, he wrote: "On the continent from 1727 to1735 there prevailed an epidemic of vampires. Thousands of people died, as was supposed, from having their blood sucked by creatures that came to their bedsides at night with goggling eyes and lips eager for the life-fluid of the victim." His mention of the goggle-eyed vampire inspired the editors of the *San Francisco Examiner* (January 26, 1896), in their publication of Bache's article, to add an illustration entitled, "The Goggle-Eyed Vampire that Frightens Rhode Islanders." It depicts a wide-eyed vampire bat hovering over a startled young woman, sitting up in her bed. Indeed, at the end his article, Bache seemed to be mimicking the *Encyclopaedia Britannica* (or, perhaps, responding to the same swell of popularity of the topic) when he devoted six paragraphs to a detailed description of vampire bats.

Did the influence of this syndicated article extend even to the creation of Count Dracula? A handful of writers have suggested that the version of Bache's article published in the *New York World* (February 2, 1896), played a significant role in the formation of Bram Stoker's novel *Dracula*, published May 24, 1897.[525] Responding to Peter Haining's assertion, in *Shades of Dracula* (1982), for example, that the newspaper article was "enormously crucial" because it "directly helped [Stoker] to create . . . Dracula," renowned Dracula scholar Elizabeth Miller wrote:

> Let's get our facts straight: "Vampires in New England" appeared in . . .1896. Obviously read by Stoker during his American tour with the Lyceum that winter, the cutting is included among his working papers. By 1896, however, what Haining refers to as Stoker's "new book" was in its final stages. Haining's assertion that this newspaper clipping was the "only incontrovertible piece of evidence" for the inspiration and the origins of Dracula, as well as "the life-blood . . . from which the vampire took his genus" is preposterous.[526]

Miller then added the coup de grâce: "The facts speak for themselves. By 1896 Stoker would have had assembled almost all the material he needed for the completion of his novel."[527]

Rene Bache (1861–1933), the fourth great grandson of Benjamin Franklin, was born in Philadelphia.[528] A journalist and author who wrote for *Scientific American* and other periodicals, his articles ranged from novel paleontological finds (*Dallas Morning News*, May 14, 1896) to photographing the human voice (*Springfield Republican*, April 18, 1896). Commenting on Bache's engagement to edit a column (to be entitled, "Lung to Lung Talks on Crabs and Caterpillars") for a new women's magazine, an announcement in the *Kansas City Star* (October 5, 1900) designated Bache "the popular and interesting writer on unpopular and uninteresting matters of science." In that light, a discussion of vampire bats would not have been a stretch for Rene Bache.

"Vampires In Exeter," V. J. Briggs, Syndicate Press, Boston (March 28, 1896)

Vernum Judson Briggs (1863-1936), of East Greenwich, Rhode Island, undoubtedly would have been aware of New England's vampire tradition years before 1896.[529] Not only was he born and raised in Rhode Island, a hotbed of vampire activity and the setting of Stetson's article, he also served as a correspondent for several state and local newspapers, including the *Providence Bulletin*, *Rhode Island Pendulum* (East Greenwich), *Wickford Standard*, *Pawtuxet Valley Gleaner*, and *Providence Journal*. Briggs was a correspondent for the latter two newspapers in the spring of 1892 when the story of Mercy Brown's exhumation was being widely examined in the press.[530] The *Providence Journal*'s coverage of the Brown family exhumations included a lengthy account from an unnamed "local correspondent," and the *Pawtuxet Valley Gleaner* published an unsigned letter describing the exhumations in some detail, during which the writer referred to George T. Brown as "our neighbor" (see chapter 10). It is unknown if Briggs was either of these two narrators, or if he contributed to any of the *Providence Journal*'s several accounts of the event.

Briggs did, however, write an article on Rhode Island vampires in early 1896. It was published under his name, with a dateline of East Greenwich and a copyright by Syndicate Press, Boston. During 1895 and 1896, Syndicate Press routinely assembled a full page of feature articles with accompanying illustrations—entertaining soft news stories without a shelf life—to which several newspapers subscribed. Between March 28 and April 2, 1896, the Briggs article was published verbatim under the headline, "Vampires in Exeter" (or "Exeter Vampires"), in at least five newspapers.[531] The article did not mention Stetson directly, but Stetson's recently published article seemed to loom in the background when Briggs wrote of the vampire practice: "It is a startling discovery that was made the other day not twenty-five miles from Providence." Briggs provided details of the 1892 Brown family exhumations, including names of participants, that Stetson had omitted. Under the heading, "A Relic of Voodooism," Briggs traced the possible origins of the tradition to the voodooism of Southern Blacks (another echo of Stetson's article). But, in the next section, "A Relic of Vampirism," he connected the New England tradition to Eastern Europe (as did Stetson). An accompanying, uncredited illustration shows three bodies near a roaring fire where, presumably, Mercy Brown's heart was burned. Briggs wrote, incorrectly, that the ashes of her incinerated heart were scattered at the gravesite.

Since Bache's article began circulating in newspapers two months before that of Briggs, it is possible that Bache's account prodded Briggs to contribute his own version of the 1892 Brown family exhumations. Bache's lengthy discussion of vampire bats also may have inspired Briggs to add a brief paragraph on vampire bats at the end of his own article.

"Signs And Omens," *New York Sun* (February 3, 1896)

A third major derivation of Stetson's vampire study was an anonymous article, "Signs and Omens," first published in the *New York Sun* (February 3, 1896), which was subsequently picked up by at least two other newspapers.[532] The *Sun* debuted in 1833, the first successful "penny press" newspaper. Inexpensive and eager to break long-standing tradition by publishing personal stories, such as suicides and divorces, the *Sun* appealed to working class people.[533] The text of "Signs and Omens" was devoted mainly to commenting on current superstitions. Near the end, the author addressed the vampire, alluding to Stetson (without naming him) as "a student of anthropology":

> Superstitions as to strange remedies for sickness hold out in various parts of the country. One of the most shocking of these has been discovered in Rhode Island by a student of anthropology. The remedy is associated with a belief in the vampire, which is supposed to be the escaped soul of a dead person that sucks the blood of living relatives until they die. Many times within the last fifty years, and several times within the last ten years, the bodies of the dead have been dug up in Rhode Island cemeteries and the heart in each case burned in order to save the life of some vampire-hunted relative. This has occurred within fifteen or twenty miles from Newport.

The influence of Stetson's academic essay continued, as at least four additional direct or indirect appropriations appeared after its publication.

"Believe In Vampires," *Boston Daily Globe* (January 27, 1896)

This article's anonymous author credited George R. Stetson for a renewed interest in vampires, writing, "The smoldering interest in the subject has been revived by the recent publication of a newspaper syndicate article over the signature of a rather well-known writer, who borrowed the article almost word for word from an essay by George R. Stetson in the Anthropologist." As we have just seen, that "well-known writer" was Rene Bache, Special Correspondent for the *Boston Transcript*. Bache did not borrow Stetson's essay "almost word for word," although he *did* include a summary of it, and, like Stetson, he excluded names of the families involved in the rituals.

A unitary thread of disgust and disapproval coursed through nearly all accounts of New England's vampire tradition in the nineteenth century. But there was a significant transformation in the 1890s, a banner decade for vampires, especially those in New England. "As the region became more secular in the nineteenth century," folklorists Kent C. Ryden and Simon A. Bronner observed, "writers profiled the representative New Englander from the sober, God-fearing Puritan into the comic Yankee. A staple of regional and national newspaper and almanac humor, the rural Yankee bumpkin appeared at times crude (as a hardscrabble farmer), at times shrewd (as a guileful peddler), and at most times awkward and naive."[534] After decades of censuring Yankee vampire practitioners, the circumstances were favorable for lampooning them. The fact that consumption was a contagious disease caused by a microbe had been settled (at least within the medical establishment, if not among the general populace) when Robert Koch announced his discovery of the tuberculosis bacterium in 1882. Deliberate exaggeration for comic effect or irony had a reasonable chance of succeeding only when a vampire exhumation lost its vigor as a menace to civilization. Transforming vampires from hard news to feature story was assisted, not only by increasing secularization in the region and the identification of the tuberculosis bacterium, but also by a coincident change in the newspaper industry, as the importance of story journalism rose at the expense of information journalism.[535] The infotainment value of vampires plaguing local yokels proved irresistible to some news writers.

An illustration in the *Boston Globe* (January 27, 1896) shows a man with a long beard, wearing comically rustic clothing and carrying a buggy whip, walking next to a cart hitched to a horse *and* a pair of oxen. It is captioned: "A member of the anti-vampire party." The illustration is a paradigm for the narrative that follows, mixing the factual and fanciful in what surely was meant to be a parody of southern Rhode Island's vampire tradition, laid bare four years earlier by the far-flung story of Mercy Brown's exhumation, and then reanimated in Stetson's just-published article. With a dateline of "Sodom, R.I. Jan 26," the headings sketched-out what was to follow: "Believe in vampires. Rhode Islanders who are sure that they do exist. Instances told of where the living have been attacked and preyed upon by these representatives of an unseen world."

The narrative's humor relies on othering its characters, beginning with geography. Odd place names tell us we have entered the hinterlands: Swamptown City, Escoheag and Usquepaugh, Noose Neck Hill and Skunk Hill, Exeter Hollow, Gomorrah, and, of course, the home of the "Sodomites." The anonymous author could not have invented a more suitable place name, a city that God destroyed because of the deviant behavior of its citizens. "Sodom is a back number," we are informed, "Too small to be on the map." There is, indeed, a Sodom, Rhode Island, and it is on some maps. I was at the end of Sodom Trail, in the town of Exeter, on November 18, 1981, to listen as Everett Peck related his family's story of Mercy Brown's exhumation (chapter 10). At the beginning of our interview, Peck presented me with a yellowed clipping (coincidentally dated 1896) that included text, photographs, and a map showing Sodom. The Sodom Mill Historic and Archeological District had been established as a National Register District the year before our interview. A topographic map included in the National Register nomination form shows the old farmhouse at the end of Sodom Trail (where Peck was living), along with the ruins of the 1814 mill, located just to the south on the mill pond created by a dam on Sodom Brook.

Backcountry people are easy targets of derision. The author used Sodom's geographical isolation to paint a scene of outdated manners, customs, and language retained by people distant from centers of cultural innovation. The marginalization, which proceeded from geographical isolation through rustic clothing and antiquated speech, was consummated with dubious kinship practices (family intermarriage) and primitive beliefs.

The vampire notion is introduced with the European vampire trope, characteristic of outsider coverage of the New England vampire belief, derived explicitly from the *Encyclopaedia Britannica*'s "vampire" entry. Thus, we get: "Now a vampire, as everybody knows who has seen one, is a blood-sucking ghost—the soul of a dead person which quits the body by night to feed upon the blood of the living, especially of its relatives and dearest friends, if it has any. When the vampire's grave is opened the corpse is always found to be fresh and rosy from the blood which it has thus absorbed; otherwise it is not a genuine vampire." We learn that a vampire can be halted by pouring boiling water and vinegar on its grave, or, if it is especially persistent, driving a stake through its heart, cutting off its head, or removing its heart and liver, burning them to ashes, and eating the ashes. The latter remedy bends the narrative back toward New England. "In Rhode Island, no one becomes a vampire after death unless he has died of consumption."

As the narrator continued interviewing the Sodomite, "slowly but surely the conversation drifted to vampires." Given what we know of the regional tradition, of the four narratives offered, two seem to be outright fabrications, one is a well-documented event, and the other appears to be a blend of fiction and folklore. The latter tale likely is based on a story published in 1888 by Rhode Island historian, Sidney S. Rider (which I discuss in chapter 19). "About 100 years ago there lived two families on the western slope of Pine Hill in Exeter," the anonymous author begins. "They were prosperous farmers for those days." The Rider story also unfolds on Pine Hill in the late 1700s. But the *Globe*'s tale quickly veers off with unusual names and outlandish events. Isaiah falls for Mehitable, but she unfortunately dies of consumption. Isaiah's heart is broken, but then, he, too, contracts consumption. One night his mother hears him groaning and, upon entering his room, sees Mehitable, who has turned into a vampire, sucking his blood. The Sodomite narrator seems to have a spate of rationality at the end of his tale, theorizing that "them two young folks had probably been kissing each other a good deal, and Isaiah caught the disease from his sweetheart. Contagion, they call it, don't they?" His scientific insight was short-lived, however, as his next two narratives revealed.

In the first, the wife of Godlove Arnold, "who lived on the southern shore of Yawgoo pond in South Kingstown," died of consumption. The couple "had not always been on the best of terms," and Mrs. Arnold returned as a vampire to plague her husband. One afternoon, "she chased him . . . all the way to Bald hill, and finally he had to give in. They found his body about a week later on the hillside, and the expression on his face was something ghastly." In another rational moment, the Sodomite speculated that Godlove "probably died of heart disease."

The second tale is even more outlandish. "Over around Kettle hole and Goose Nest spring, in the Pork hill district of North Kingstown, there once lived a man by the name of Isaac Harvey." When Ike, as he was known, died of consumption, his wife seemed relieved, since "he had seldom contributed anything but advice to her support." But Ike returned as a vampire and "tormented her by night and by day, following her around in the shape of a ball of fire, until she finally hit upon the happy thought of wearing a horseshoe around her neck. It cured Ike completely. The horseshoe was rather heavy and cumbersome, but it was better than being singed by a ball of fire." Rationality must have eluded the Sodomite, because "for this legend the Sodomite had no explanation."

But rationality returned in the final narrative. What does not return, at least in corporeal form, is the vampire/revenant of European tradition. "Coming down to historic events, which are matters of record, and omitting a score or more authentic cases within the memory of any middle-aged man now living," the author began, "the most important vampire incident of recent years was the celebrated Brown case." In a detailed and mostly accurate narrative, the author reprised the now-familiar story of George T. Brown and his family's unfortunate encounter with consumption.

We are informed that, since the Brown family event, "the belief of the community in vampires has been rather wavering. A great many of the leading men in Exeter do not believe in the theory at all." As is the case with all the named people in this narrative, some have a documented existence while others do not. The "Hon. Edward P. Dutemple, state senator from this town, who is a good legislator and a still better blacksmith," for example, apparently does not believe in the vampire theory, although "his private opinion" about "the vampire question" is known only among his friends. Edward P. Dutemple was an actual person, born in Exeter on September 30, 1848. He was "educated in the public schools" and was, in 1895, the town's coroner and a member of the Republican State Central Committee. His occupation was, indeed, blacksmith.[536]

Also among the anti-vampire party is Reynolds Lillibridge, "the successful farmer, gunner and trapper of Pine Hill," who is "much more interested in minks and otters, and the trout in his fine pond, than in the vampires." After Lillibridge asserts, "When a man's underground, he hasn't anything more to do with anybody that's above ground—that's my theory," he appears to be the unaffected Yankee. "Still," he reasons, "I can understand how a man like Brown must have felt. When you are in trouble you will grab at a straw, and when you are in a good deal of trouble you will grab at a whole bundle." Several vital records show that Reynolds Lillibridge (1837-1900) was an Exeter farmer who served in the Civil War.

"Then there is the good elder Edwards," the narrator continues, "town clerk, librarian of the public library on Pine Hill, farmer and preacher. He is one of the most pronounced of the anti-vampirites. Among the laity, the hard-headed farmers of the town who work early and late to coax a living from the reluctant soil, there are plenty who are outspoken in their disbelief in vampires." Reverend John H. Edwards did indeed wear many hats. He was pastor of the Chestnut Hill Baptist Church from 1882 to December of 1892, when he resigned due to poor health. Notably, he was the pastor when the Brown family exhumations took place in the cemetery behind the church to which they belonged. But, apart from the anonymous author's assertion, we have no information regarding Edwards' disbelief in vampires.[537]

The author presents us with a riddle that has baffled most who have investigated New England's vampire tradition: "As to the origin of the belief there is no satisfactory explanation given. How it could have been transplanted from the old world and found a lodgment only in Rhode Island, among an otherwise very intelligent and enterprising and wideawake population, is a mystery. It is not an English superstition, and yet the settlers of this region were all English." Of course, we now know that the tradition was diffused throughout New England, and we have clues regarding how it arrived and was disseminated in the region.

Finally, I grieve that Madame Douglas was among the people cited in this narrative whose existence has not been authenticated: "Mme. Douglas, the lone clairvoyant and business medium, who lives on the Ten Rod road, hasn't any doubt about the existence of vampires and lots of other things, seen and unseen." If she *had* existed, her narrative would have been a most welcome addition.

The anonymous author of the *Globe* article seems to have been tapping into a journalistic trend of his time. Michael Schudson, in *The Power of News* (1995), discussed ways in which a journalist could manipulate the interview form in the late nineteenth century, writing, "The reporter could determine what he wanted the news to say and then choose an interviewee he believed would be willing to say it." In his work, Schudson observed that "interviews were routinely faked. If reporters sent to hotels in search of newsworthy visitors found the pickings slim, they sometimes simply conjured up quotable characters."[538] While several of the characters in the *Globe* article were actual people, it seems likely that the quotable key informants, who provided the outlandish vampire tales, were, indeed, conjured up. In "Believe in Vampires," the author was following the worn path of the comic Yankee, whose evolution during the nineteenth century began with the likes of Brother Jonathan and Seba Smith's shrewd but homely Major Jack Downing.[539] Folklorist Richard

M. Dorson dubbed such "folksy" narratives "subliterature," presuming that they do not meet the standards of "real literature." Dorson argued, "The dark and somber theme of supernatural legends merits as much attention as the current of humorous exaggeration in our literary and subliterary history."[540] The *Globe* article, while not unique in combining the supernatural with the humorous, stands out in that regard. The humor arises from burlesque of the backwoods bumpkin. To create the necessary broad appeal, the stereotypes displayed had to be accessible to the general reader. The criteria that shape historical facts to fit a predetermined mold apply to both yokel and vampire. Since the "joke" is shared implicitly with the reader—an evident assumption in this article—there is no need to label the narrative as satire and not news. Newspaper readers, unlike scholars, were not always concerned with sorting out history from hokum.

Folklorist MacEdward Leach's summary of how writers fail as they attempt to incorporate folklore into American regional literature is more than apt for the *Globe's* caricatures: "In a number of instances, the regional writer found a scrap of story in folk tradition or a vague memory of a folk character incentive enough on which to build a whole system of stories in a pseudo-folk style. . . . The result is self-consciousness, straining for effect, gross exaggeration, synthetic lustiness."[541] In the same article, Leach gave us a stake to drive through the heart of the *Globe's* parody: "Many writers of regional literature are content with a surface picture. Many sense the need of going deeper into the fundamentals of human nature and, with little knowledge of the underlying folk life, try to create it. The result is usually manifestly phony."[542]

"Talk Of the Day," *Boston Journal* (January 29, 1896)

An article in the *Boston Journal* (January 29, 1896) was, credibly, a spoof of the *Boston Globe's* parody, which had appeared in the city two days earlier. While there was no mention of Stetson or his article, the text included a summary of Rhode Island's vampires, in the context of a story from Bohemia. The article's punchline was: "Are these Rhode Island vampires of the quiet species, or do they howl and shreek [sic] dismally?"

"The West Leads," *Wichita Daily Eagle*, February 1, 1896

Another iteration, "The West Leads," published in the *Wichita Daily Eagle* (February 1, 1896), made no pretense that it was using the *Boston Globe's* burlesque to turn the tables on the snobby East. The article's opening set the tone: "The more intelligent people of the east are not a little perturbed over the superior intelligence of the west. Supposed centers of culture, like Boston and her imitators, are green-eyed." Yet again, the vampire ritual is a marginalizing cudgel: "The Boston Globe is responsible for the statement that in a certain locality in Rhode Island there is still a prevalent belief in vampires. This horrible superstition is that those who die of consumption come again to earth, where they fall upon their relatives and friends and suck their life blood. . . . If all that is true, it is another proof that in parts of the United States in this enlightened age men live who are as much in need of conversion to sanity as were the men of Salem many years ago."

"The Vampires of Rhode Island," *New York Press* (December 28, 1902)

The *New York Press*, which began publishing in New York City in 1887, often included articles written by well-known authors, notably Stephen Crane. It seems possible that a remarkable article like the "The Vampires of Rhode Island," published in the *New York Press* (December 28, 1902), may have been written by a renowned author whose identity remained undisclosed. (It was not Crane, for he died of tuberculosis in 1900, at the age of twenty-eight.) The article is lengthy by newspaper standards, running to nearly 3,000 words. Its author—who disclosed that he visited Rhode Island in the Autumn, accompanied by a local informant or "guide" as he was dubbed—was not intimately familiar with the state. His continual reference to North Kingstown (one town of the state's thirty-nine cities or towns) as North Kingston (Kingston is a village in the Town of South Kingstown) is telling. But his detailed description of landscape elements, both natural and built, discloses the eye of a critical observer. His writing, too, is well crafted.

The writer (as he dubbed himself) seemed at a loss to explain the existence of the vampire superstition among "clear eyed, level headed, well informed people" residing in a "community of educated, well-to-do farmers and mechanics, who send their children to the best schools, go regularly to church, take the daily papers and live in ancestral homesteads surrounded by orchards and broad and fertile acres." After giving a succinct account of a recent exhumation, the writer provided an unusually good summary of the vague theory underlying the belief:

> They do not seem to think that it is exactly the spirit of the dead person, either, which does the mischief—the real soul of the deceased they have consigned to heaven or the other place, according as they held him in esteem, like all other Christians—but a vague 'something' which has its residence in the heart and liver of the dead—a sort of astral body, as it were, projected from a corpse. When questioned closely on the subject the believers in the vampire superstition are rather hazy as to just what the influence is that draws the blood of the living to the body of the dead.

His verbatim quotation of a doctor (almost assuredly Harold Metcalf, of Saunderstown, who attended the Brown family exhumations in his capacity as Medical Examiner) suggests that the writer interviewed him in person: "A doctor of the neighboring village of Wickford (adjacent to Saunderstown), whose practice extends among these people, and who himself attended the last vampire burning in his professional capacity, said: 'They believe it and they don't; just as some people will not sit at table with thirteen, or walk under a ladder, though acknowledging that it is all foolishness.'" This was followed by a quote from an old man, whose appraisal accords with the local attitude that I documented regarding the Brown case: "It may be all foolishness, but even if it is a man is bound to do everything he can, no matter how absurd it may seem, to stop the ravages of disease in his family. I have seen many burnings in my time, and in every case the rest of the family escaped the disease, or got well, except in cases where the sick ones were already nearly dead of consumption."

Some phrases that the writer included in a summary of the European tradition suggest he, as did other vampire writers, made use of the "vampire" entry from the *Encyclopedia Britannica*, either directly or, possibly, indirectly by way of the *Boston Globe* article (January 27, 1896): "To put a stop to the ravages of the vampire the heart is taken out and the body burned, or a stake is driven through the corpse, or the head cut off. Boiling water and vinegar is also poured into the grave. . ... In the East it is believed that those who turn vampires after death are wizards, witches, suicides or persons who have been cursed by their parents or the Church." Toward the end of his narrative, the writer returns to the European vampire tradition, again echoing the *Encyclopedia Britannica*'s entry: "Among the Slavonia peasants it is believed that any person over whose body a bird flies or an animal (especially a cat) jumps becomes a vampire."

The writer was told that five exhumations had occurred in the neighborhood in the past twenty-five years, with a physician in attendance at each one. He wondered how many others had been performed beyond official oversight. Then, almost casually, he delivered an unexpected and stunning revelation, providing surnames of four families who performed the ritual (one of them twice): "Of the well-known cases one burning each took place in the families of the name of Thomas, Brown and Green, and two in a family named Gardner. In two cases it was a daughter's body which was exhumed, in one case a wife's and in another the bodies of a wife and son." If these disclosures are accurate, we can add another four or five cases to our inventory of rituals that were performed in that locale during the last quarter of the nineteenth century. The next paragraph describes the by-then fabled Brown family event. The writer named both "Mr. Brown" and "Dr. Metcalf" but mistakenly wrote that Brown and *two* daughters survived the family's consumption epidemic. The recently married Myra Frances Brown Caswell died on June 25, 1899, at the age of eighteen—three-and-a-half years before the *Press* article was published (see part IV).

The writer continued with a summary, including apparent verbatim transcriptions, of his interview with an eighty-year-old man who claimed that his grandparents spoke of the heart-burnings. The man said that, although he had heard of many, he had attended only one, himself—which he described (quotation marks in original):

"In that corner of that field over beyond the meadow there is an old graveyard where Mrs. H— was buried. She died of consumption, and soon after another member of the family began to grow sick, and wasted away until she died. Then another was attacked. They seemed to fade away from lack of blood. When the third was attacked with the same disease it was determined to open the grave of Mrs. H—, though she had now been dead for fourteen years. A doctor was sent for and came. The body was as black as your coat when it was taken up, but when the doctor took out the heart it was well preserved and full of blood. There was so much blood in it that it ran out over the stone on which it was burned. Strange, wasn't it? And she dead fourteen years! If I hadn't seen this I would not have believed it. And, more singular still, no member of the family died of consumption after that. Explain it? I can't. It is an old, old belief about her [here?]. In all the cases I have known of the burnings stopped the ravages of the disease in the family except in one case, when a daughter was already 'most dead. Coincidence very likely, but it is an old, old belief."

The man led the writer across a meadow to the old burying ground in its corner. He pointed out the rock where the woman's heart had been burned. From a distant schoolhouse, in the "still October air," the two men could hear the "hum of the children reciting their lessons." In a farm wagon, the pair continued along a country road to a hill overlooking Narragansett Bay (another sign that they were in North Kingstown, not land-locked Kingston). There, in an overgrown, neglected cemetery, the old man pointed to the spot where two bodies—a mother and son—had been exhumed and their hearts removed and burned.

To this point in the narrative, events and participants are credible, if not completely verifiable. The clues are minimal, but unyielding research might add even more cases to the growing inventory of vampire rituals in New England. But then the writer offered this:

One phase of the vampire belief in this region is that while a burning is in progress any member of the family upon whom the vampire has been feeding suffers excruciating torture, even though he or she may be entirely ignorant that the cutting and burning is taking place. They tell of a girl stricken with consumption who, while the heart of her dead sister was being taken out, kept exclaiming: "What is the matter? Some one is trying to cut my heart out!" although she did not know that a vampire hunt was in progress.

I have found nothing in the regional tradition that corroborates this motif. While the transference of pain (as well as disease) through sympathetic magic is a known folk belief,[543] I think here it is narrative license. Did the writer insert this theatrical departure to captivate our attention? If so, he got mine.

Part VIII Community-Based Storytellers

The five authors discussed in the next two chapters crafted vampire narratives based on documented events. These writers were enmeshed in the community that provided the setting for their nonfictional narratives; they formed connections to the surrounding places, people, and events that became central to their work as published authors.

The vampire story by Adaline M. Tirrell (1861-1937) appeared in the *Springfield Republican*, the first of a ten-installment series entitled "Legends of Hampshire Hills." The series, published from 1929 to1932, was composed primarily of tales that Tirrell apparently heard and remembered as she was growing up in town of Chesterfield in Hampshire County, Massachusetts.

Joseph R. Chandler's (1792-1880) vampire story first appeared in several newspapers in 1822. This article was based on stories he recalled from his hometown of Kingston, a small village on the shore of Plymouth Bay, Massachusetts. Since he was a young man when the exhumation ritual was performed, Chandler probably heard the story from other residents in this close-knit community. And, perhaps, as both his narrative and the first line of a poem he composed about the event intimates, he was also an eyewitness: "I saw her, the grave sheet was round her."

John W. Corbett (1854-1919) wrote about the vampire of Seneca Lake in his book, *The Lake Country: An Annal of Olden Days in Central New York* (1898). Corbett was born and raised in the town of Reading, located along the western shore of Seneca Lake. After selling the family farm, he turned to the newspaper business; many of the stories he wrote, edited, or published centered on the region around his hometown in Schuyler County, New York. Corbett did not disclose the origin of his vampire story, but given his personal and professional biography, it seems likely that he heard it through an oral report or encountered it in a yet-to-be-discovered newspaper article.

The historical writings of Casey B. Tyler (1819-1899) focused on the Pawtuxet Valley in western Rhode Island, especially his hometown of Foster. The story of Nancy Young's exhumation in 1827 first appeared in Tyler's serialized "Reminiscences of Foster," published in the *Providence Journal* from 1858 to 1859. In a town whose population in 1827 was not quite 3,000, Tyler, a lad of seven years, probably heard about the exhumation through word-of-mouth communication.

Although Sidney S. Rider (1833-1917) was born in Rensselaer County, New York, he relocated to Providence, Rhode Island, at the age of twelve with his family. At that early age, he embarked on a career that led to his publishing of pamphlets on Rhode Island history and culminated in his establishment of *Book Notes*, a periodical devoted to book reviews and articles connected to Rhode Island history. Rider's story of Sarah the vampire appeared in *Book Notes* in 1888.

Community-based narratives easily veer into fiction. Stories that can be improved get improved. What "improved" means, of course, depends on one's perspective. Historians probably would not consider a narrative improved if it moved away from accepted historical fact. A newspaper publisher (and, by extension, writer) might favor a narrative that had broad appeal, even at the expense of (complete) accuracy. One could get lost pursuing the categories and fine details of creative nonfiction and historical fiction (not to mention memoir). The community-based narratives we consider in this part show varying degrees of both artistic merit as well as fidelity to historical accuracy.

Gravestone of Thankful Robinson, Robinson Hollow Cemetery, Chesterfield, Massachusetts. *Courtesy of Cyril Place.*

18 A Strange Superstition Was Whispered in Robinson Hollow

In chapter 12 we encountered the dramatic story of a dying girl extracting from friends "a solemn promise" that after she died, her heart would be removed and burned so that her sisters could inhale the smoke and be spared. Now we have a similar tale of a dying girl begging her family to perform this remarkable preemptive ritual. Titled "vampire vitals," it was published in the *Springfield* [MA] *Republican* (August 25, 1929), about eighty years after the described event:

> From Robinson Hollow, known later as "The Other Hollow," comes another story that has an element of weirdness. One wonders how such a superstition crept into Puritan New England. It sounds more like vampire stories from Russia. Generations ago, when "T.B.," then called consumption, had so many victims in New England communities, the children of a family were sometimes all taken by its ravages. So many of the children of the family of Dr. Robinson had succumbed to the disease that, when one of the daughters fell ill, it began to look as if all the Robinsons would go in the same way. No one knew why, in so healthful a locality, this particular family had been stricken.
>
> When a strange superstition was whispered in Robinson Hollow, it is no wonder that this girl should have given the horrible story credence. One can marvel only at the unselfish devotion to her loved ones that made her worry about them in the midst of her sufferings and be anxious to save them from her own fate. She had heard the story of the sinister work of "vitals." The story said that, when one member of a family died with consumption, his or her "vitals," meaning by that term the lungs, heart, and liver, became animated after burial and came back to earth in invisible form to prey upon the "vitals" of others in the family until they, in turn, wasted away with the same disease. The sick girl brooded over this story until she determined that her own "vitals" should never become ghoulish creatures feeding upon the life blood of nearest kindred. She had also heard of the charm, the directions for which stated that the only way to prevent such vampire work by "vitals" was to remove them from the dead before interment and burn them.
>
> Although no one in the hill towns had ever tried this charm, cremation not then being in vogue, she called the family to her bedside one day, a short time before she died, and begged them to have the doctors cut out her own "vitals," after her death, and burn them. Her parents consented reluctantly and, before her burial, her "vitals" were taken from her body and burned. There may have been too many other family "vitals" lingering about the Robinson farmhouse: for the family continued to die off with consumption until the two or three left sold the estate at auction and moved west. Out upon the open prairies there

> seems to have been no lurking place for "vitals": at least, we never heard of "T.B." attacking any more of the Robinson family.
>
> In the old family burying-ground, stands to this day the monument of the Robinson family. Around it are broken and sunken headstones. Occasionally, a visitor parts away "from the graves of old traditions, the blackberry vines," but out of the silence, nothing tells of the tragedy that, long ago, broke up a happy home, or of the devotion of the dying girl, who begged that her "vitals" might be offered in that useless charm.[544]

The article was the first in a series of ten installments, authored by Adaline M. Tirrell (1861-1937) and published under the general heading, "Legends of Hampshire Hills," that appeared in the *Springfield Republican* from 1929 to 1932. The series is comprised primarily of tales that Tirrell, a native and long-time resident of Chesterfield (the town in Hampshire County, Massachusetts, where Robinson Hollow is located), apparently heard and remembered; she noted that she was "fascinated by these stories" as she was growing up.[545] While the anecdotes of specific incidents with named characters and places give them an aura of authenticity, Tirrell cautioned that the relationship between oral tradition and history can be tenuous. Recounting a story about a house that burned, she wrote: "Many times a legend or myth grows out of something that really has happened. … [T]here is no doubt concerning the burning of the house. What was told afterwards in connection with the fire is less authentic."[546] Evidence unpacked below suggests 1847 as the probable year of the exhumation. My hunch is that the narrative began as a personal-experience story, but by the time Tirrell heard it—she was born fourteen years after the event—it may have been circulating as a local legend.

Adaline Maria Tirrell was born in Chesterfield in November 1861. She studied at Smith College, located in her native Hampshire County, for two years, then at Boston University for two more, graduating with a Bachelor of Arts. A letter of recommendation described Tirrell as "a faithful student in all departments of her work . . . who took special interest" in English literature. The letter writer noted, "Her powers of original thinking are excellent."[547] In 1895, she graduated from the Boston College of Oratory, the precursor to Emerson College,[548] then did post-graduate work in elocution at Harvard University.[549] Tirrell devoted her professional career to teaching, in New England as well as the South.[550] After retiring, she returned to Hampshire County. The 1930 federal census shows her boarding with a family in Williamsburg, Massachusetts. Tirrell, never married, died in Northampton, Massachusetts, on August 12, 1937.[551]

Robinson Hollow Cemetery ("the old family burying ground" mentioned by Tirrell) is in West Chesterfield, Hampshire County, Massachusetts. Even for the hill country of the Connecticut River Valley, it is an isolated place. But, according to Tirrell, it was not so in the early nineteenth century:

> In the early years of the preceding century, the population of the western part of Chesterfield shifted from the mount to the valley below, to Robinson Hollow, so called from a man by the name of Robinson who lived there. The Westfield river runs near the former home of Mr. Robinson, wending its way between great rocks in a little canyon whose rugged scenery . . . now draws hundreds of visitors every summer. In the palmy days of Robinson Hollow, when the place boasted of a church, school, store, and other business interests

connected with the tannaries and shoemakers' shops, there was a long bridge spanning the river at the 'Gorge.'[552]

The Robinson family monument is a tall stone obelisk, set in the center of a small, overgrown plot just off a narrow, winding asphalt road running along the Westfield River. When I visited the Robinson cemetery in the Summer of 2012, the eight visible gravestones surrounding the obelisk incorporated three generations of Robinsons: Zebulon (1770-1849) and his wife, Eunice Josslyn Robinson (1770-1850); two of their children, Roxana Robinson Dodge (1791-1850) and Asa Robinson (1795-1852); Asa's wife, Thankful Taber Wood Robinson (1798-1848); and three of Asa and Thankful's children, Mary Ann (1820-1847), Sarah (1822-1848), and Walter (1825-1846). The Robinson Hollow gravestones were transcribed by Inez Stevens Lederer and published (along with her transcriptions of other Chesterfield cemeteries) in 1940. Lederer was probably unaware that her matter-of-fact statement, "Eliza stone not found," hinted at Eliza's central role in this family's tragedy.[553]

Their gravestones are not merely mute artifacts; in context, they paint a grim picture. Bits and pieces of mundane data, including genealogical records, census tracts, and vital statistics, combine with the gravestones to reveal the progressive dissolution of a single family. In a six-year span, from 1846 to 1852, nine members of the Robinson family died; seven of those deaths were attributed to consumption, destroying virtually an entire generation.

Four of Asa and Thankful Robinson's children were the first to die of consumption: twenty-one-year-old Walter on September 22, 1846; twenty-six-year-old Mary Ann five months later, on February 6, 1847; seventeen-year-old Eunice E. (known as Eliza—the one without a gravestone), on June 28, 1847; and twenty-six-year-old Sarah (known as Sally) on September 20, 1848. Their mother, Thankful, died three months later, on December 29, 1848, at the age of fifty. A month later, Zebulon, the family patriarch, succumbed to "old age" (according to the death record) at the age of seventy-eight. His wife, Eunice, followed him to the grave in March 1850. The Federal Census Mortality Schedule for 1850 listed her cause of death as consumption, noting that she had been ill for ten years. The married daughter of Zebulon and Eunice, Roxana (known as Roxany) Dodge, died July 18, 1850, just ten days shy of her fifty-ninth birthday. Finally, Asa, died of consumption on August 10, 1852, at age fifty-seven.[554]

Vital records show that Asa and Thankful bore three other children who, like Eliza, have no visible gravestones in the Robinson Hollow Cemetery: Horace, Pauline Ellen (known as Ellen), and Augusta. Horace, born February 23, 1827, married Ann E. Warner in Chesterfield on November 18, 1851. No doubt hoping to leave the family's consumption scourge behind, the young couple relocated to Milo, Bureau County, Illinois, where a daughter, Mary A., was born about 1858. But Horace could not outrun the bacterium that almost certainly had already possessed him. He died of consumption on August 22, 1864, at the age of thirty-seven. His son, Eddy (Walter Eddie), died less than two months later at the age of nine months.[555] The fates of Pauline Ellen Robinson, born June 6, 1833, and Augusta T. Robinson, born September 27, 1840, exemplify the family curse that descended even on those who managed to escape consumption's grasp. Following the death of their father, Asa, in 1852, nineteen-year-old Ellen and twelve-year-old Augusta, bereft of both parents and four older siblings, relocated about twenty-five miles west to live in Pittsfield with their cousin, Elijah

Dodge, and his wife and young child.⁵⁵⁶ Elijah's mother (Aunt Roxany to the Robinson children) had passed away two years earlier.

When Inez Stevens Lederer noted "Eliza stone not found," she was perhaps aware that Eunice E. was known in her community as "Eliza" (a nickname for her middle name of Elizabeth)—probably because she shared the name Eunice with her grandmother.⁵⁵⁷ The daughter who begged that her "vitals" be removed and burned could have been Mary Ann, Eliza, or Sarah. Tirrell's narrative indicates that the therapeutic excision occurred around 1850. Because the grave of Eliza was the only one not found, I have a hunch it was her "vitals" that were removed and burned. I make this guess, in part, because often I have been unable to locate the grave of a suspected "vampire." I'm not introducing some supernatural element into these events. One reason the vampires and their families are difficult to track down is because many died young and unmarried and, therefore, left no direct descendants. People filling in family trees care little about dead branches.⁵⁵⁸

How should a folklorist evaluate Tirrell's writing in relation to authentic folk material? Richard Dorson, a pioneer in investigating the use of folklore by American regional writers, argued for three "principal kinds of evidence" that can "establish the relationship of a given work to folk tradition." Dorson's first kind of evidence is that "an author may be shown through *biographical evidence* to have enjoyed direct contact with oral lore."⁵⁵⁹ Tirrell's biography checks this box. Secondly, there must be "*internal evidence* in the literary composition itself, that indicates direct familiarity of the author with folklore. This evidence includes the alleged folktales, folksongs, folk sayings, or folk customs embedded in literature, as well as their settings."⁵⁶⁰ Tirrell named the local storytellers—such as the old teacher who called himself "Ye Auld Pedagogue"—and described their styles as well as the storytelling situations. She was aware of the folk process, as evidenced by her recounting of a "story based upon facts, but colored by subsequent retellings."⁵⁶¹ In another example of her familiarity with the folk process, Tirrell wrote: "The following is recalled from a conversation of elderly people, many years ago, who were indulging in reminiscences of the pastor whom they knew in childhood, an old story of a peculiar ride that Mrs. Hallock took one early morning."⁵⁶² Finally, Dorson argues that the author "must prove that the saying, tale, song, or custom inside the literary work possesses an independent traditional life. . . . [O]ur critic must present *corroborative evidence* to supplement his proofs from biographical and internal evidence."⁵⁶³ Part of the corroborative evidence for Tirrell's narratives having an "independent traditional life" includes the folk motifs that appear in her written narratives, such as the hair of one of her characters turning white overnight from fright (Motif F1041.7 - Hair turns gray from terror).⁵⁶⁴ Although Tirrell wrote that "no one in the hill towns had ever tried this charm," she acknowledged that the consumption ritual was circulating orally, writing that "a strange superstition was whispered in Robinson Hollow" and that the Robinson girl "had heard the story of the sinister work of 'vitals.'" We know now that the practice had, in Dorson's words, "an independent traditional life" in the Berkshire Hills of Massachusetts, with documented cases from Berkshire, Hampden, and Hampshire Counties. Tirrell was not alone when she pondered "how such a superstition crept into Puritan New England." Acknowledging European connections, she noted," It sounds more like vampire stories from Russia."

The Robinson family's existence in Robinson Hollow is certain. But what about the story of the young woman who extracted the promise to have her vitals removed and burned at her death? This version of the dying promise motif is so like the one from Harvard, Massachusetts—separated by about fifty years and sixty miles—that one must wonder if there is a connection, through narrative motif, if not practice. The deathbed promise made to the dying Anna Bowles that Henry Nourse wrote about in his history of Harvard, Massachusetts (chapter 12), is an extraordinary motif in New England's exhumation narratives (M250 - Promises connected with death; M251 - Dying man's promise will be kept). While the dying-promise motif is ancient and widely distributed in narratives of all categories—folk, literary, and popular—it seemed unlikely to turn up in another vampire narrative from New England. Yet, it did.

After the Robinsons had died or relocated, Robinson Hollow became known as West Chesterfield. The Town of Chesterfield was fortunate to have several residents who, throughout the nineteenth century, recorded deaths (and other matters, such as strange occurrences) in their community. These records have been preserved in the Town Clerk's safe and made available online to researchers by the Chesterfield Historical Commission. Contained in the following four documents are numerous details that flesh out the dry bones of numerical data and corroborate much that is found in Tirrell's stories:

- Record of Joseph Burnell, Jr., maintained from 1804 through 1911, consisting of an index and three parts.[565]
- Engram's Book, kept from 1834 to 1902.[566]
- Record of Deaths, 1801-1947, recorded by: Bela Stetson, 1801 through 1855; his son William Lyman Stetson, 1856 to 1872; William Bancroft, 1872 to 1892; Ellen J. "Nellie" Bancroft, 1892-1928; Mary Adelle Damon Bryant, 1929 to 1947.[567]
- Olive Cleveland Clark, *Things that I Remember at Ninety-Five: Memoir of Life on the Mount*, 1881.[568]

These memory-based records depict a community based on close social networks in which oral communication, from firsthand accounts to rumor and legends, played a prominent role in connecting people to each other and their shared past. The supernatural accounts scattered throughout the memoirs show an unobtrusive mingling between ordinary and extraordinary events. The matter-of-fact tone evident in the Chesterfield Historical Commission's documents that reported on Joseph Vinton digging up and burning the corpse of his wife serves as a good example:

- "1804, January 17, Mrs. Vinton, wife of Joseph Vinton, corpse burned by husband after burial."[569]
- "1804, January 17, Mrs. Vinton the wife Jos Vinton corpse burned after buried by husband."[570]
- "January 17, 1804 Mrs. Vintin [sic] the wife of Joseph Vinton corpse burnt after burial by husband."[571]

The same nonchalance was echoed some twenty-two years later when Joseph Vinton's death was noted:

- "February 9, 1826 old Joseph Vinton died (who dug & burned his wife's corpse)."[572]

- "February 9, 1826 Joesph Vinton dug up & burned his [illegible][death date]."[573]
- "February 9, 1826 Joseph Vintin [sic] he dug up and burnt his wife's corpse."[574]

Who was Joseph Vinton and, bearing in mind that cremation was not a common mortuary practice, why did he exhume and burn his wife's corpse? These questions remain unanswered, but I am convinced that consumption played a role.

The tales of Adaline M. Tirrell are grounded in the fate of all who live—death. Rituals provide means to cope with this rupture in the social fabric, but death's uncertainties are also bridged with stories, prosaic and strange. Tirrell, herself, met her own fate at an advanced age, in 1937, in that part of the Connecticut River Valley where she was raised, hearing and remembering tales of "weirdness" and "superstition."

※※※

Joseph R. Chandler (1792-1880) wrote for a national audience, targeting the perceived tastes of the mid-nineteenth century woman. His vampire story, set in his hometown of Kingston—a small village on the shore of Plymouth Bay, Massachusetts—was incorporated into an article in *Godey's Lady's Book* about superstitions in New England. Chandler was a young adult when the exhumation took place, opening the possibility that it came to his attention as a family-experience narrative or, perhaps, a community-based tale of several variants. As we see below, it is also possible that he was an eyewitness.

In *Food for the Dead*, I discussed a case from Plymouth County that appeared in several newspapers in 1822.[575] The narrative—which included a poem with the line, "The living was food for the dead!"—contained sufficient information to encourage further research, but not enough to achieve closure. The people in this drama were unnamed, and the article was cryptically signed with just the letter *C*. Fast forward through a decade of enormous growth in the amount of textual material electronically formatted and available online and—voilà! —we have an author: Joseph Ripley Chandler.

The same article, with Chandler's name as author, was published in *The Lady's Book* more than ten years after appearing in newspapers. Chandler's vampire narrative, like Tirrell's, was one of a series of articles featuring folklore narratives ostensibly recalled from experience. The serialization of writings grouped around a specific topic, and gathered from previously published narratives, is not unusual, especially in the case of authors who have a history of successful work. The vampire story was among Chandler's first successes. The initial installment of his series, "Superstitions of New England," appeared in Volume 6 of *The Lady's Book* (January 1833). Five installments followed in the same volume. In the third, Chandler wrote: "I derive pleasure from a remembrance of the many scenes to which the superstitions of my native village gave origin."[576] Chandler was well into this installment before he revealed that his native village was Kingston, in Plymouth County. In his fourth article, as he contrasted religious beliefs with "those unnatural impressions," Chandler seemed to be to casting for a way to explain the survival of superstitions in his community:

> In a land almost entirely influenced by religious feelings, where the mind was early imbued not only with the general principles of piety, but equally tinctured with those peculiarities of beliefs, or, perhaps, made susceptible of those unnatural impressions, which are common to a people who do not separate the ordinary occurrences of life from the extraordinary events

which characterized the first promulgation of Christianity. In such a community it is not strange that some peculiarity should distinguish the opinion held in common by its members, and it is even less to be admired that, where the mind of a single individual should by any means be rendered liable to aberration, its wanderings should be characterized by some leading feature of the general superstition.

> Education may give force to misconceived opinions, or tinge the whole with some classic hue—but custom and early imbibed creeds will finish the groundwork of public or private hallucination, and the progress of society may be almost as easily traced in the conversation and conduct of the deranged, as in the speculations and pursuits of the sane.[577]

I welcomed Chandler's opinion that the conduct of the "deranged" should not be ignored, for it compelled him to preserve the following vampire narrative:

> In that almost insulated part of the State of Massachusetts, called *Old Colony* or *Plymouth County*, and particularly in a small village adjoining the shire town, there may be found the relicks of many old customs and superstitions which would be amusing, at least to the antiquary. Among others of less serious cast, there was, fifteen years ago, one which, on account of its peculiarity and its consequence, I beg leave to mention.
>
> It is well known to those who are acquainted with that section of our country, that nearly one half of its inhabitants die of a consumption, occasioned by the chilly humidity of their atmosphere, and the long prevalence of easterly winds. The inhabitants of the village (or town as it is there called) to which I allude were peculiarly exposed to this scourge; and I have seen, at one time, one of every fifty of its inhabitants gliding down to the grave with all the certainty which characterises this insiduous [sic] foe of the human family.
>
> There was, fifteen years ago, and is perhaps at this time, an opinion prevalent among the inhabitants of this town, that the body of a person who died of a consumption, was by some supernatural means, nourished in the grave of some one living member of the family; and that during the life of this person, the body retained, in the grave, all the fullness and freshness of life and health.
>
> This belief was strengthened by the circumstance, that whole families frequently fell a prey to this terrible disease. Of one large family in this town consisting of fourteen children, and their venerable parents, the mother & the youngest son only remained—the rest within a year of each other had died of the consumption.
>
> Within two months from the death of the thirteenth child, an amiable girl of about 16 years of age, the bloom, which characterised the whole of this family, was seen to fade from the cheek of the last support of the heartsmitten mother, and his broad flat chest was occasionally convulsed by that powerful deep cough which attends the consumption in our Atlantic States.
>
> At this time as if to snatch one of this family from an early grave, it was resolved by a few of the inhabitants of the village to test the truth of this tradition which I have mentioned, and, which the circumstances of this afflicted family seemed to confirm. I should have added that it was believed that if the body thus supernaturally nourished in the grave, should be raised and turned over in the coffin, its depredation upon the survivor

would necessarily cease. The consent of the mother being obtained, it was agreed that four persons, attended by the surviving and complaining brother should, at sunrise the next day dig up the remains of the last buried sister. At the appointed hour they attended in the burying yard, and having with much exertion removed the earth, they raised the coffin upon the ground; then, displacing the flat lid, they lifted the covering from her face, and discovered what they had indeed anticipated, but dreaded to declare.—Yes, I saw the visage of one who had been long the tenant of a silent grave, lit up with the brilliancy of youthful health. The cheek was full to dimpling, and a rich profusion of hair shaded her cold forehead, and while some of its richest curls floated upon her unconscious breast. The large blue eye had scarcely lost its brilliancy, and the livid fullness of her lips seemed almost to say, "loose me and let me go."

In two weeks the brother, shocked with the spectacle he had witnessed, sunk under his disease. The mother survived scarcely a year, and the long range of sixteen graves, is pointed out to the stranger as an evidence of the truth of the belief of the inhabitants.

The following lines were written on a recollection of the above shocking scene:

> I saw her, the grave sheet was round her,
>   Months had passed since they laid her in clay;
> Yet the damps of the tomb could not wound her,
>   The worms had not seized on their prey.
>
> O, fair was her cheek, as I knew it.
>   When the rose all its colours there brought;
> And that eye,—did a tear then bedew it?
>   It gleam'd like the herald of thought.
>
> She bloom'd, though the shroud was around her,
>   Her locks o'er her cold bosom wave,
> As if the stern monarch had crown'd her,
>   The fair speechless queen of the grave.
>
> But what lends the grave such a lusture?
>   O'er her cheeks what such beauty had shed?
> *His* life blood, who bent there, had nur'st her,
>   *The living was food for the dead!*[578]

Joseph Ripley Chandler was born August 22, 1792, in Kingston, "a small village adjoining in the shire town" of Plymouth. After attending "the common schools" and doing commercial work in

Boston, he relocated to Philadelphia, in 1815. There, he founded a seminary for young ladies and, from 1822 to 1847, edited the *United States Gazette*, a periodical that had merged with *The Union*. He served as a city councilman in Philadelphia for sixteen years, and then was elected to Congress as a member of the Whig party, where he served from 1849 to 1855. For two years, he was Minister to the Two Sicilies, appointed by President Buchanan in 1858. He also took an active interest in prison reform and was president of the Board of Directors of Girard College.[579] In their history of Philadelphia, Scharf and Westcott provide a glimpse of Chandler's initial success in writing and editing, noting that "the fortunes of *The Union* were at a low ebb, when in 1822, Joseph R. Chandler offered a story for publication in its columns. The acceptance of this story proved to be the starting-point of the subsequent great success of the paper." Readers apparently viewed Chandler's writing to be "fresh and vigorous," leading to his editorship.[580] Chandler's first story in *The Union*, which attracted such favorable attention, likely was the Old Colony vampire tale, entitled, "Superstitions of New England."

Knowing that this incident probably occurred in Kingston narrowed the search for graves considerably: the sole possibility seemed to be the Ancient Burial Ground at Kingston. Fortunately, the gravestones in this cemetery had been transcribed in 1895, and these transcriptions are available on-line, organized by surname. Unfortunately, none of the family groups showed "the long range of sixteen graves," dating about 1807, that should have been there if the story was genuine and I had, indeed, found the right cemetery. A check of the death records for Kingston, conveniently organized by surname, also failed to identify a family that corresponded to the narrative. For now, one may add this to the list of cases only partially solved.[581]

A physician from Plymouth took issue with some of Chandler's statements in his *Old Colony Memorial* article. The next issue of this weekly newspaper (May 11, 1822) included the following "communication" from the unnamed physician:

> Sir—It is highly gratifying to observe, that your first essay towards the establishment of the *Old Colony Memorial* is so auspicious to its eventual success. Your first number, as respects the quality of the paper, the type and the matter, cannot, in my opinion, fail of proving satisfactory to every reader. If any portion of its contents can be deemed objectionable, I should point to the communication taken from the Philadelphia Union, headed "Superstitions of New England." The writer indulges his imagination in ranting about the superstitious customs which specially prevail in the Old Colony, or Plymouth County, and this he fantastically locates an insulated part of the State of Massachusetts. His first assertion is too extravagant to require refutation. If true it would imply a phenomenon which has never occurred in any part of the habitable world, "that nearly one half of the people die of consumption." Nor will it be credited, that one of fifty of the inhabitants fall a prey to this inexorable disease. It is impossible to conceive the motive by which the writer could be actuated in advancing a position so glaringly preposterous.
>
> The fact, however, is notorious that consumption in all countries forms one of the most crowded avenues to the tomb, but it is not admitted that the disease is peculiarly destructive in the district. Had the writer been ingenious enough to have consulted the bills of mortality, he might have informed his readers, that the section of Massachusetts in question will bear comparison, in point of exemption from consumptive cases, with any of

our Atlantic situations, not excepting New York, or the more salubrious southern regions of our Union. It has been calculated that about one seventh of all the deaths in the State, is to be attributed to this fatal malady. A medical friend, in whose accuracy I have the most perfect confidence, has, with laudable views, devoted a share of attention to the investigation of this subject for several years past. The result of his estimation is, that the average proportion of consumptive cases is about one to every five and a half deaths. This estimate applies to New England and the intermediate States, extending to the Carolinas. Instances, it must be conceded, too frequently occur of whole families having a constitutional predisposition to consumption, by which parents are bereft of children in the early periods of life. But the writer in the Philadelphia Union, cites one instance of a family in this vicinity consisting of sixteen persons, all of whom were victims of consumption. The number specified, as well as his fanciful ideas relative to the superstitious belief in the salutary effects to be derived from touching the entombed corpse, bear evident marks of great exaggeration. During a residence of nearly forty years in the district referred to, and favoured with opportunities of correct observation respecting this subject, the writer of this reply, has not been made acquainted, with but one solitary instance of raising the body of the dead for the benefit of the living: and this was done purely in compliance with the caprice of a surviving sister, reduced to the last stage of hectic debility and despair. Although the family and connections entertained not the smallest hope of beneficial consequences, they could not in duty and tenderness refuse to indulge a feeble minded and debilitated young woman in a mean, on which she had confided her last fallacious hope. Plymouth, May 10, 1822. A Physician.[582]

In his argument denying the extent of the vampire tradition—accusing Chandler of "great exaggeration"—the physician ironically boosts the counterargument by adding yet another vampire incident. Internal evidence in his text indicates that the "raising of the body for the benefit of the living" must have taken place sometime between 1782 and 1822, a convenient midpoint being about the turn of the century. So, add another to the list of "at this time dead end" cases.

The physician had a good argument against some of Chandler's details, particularly his fifty-percent figure for consumption deaths, which seems much too high. Even with a generous extension of consumption's scope to include other respiratory ailments as a cause of death, the database of Kingston deaths prior to 1860 does not support that number.

19   Sarah Came Every Night, Causing Great Pain and Misery

President Harry Truman is supposed to have said, "The only new thing in the world is the history we have not learned." George Alfred Hazen wrote that, until the spring of 1950, he "was unaware of any stories about vampires in New York State."[583] No shame is attached to Hazen's unfamiliarity with New York's vampire history, consisting of more than a dozen incidents beginning in the late 1700s. Vampire incidents anywhere in New England were frequently characterized as unprecedented. We don't know why Hazen, a native of upstate New York and well-educated teacher, decided to share, verbatim, John Corbett's vampire story in the *New York Folklore Quarterly* more than fifty years after its initial publication.[584] In his lean commentary, Hazen writes that "the details of this story are reminiscent of" European vampire lore, which argues that he was unaware of New England's own tradition.

When John Weldon Corbett (1854-1919) wrote about the vampire of Seneca Lake, in his 1898 book, *The Lake Country: An Annal of Olden Days in Central New York*, his relationship to that community was both personal and professional. He was born in Reading, a small town in Schuyler County, which incorporates the southern portion of Seneca Lake. He eventually sold a large portion of his family's farm, located along the western shore of the lake, and devoted himself to editing, publishing, and writing for newspapers. Corbett did not provide a source for the following:

> The superstition of the vampire, that horror of the grave which was supposed to harbor the dead yet derive its sustenance from the living, had one illustration at least about Seneca Lake. Down the western shore not many miles from its head, in the early years the corpse of a young woman was exhumed, and the heart and other vital parts committed to the flames. The grewsome tale comports in a remarkable manner with the general sayings in regard to vampires. Of several sisters, all in succession had wasted away, until one remained and she was ill. Though in the grave for many months, the burned portions of the body were fresh in appearance. The living sister, undoubtedly from mental relief, recovered her health after the event.[585]

In the book's dedication, Corbett evokes his ties to the Lake Country:

> The Annal of Olden Days is the outgrowth of several years of endeavor as a newspaper writer in the local field of the Lake Country, and the research for the facts presented was pursued with great care and diligent application. The work portrays the period of the pioneers of Central New York, and is designed to be a correct chronicle of the time. To those whose eighty years have been passed amid the scenes depicted, and another whose life has been about the lakes, these sketches are inscribed—Otis R. Corbett and Adelia B. Corbett, the parents of the author.[586]

Otis R. Corbett (1818-1908) was the beneficiary of a glowing biography in 1908; it stated that John, one of Otis and Adelia's ten children, had "somewhat of a reputation as a newspaper writer on local historical subjects and is now with the Elmira Advertiser."[587] The biographer wrote,

"From the heights of Seneca, near the family dwelling, the shores and surfaces of this peerless lake are visible for a score of miles."[588]

Near the end of his book, Corbett offers clues about sources: "The Topics of the foregoing pages are neither treated at length nor in attempt at exhaustive consideration. Each subject could be amplified, but the intent of the work is rather of a cyclopedic character than an extended narration of the past. In gathering the facts presented, gazetteers, session laws, local histories, newspaper files and old residents' recollections have been consulted."[589] Most of New England's consumption narratives eventually turn up in popular periodicals. Given Corbett's ties to the newspaper business, it seems likely that he may have discovered the Seneca vampire story in one of the newspaper files he consulted. Even so, directly or indirectly, the story probably was rooted in "old residents' recollections."

Corbett's chapters on folklore and religion illustrate why that region of New York was labeled the "burned-over district," as one cult after another took root, flourished, and then vanished. Here is how Corbett expressed it:

> The vagaries of religious belief have had striking illustrations in Central New York, not however to prosper long at the place of inception. The Friends whose deeds about Lake Keuka Outlet are now ancient annals, had faith that Jemima Wilkinson was controlled by the Divine Spirit in propagating the tenet that celibacy was indispensable to a pure life. Mormon Hill near the north line of Ontario County, is the pretended place of discovery by Joseph Smith in 1827, of the golden plates of the Book of Mormon, and Brigham Young, after living for a time west of the head of Seneca Lake, resided long at Canandaigua. The Oneida Community, established by John H. Noyes in 1847, held all things in common up to 1879, when their peculiar family relations were abandoned.[590]

This district accommodated magical thinking. It would be no surprise if Jemima Wilkinson and her followers, who settled in the vicinity of Corbett's vampire story, knew about, and perhaps even performed, the consumption ritual.

Jemima Wilkinson was born in 1752 to Quaker parents in Cumberland, Rhode Island, a town that, in 1796, officially gave Stephen Staples permission to exhume the body of his daughter "in order to try an experiment" that he hoped would spare another daughter, dying of consumption.[591] Could she and her followers—many also from Rhode Island—have taken the vampire tradition with them to Central New York? According to Harold Wisbey, one of her several biographers, Wilkinson became ill in 1776 while she was "troubled by religious conflicts." "During the course of her sickness," Wisbey wrote, "she experienced a vision that convinced her that she had died, gone to heaven, and had been sent back to preach to a dying and sinful world. She rose from her sick bed declaring that Jemima Wilkinson was dead and that she was the 'Publick Universal Friend' whose mission was to work for the salvation of sinners." Until her death forty-three years later, "this resolute woman never faltered in her role of a reborn spirit called to preach to an unregenerate world."[592]

By all accounts, Wilkinson was charismatic. Even detractors acknowledged her good looks and speaking ability. Wisbey noted that she "always spoke extemporaneously, as the spirit moved her, using few gestures and seldom raising her voice. Yet she attracted large crowds in Rhode Island,

Massachusetts, Connecticut, and even in the city of Philadelphia."[593] Wilkinson left divided communities in her wake, as few were neutral regarding the "Publick Universal Friend." Her followers adored her while the rest designated her a fraud. Eventually, in 1794, she and her devotees founded a settlement, dubbed New Jerusalem, just west of Seneca Lake. In 1909, when Corbett was editor of the *Schuyler County Chronicle*, he wrote a story about how the Town of Reading was greatly reduced in size, noting that at one time, its northern boundary was "the southern line of the Settlement of the Friends, as were called the followers of Jemima Wilkinson."[594]

Naturally, oral traditions, both laudatory and derogatory, surround Wilkinson. Some legends recount the many gifts attributed to her, including the ability to foretell future events, heal the sick, see into people's hearts, raise the dead, and even walk on water. As we saw in chapter 8, one of the negative legends about Wilkinson circulated among the Rose family of southern Rhode Island, a branch of which had its own vampire problem in 1872. Once again, Wisbey is our source:

> A story that has never appeared in print before was told for many years in the Rose family of South Kingstown, Rhode Island. It was recounted many years ago by a member of the family who lived not far from the site of the old Potter mansion in Kingston. Mrs. Rose, the story went, had some suspicion that Jemima was interested in her husband, and when Jemima sent her a cake she did not eat it but put it on a high shelf. One of the children got to it, ate some, and became sick (or died in another version). When the story teller, at least five generations from the event, was questioned whether the sickness might not have been from some other cause, she replied with great feeling, "I tell you that cake was poisoned!"[595]

Was Jemima Wilkinson sent by God to prepare mankind for the impending rapture? Or was she delusional, genuinely believing that her body had become animated by "the Spirit of Life"? Or was she merely a fraud who used her attractive features and extemporaneous speaking ability to create a community of followers eager to do her bidding? The only person who can truly answer these questions has yet to return. We are also left to wonder if she or her followers from Cumberland, Rhode Island, introduced the consumption ritual to Central New York.

***

Casey B. Tyler (1819-1899) directed the major portion of his historical writings to the residents of the Pawtuxet Valley in western Rhode Island. He wrote as though readers knew and cared about the people, places, and incidents that he related in generous detail. His serialized "Reminiscences of Foster"—first published in the *Providence Journal* (seventeen installments from 1858-1859) and then republished in the *Pawtuxet Valley Gleaner* twenty years later—included his story of about the exhumation of Nancy Young in 1827 (*Pawtuxet Valley Gleaner*, June 22, 1878):

> The Gleaner to-day presents its readers with the fourth paper on the reminiscences of Foster, as originally published.
>
> Sixty years ago, or more, Capt. Levi Young of Sterling, Connecticut., who married Anna Perkins, bought the extreme southern portion of the original "Dorrance purchase" and erected a house thereon, which is now the southwest corner house in the town of Foster, and commenced life as a farmer. His oldest daughter, Nancy, a very bright and intelligent girl, at an early age became feeble in health and died of consumption on April 6th, 1827, aged 19 years. Previous to the death of Nancy, the second daughter, Almira, a very

sprightly girl, commenced a rapid decline in health with sure indications that she must soon follow her sister. The best skill of the most eminent physicians seemed to be all in vain. There was a large family of children and several of them were declining in the same manner. Mr. Young was a very worthy and pious man and wished to do everything possible to benefit his family, and he had the sympathy of all his friends and neighbors.

There seemed to be a curious idea prevailing at that time, in some localities, that by cremating or burning the remains of a departed friend or relative, while the living relatives stood around and inhaled the smoke from the burning remains, that it would eradicate the disease from the systems of the living and restore them to health.

A short time after the decease of Nancy, in the summer of 1827, the neighbors and friends, at Mr. Young's request, came together and exhumed the remains of Nancy and had her body burned, while all the members of the family gathered around and inhaled the smoke from the burning remains, feeling confident, not doubt, that it would restore them to health and prevent any more of them falling a prey to that dread disease, consumption. But it would seem that it was no benefit to them, as Almira died August 19th, 1828, aged 17 years. Olney, a son, died December 12th, 1831, aged 29 years. Huldah died August 26th, 1836, aged 23 years. Caleb died May 8th, 1843, aged 26 years, and Hiram died February 17th, 1854, aged 35 years. Two other sons lived to be older but are now dead. The youngest daughter, Sarah, is the only one now living of the family. She seems as yet to escape the disease of consumption.

Some scientific persons thought perhaps the water in the well contained impurities which caused the disease, as the whole family were of exemplary habits and very much respected; but it seems to have not been so, as no disease of that kind has visited the people who have since occupied the same premises. Possibly this [is]the only instance of cremation in Rhode Island for the purpose of curing or preventing disease.

Kc.

When the story was republished in the *Gleaner* (October 14, 1892) fourteen years later, Tyler updated it with the following: "Sarah is now dead but lived many years after removing from the old place. Dwight B. Jenks now owns and occupies the farm and has done so for years with a healthy family and no death in the house to the knowledge of the writer since the Young family passed away."

The dates and ages at death provided by Tyler match those on the family gravestones in Foster Historic Cemetery Number 142; Tyler likely visited the cemetery and transcribed the dates directly from the stones. This cemetery—appropriately named the "Young Lot"—contains only nine graves, eight of which are the nuclear family of Levi Young and his wife, Anna Perkins Young.

Fewer than 3,000 people lived in Foster at the time of the exhumation, so it's likely that seven-year-old Tyler learned of Nancy Young's exhumation through word-of-mouth communication not long after the event. A possible source is Hiram Young, the last of Levi and Anna's children to be interred in the Young Lot. Tyler and Hiram, who died in 1854, were cohorts (both born in the year 1819) and close neighbors (the Tylers and Youngs are two of eleven families listed on the same page in the 1850 Federal Census for Foster[596]).

Casey Tyler was born, raised, worked, and died in Foster, Rhode Island. Even though he earned his living as a storekeeper, he was proficient enough at educating himself that he taught school. Tyler's true passion was the history of his hometown. At his death, at age eighty on March 24, 1899, his large collection of books, many concerned with history, was donated to Foster, and became the centerpiece of the town's public library, appropriately named the Tyler Free Library.[597]

***

The story of Sarah Tillinghast's exhumation, published by Sidney S. Rider (1833-1917) in his biweekly journal, *Book Notes*, bears the marks of an experienced storyteller. When juxtaposed to the known historical facts of the event—which occurred in Exeter, Rhode Island, at the end of the eighteenth century—however, Rider's tale appears to have been reshaped; this vampire behaves like she crept out of a nineteen-century novel, and some of the marvelous motifs seem *too* marvelous:

> At the breaking out of the Revolution there dwelt in one of the remoter Rhode Island towns a young man whom we will call Stukeley. He married an excellent woman and settled down in life as a farmer. Industrious, prudent, thrifty, he accumulated a handsome property for a man in his station in life, and comparable to his surroundings. In his family he had likewise prospered, for Mrs. Stukeley meantime had not been idle, having presented her worthy spouse with fourteen children. Numerous and happy were the Stukeley family, and proud was the sire as he rode about the town on his excellent horse, and attired in his homespun jacket of butternut brown, a species of garment which he much affected. So much, indeed, did he affect it that a sobriquet was given him by the townspeople. It grew out of the brown color of his coats. Snuffy Stuke they called him, and by that name he lived, and by it died.
>
> For many years all things worked well with Snuffy Stuke. His sons and daughters developed finely until some of them had reached the age of man or womanhood. The eldest was a comely daughter, Sarah. One night Snuffy Stuke dreamed a dream, which, when he remembered in the morning, gave him no end of worriment. He dreamed that he possessed a fine orchard, as in truth he did, and that exactly half the trees in it died. The occult meaning hidden in this revelation was beyond the comprehension of Snuffy Stuke, and that was what gave worry to him. Events, however, developed rapidly, and Snuffy Stuke was not kept long in suspense as to the meaning of his singular dream. Sarah, the eldest child, sickened, and her malady, developing into a quick consumption, hurried her into her grave. Sarah was laid away in the family burying ground, and quiet came again to the Stukeley family. But quiet came not to Stukeley. His apprehensions were not buried in the grave of Sarah.
>
> His unquiet quiet was but of short duration, for soon a second daughter was taken ill precisely as Sarah had been, and as quickly was hurried to the grave. But in the second case there was one symptom or complaint of a startling character, and which was not present in the first case. This was the continual complaint that Sarah came every night and sat upon some portion of the body, causing great pain and misery. So it went on. One after another sickened and died until six were dead, and the seventh, a son, was taken ill. The mother also now complained of these nightly visits of Sarah. These same characteristics were present in every case after the first one. Consternation confronted the stricken household. Evidently something must be done, and that, too, right quickly, to save the remnant of this family. A

consultation was called with the most learned people, and it was resolved to exhume the bodies of the six dead children. Their hearts were then to be cut from their bodies and burned upon a rock in front of the house. The neighbors were called in to assist in the lugubrious enterprise. There were the Wilcoxes, the Reynoldses, the Whitfords, the Mooneys, the Gardners, and others. With pick and spade the graves were soon opened, and the six bodies were found to be far advanced in the stages of decomposition. These were the last of the children who had died. But the first, the body of Sarah, was found to be in a very remarkable condition. The eyes were opened and fixed. The hair and nails had grown, and the heart and the arteries were filled with fresh red blood. It was clear at once to these astonished people that the cause of their trouble lay there before them. All the conditions of the vampire were present in the corpse of Sarah, the first that had died, and against whom all the others had so bitterly complained. So her heart was removed and carried to the designated rock, and there solemnly burned. This being done, the mutilated bodies were returned to their respective graves and covered. Peace then came to this afflicted family, but not, however, until a seventh victim had been demanded. Thus was the dream of Stukeley fulfilled. No longer did the nightly visits of Sarah afflict his wife, who soon regained her health. The seventh victim was a son, a promising young farmer, who had married and lived upon a farm adjoining. He was too far gone when the burning of Sarah's heart took place to recover.

> The conditions here narrated are precisely similar to those alleged to have taken place in the Danubian provinces and the remedy applied the same. But in those countries certain religious rites were observed, and occasionally, instead of burning a part or the whole of a body, a nail was driven through the centre of the forehead. At the period when this event took place, religious rites were things but little known to the actors in the scene, and fire in their hands was quite as effective an agent as an iron nail. Those from whom these facts were obtained little suspected the foreign character of the origin of the extraordinary circumstances which they described; but extraordinary as they are, there are nevertheless those still living who religiously believe in them.[598]

At this point, Rider noted that he had sent the story to the *Providence Journal* for their consideration, but it was rejected for being "too sensational." Defending the integrity of his article, Rider then cited two additional vampire examples, one, in Wakefield, apparently already undertaken, and the other under consideration:

> Regardless of the opinion of the *Journal*, I maintain that the tale which I have told deserves and will receive the consideration of thinking men. It details an extraordinary belief, considered in connection with the supposed enlightened intelligence of the American people. Since it was written another similar case in Wakefield, Rhode Island, has come to my knowledge; and still another now is in contemplation in a family of respectable surroundings, several of the members of which have recently died. Such delusions ought to be obliterated, and the way to obliterate them is to expose them to the light of reason and to educate men to better beliefs.[599]

Was the performed ritual that of William R. Rose's family (the cemetery was near Wakefield)? Was the one "in contemplation" that of George Brown's family? Rider's view of "such delusions," however, is not an open question.

As to the extraordinary motifs that do not fit the traditional New England pattern, some are more suited to the "classic" vampire, while others are more characteristic of folktales (fairy tales) than legends. The former includes Sarah as an actual revenant, returning from the grave in corporeal form to harass her family during the night; that she was the first, not the last, to die; and that, upon her exhumation, the signs of her undead status included the growth of her fingernails and hair. Perhaps the most striking folktale motif in Rider's narrative was Stukeley's prophetic dream. Many folktales include the motif "future revealed in dream" (D1812.3.3). A logical inconsistency reinforces the conclusion that Sarah's nocturnal visits is an intrusive motif. If Sarah's siblings complained of her visiting them at night—sitting "upon some portion of the body, causing great pain and misery"—then why were the corpses of *all* six of the deceased siblings exhumed and examined? Clearly Sarah was the vampire. Absent these nonconforming (undeniably startling) motifs, Rider's narrative presents a credible instance of New England's vampire tradition.

Sidney Rider was a prolific chronicler of Rhode Island history, yet his personal history is elusive, as Erik J. Chaput and Russell J. DeSimone noted in their article, "Sidney Rider and the Business of Rhode Island History":

> For those who have set out to write about Rhode Island's rich history, Rider is a familiar name. The size of the 'Sidney Rider Collection' at Brown University's John Hay Library is extensive, often overwhelming those who set out to sift through it. Researchers will encounter more than 15,000 items including books, pamphlets, manuscripts, and newspaper clippings chronicling the founding of the colony in 1636 to the post-Civil War era. However, while researchers may spend months with this material, most know nothing about the man who assembled it.[600]

As I recounted in *Food for the Dead*, I spent some time with the Rider Collection at the John Hay Library in early 1983, attempting to get to the bottom of Rider's astonishing narrative.[601] In a small, spiral-bound notebook labelled "Exeter Notes," Rider's handwritten entry provides the family name of Tillinghast and a one-sentence summary of the event: "Story of Stukeley Tillinghast & his digging up his six dead children—Snuffy Stuke." The notebook also includes pasted-in newspaper clippings. Did Rider hear the story directly from an informant, or did he find it in print, possibly in a newspaper? I lean toward the former, based on Rider's statement near the close of his narrative: "Those from whom these facts were obtained little suspected the foreign character of the origin of the extraordinary circumstances which they describe."[602] Reinforcing this conclusion is that the only mention of this story in Rider's notebook is handwritten. If he had taken it from a published source, such as a newspaper, he probably would have pasted it into the notebook, as he did with other printed articles. In any event, somewhere in the chain of this story's transmission it was apparently "improved" by the addition of Snuffy's dream that half of his orchard died. For the dream to be entirely prophetic, half of Snuffy's fourteen children had to die. Yet, my research showed that no more than five of the fourteen Tillinghast children died during the short time span.[603]

Apart from the several out-of-place narrative motifs, the structure of Rider's vampire essay follows a conventional pattern: (1) definition of vampire; (2) historical and geographical distribution of the belief and practice; (3) the local narrative under consideration; (4) speculation concerning how the tradition came to New England; (5) an attempt to contextualize the incident and summarize its meaning. Rider identified two European vampire traditions. The first was an earlier form that originated in Eastern Europe, which, today, we might view as the "classic" vampire who leaves the grave at night to suck the warm blood of living persons; the second was the werewolf tradition. According to Rider, the first form "came to this country, and seems to have been prevalent at one time here in Rhode Island. In fact, it may even at this day be held in her remote regions, if, indeed, that term be not inapplicable within the narrow confines of this little State. Strange, even incredible is it that anybody should believe in such absurd superstitions. It is true, nevertheless. There were, and there are now, those who do believe them, and the purpose of this paper is to narrate a case which took place here in Rhode Island at no very remote period. It was of a genuine vampire.[604]

Sydney Rider was born in Rensselaer County, New York, but relocated to Providence with his family in 1845 at the age of twelve. He began working as an apprentice in a Providence bookstore that same year, and, in his twenties, became a partner in his own bookselling business, eventually assuming sole proprietorship. After the Civil War, he began publishing pamphlets on Rhode Island history, which led to his establishment of *Book Notes* in 1883. Its masthead described the fortnightly publication as "consisting of literary gossip, criticisms of books and local historical matters connected with Rhode Island."

Was Rider the kind of author who would distort the historical veracity of a narrative for the sake of creating a story that would appeal to a wider audience and, thus, generate more income? Or was he too much of a stickler for facts to deliberately change them? A strong argument can be made for the latter. When he was awarded an honorary Master of Arts degree from Brown University in 1880, he received the following acknowledgement:

> Mr. Rider, though lacking in early life the advantages of a classical education, has by patient labor through a long series of years, become one of the most accurate students in the State in matters pertaining to Rhode Island. As the writer of many historical articles in the current newspapers of the day; as publisher of the series of Rhode Island Historical Tracts; as well as the Editor, or author of many memorial sketches of prominent Rhode Islanders, Mr. Rider has added not a little to the historical literature of the State.[605]

Reinforcing the case for Rider's adherence to facts is this notice that appeared in the *Boston Globe*:

> Mr. Sidney S. Rider, whose mind is a cyclopedia for everything connected with the past in Rhode Island, is continually on the lookout for historical articles in the daily papers, and woe betide the unlucky wight who slips up on his facts; many a delightful romantic story has he spoiled by knocking the bottom completely out of it with his cold facts. ... He is terror to reporters and to others who write historical articles, ... (but he) has secured to the newspaper reader more accurate bits of local history."[606]

Despite Rider's reputation for assembling hard facts, he failed to disclose "those from whom these fact were attained," so we may never know who told the vampire story as it was related by Rider.

Part IX   Authors

Storytellers have deployed elements of the vampire ritual in many contexts, from personal correspondence and local histories to newspapers and medical journals. A ritual's script is created by people performing it over time in different communities, which creates variability, a hallmark of folklore. Yet, folk traditions also have boundaries that mark the limits of acceptable variation. Afflicted families were under pressure to perform the ritual as prescribed in their communities, for they hoped it would succeed. When these boundaries are violated in a story, its authenticity is called into question. Fortunately, if an author got it wrong, no one (actually) died.

The nine authors who appear in this part range from an anonymous newspaper storyteller of the early nineteenth century to a late twentieth-century writer who crafted tales of history and mystery. Some have been celebrated and others have all but vanished from the literary landscape. A portrayal of New England's vampire tradition, with varying degrees of authenticity, is their commonality.

The three vampire stories penned by Francis Gerry Fairfield (1836-1887), published within a five-year span, tread between fact and fiction. Yet all have core elements that accurately portray the New England consumption ritual. Fairfield was born and raised in the town of West Stafford, Connecticut, where his vampires lived and died—and died again. His ancestors were tethered to the land, built into the landscape. Trees and houses had memories, but as we shall see, not all were tranquil.

Like Fairfield, Walter Prichard Eaton (1878-1957) wrote articles for literary journals and popular periodicals. These vampire narratives inhabit the cusp between nonfiction and fiction, although their presentation is not self-consciously fictional. Both authors used local knowledge in their place-based narratives. How that esoteric understanding is acquired, however, makes all the difference. Fairfield's writing about West Stafford is imbued with an intimate understanding of the place. Eaton's travel article discloses a casual acquaintance with South County, Rhode Island, that undermines his vampire narrative's credibility. But storytelling is not always an ally of history or fact. Narrative often subverts one truth while advancing another. Sorting this out is an ongoing preoccupation.

When Owen Wister (1860-1938) inserted a vampire story into his 1915 essay, "Superstition and the Doctor," he was a celebrated author, having written the best-selling, quintessential Western novel, *The Virginian* (1902). How he came to learn of a consumption ritual in Rhode Island is a fascinating and intricate tale of its own, with a cast of familiar characters, including Dr. Harold Metcalf and Edward Everett Hale.

An early, anonymous, piece of romantic fiction, published in a Rhode Island newspaper in 1829, illustrates the jarring disjunction of a consumption narrative when its core ritual disregards traditional boundaries. Although the vampire story of Mrs. S. E. Farley (ca. 1808-1880) was written in memoir format, its veracity is suspect. Most of the details of the ritual, itself, have credible standing in the New England tradition; context rather than content betrays this incident, as well as many of the other "popular prejudices" that she claims were practiced by her "comparatively cultivated and enlightened" neighbors. If her story is indeed fiction, it is, as far as I am aware, the first to use New England's *authentic* vampire tradition. Moreover, in 1848, she was the first to affix the vampire label.

Mary A. Denison (1826-1911) wrote some eighty popular novels that sold more than a million copies, including her best-selling, *That Husband of Mine* (1877). Her work touched upon the diverse issues of the day, including slavery, temperance, and class antagonism, with unexpected twists and turns. Her third novel, *Home Pictures* (1853), was, in the words of her editor (who was also her husband), "a series of Pictures of Home, as supposed to be viewed by a country girl, . . . sketched by her own hand from real life." In the chapter, "Old Superstition," Denison, like Farley, places her vampire narrative in the mouth of a neighbor: "One learns many a curious little thing in a village like this. I listened to the narration of a most singular incident yesterday at the house of a neighbor." Whether it is entirely fictional or based on an actual event is not evident, but every motif in this story accords with the New England vampire tradition. Denison undoubtedly had heard of or read about at least one of these exhumations.

Amy Lowell (1874-1925) was raised in a family of wealth and fame in the Boston area. Educated at home and in exclusive schools, she confessed that she knew little of the rural Yankee culture featured in her lengthy vampire poem, "A Dracula of the Hills" (1925). In a letter to the editor of *Century Magazine*, Lowell wrote that she learned of the exhumation ritual in the *Journal of American Folklore*, referring almost certainly to the article by Jeremiah Curtin that was published in that journal in 1889 (see chapter 15). Although Dracula is featured in the poem's title, Lowell's story within a story is firmly situated in the hills of western Massachusetts and unfolds according to the tradition as practiced in New England, not Europe.

While the words *consumption* and *vampire* never appear in Edith Wharton's (1862-1937) "Bewitched" (1926), her short story is almost certainly based on the New England ritual to stop the spread of consumption by counteracting the power of an undead kinsman. A troublesome corpse in this story, however, is dispatched by driving a stake through its breast, as if the mountain folks in this isolated New England backwater were living in Eastern Europe.

"Chastel" (1979), a short story by Manly Wade Wellman (1903-1986), takes the 1854 exhumations of the Ray family, in Jewett City, Connecticut (see chapter 6) into the late twentieth century. The narrative unfolds as readers learn that a descendant of that family remains undead. Following a now-familiar path, Wellman weds New England's authentic vampire tradition to the classic vampires of Gothic fiction and popular culture. Wellman's impressive knowledge of New England's vampires was on display in his nonfiction essay, "The Vampire in America," first published in 1973. Yet, in "Chastel," the vampire is corporeal: rising from her tomb, she travels through decades to infect others and finally must be destroyed with a spike through her heart.

Gravestones of brothers Lemuel B. Ray and Elisha H. Ray, Jewett City Cemetery, Griswold, Connecticut. *Courtesy of Cyril Place.*

20   The Dead Gnawing and Feeding upon the Vitality of the Living

The consumption ritual commemorated by Francis Gerry Fairfield's (1836-1887) vampire narrative, published in 1876, may have occurred during his lifetime, but his contextual commentary sheds little light on the story's historical provenance: it was a "strange tale," he wrote, that happened "many years ago." Fairfield's tale was published in *Appleton's Journal*, a short-lived (1876-1881) but popular national monthly focused on literature, science, and the arts. He was interpreting the local traditions of his small, rural hometown of West Stafford, Connecticut, for a national—perhaps international—readership.

The exhumation in West Stafford—which I first encountered in J. R. Cole's 1888 history of Tolland County—was difficult to document. My several visits to the town's three cemeteries failed to reveal gravestones whose progression of family deaths matched the description related by Cole. Lacking a surname blocked the investigation, which, without the intervention of startling new evidence, had reached a dead end. Then, years later, I found a variant of Cole's text. In 1915, the *Hartford Courant* (March 28) published an article entitled "Tragic legend of West Stafford Cemetery." A photograph above the article showed oblique rows of gravestones along a hilltop cemetery. The subscript read: "In this old cemetery the weird ceremony is said to have taken place." The text of this narrative was labelled "special to the Courant":

> On the hill north of the village of West Stafford is an old cemetery well worth visiting. It is the burial place of many of the early settlers of the town who, following the custom of early times, selected an elevated situation on which to build their homes, and the hilltop became the seat of a thriving village. In the olden days two churches stood nearby and the county turnpike passed the place. Near at hand a tavern did a thriving business and the stage drivers changed their horses there. On training days the state militia assembled here and the place was the center of the social activities of that section of the town.
>
> With the passing of years, villages sprang up in the valleys and the churches were moved away, one to West Stafford and the other to Stafford Hollow. The tavern long since closed has been torn down and a farmhouse stands on its former site. Some of the old homesteads have been burned and others have gone to decay, until the old cemetery is all that is left to remind one that once there was life here.
>
> A strange tale is told of a tragic scene enacted in this old cemetery many years ago. Of a family of six sisters living in the village, five had died of consumption. The sixth seemed doomed to follow the others. There was an old superstition in such cases that the vital organs of the dead still retain a flicker of vitality and by some strange process absorb the vital forces of the living. Instances were cited where dead bodies had been exhumed and the vital organs burned, after which a living relative apparently about to die had suddenly and miraculously recovered. In the hope that this might prove true and bring about the recovery

of the dying girl, it was determined to exhume the body of the sister last to die and perform the strange rite. The superstition held that to secure the desired results the ceremony must be conducted at night at the open grave by a single individual. No one was willing to undertake the gruesome task, but finally the lover of the sick girl volunteered to do it. He went to the graveyard in the dead of night and dug up the body. Silently he performed the weird autopsy and carried out the strange program in every detail. The story goes that the girl recovered and lived to be a very old woman.[607]

This newspaper text provided the specific location of the cemetery (absent in Cole's text), but raised other mysteries: At night? Alone? In silence? Exhuming a corpse unassisted would be a difficult feat, a fact made clear by the accounts of bodysnatching "resurrections." Adding to that improbability is the decision of the ailing girl's lover to dig up the corpse of her sister. These motifs, capped off by the girl living happily ever after (or at least to a ripe old age), lend an air of folk tradition. Of course, it would have been simple for someone to add these elements precisely for the purpose of "improving" the story. The *Courant's* text did little to move my investigation forward, even though it identified the cemetery in which the five sisters were allegedly interred.

J. R. Cole was a chimerical, professional writer who specialized in local history and biography. Aside from his *History of Tolland County, Connecticut* (1888), which includes the exhumation narrative, Cole also wrote histories of Greenbrier County and Preston County, West Virginia, and the Red River Valley. Cole, "a gentleman of long experience in the compilation of biographic cyclopedias," summarized the sources that he used in compiling the text for his *History of Washington and Kent Counties, Rhode Island*: "Records of every kind, town, church and court, unpublished manuscripts, standard histories, private diaries, letters and local traditions have furnished the material, which has been sifted, collated and arranged according to the writer's ability."[608] This knowledge led me to infer that his exhumation narrative was embedded in a previous publication, as neither primary research nor source citation were central to Cole's *modus operandi*. And it seemed reasonable that the hypothetical text might have been the source for both Cole and the newspaper.

A tedious search through periodical indexes validated my hunch. The 1876 edition of *Appleton's Journal* contained an article entitled, "A Century Ago in New England," authored by Francis Gerry Fairfield. Cole appears to have taken verbatim (without attribution) Fairfield's account of the exhumation, which follows:

> In the old West Stafford graveyard the tragedy of exhuming a dead body and burning the heart and lungs was once enacted—a weird night-scene. Of a family consisting of six sisters, five had died in rapid succession of galloping consumption. The old superstition in such cases is that the vital organs of the dead still retain a certain flicker of vitality, and by some strange process absorb the vital forces of the living; and they quote in evidence apocryphal instances in which exhumation has revealed a heart and lungs still fresh and living, incased in rottening and slimy integuments, and in which, after burning these portions of the defunct, a living relative, else doomed and hastening to the grave, has suddenly and miraculously recovered. The ceremony of cremation of the vitals of the dead must be conducted at night, by a single individual, and at the open grave, in order that the result may be decisive; and most old graveyards could mention nights when they have been thus illuminated; for, no

longer since than 1872, the Boston Health Board reports describe a case in which such a midnight cremation was actually performed during that year.[609]

My research showed that Fairfield was born in Stafford, Connecticut, August 18, 1836. His article, which focused on his hometown, located in northeastern Connecticut, was derived from stories he recalled hearing from older generations. Looking back, at the age of forty years, Fairfield began his narrative:

> Ah, days of tokens and omens and revelations! How little our more fastidious civilization comprehends of the wild, stern, and daring psychic lives, of the largeness and heroism, of the gloomy and fantastic religious enthusiasm, that were nurtured in those geometrical old houses, so few of which are left as reminders of the last century! Grand men—large and able men—a little superstitious, perhaps, but all the more picturesque and manly for it! My great-grandmother Washburn had the reputation of being the most accomplished sorceress in all that region, and old people even now tell the legend of her having turned over a heavy oaken sled, loaded four feet high with heavy timbers, by just wishing it, simply because she was offended with amiable Captain Washburn.[610]

After several paragraphs devoted to stories of old houses and their now withered or departed occupants, Fairfield returned to his great-grandmother, the "sorceress" and his primary informant:

> But I am wandering—for memories of the old legend-life when my great-grandmother trotted me on her knee, sorceress though she was, and told me stories of ghosts and wizards, and of the goblins that lived down-cellar in the dark, and the strange voices that had been heard up-garret, while the flames in the huge fireplace crackled and laughed, and made flickering pictures on the wall, by way of illustrating her remarks—such memories as made me afraid of the dark when I was a boy, and caused me to shiver when the wind shook the garret-door of a night, come thronging from every nook in my brain, where they have drowsed so many years that I have deemed them utterly effaced.[611]

Fairfield's career was not exactly a straight line, as Appleton's *Annual Cyclopedia and Register of Important Events* recorded in 1887, the year of his death:

> Fairfield, Francis Gerry, an American clergyman, physician, and author, born in Stamford, Conn., Aug. 18, 1844; died in New York city, April 4, 1887. He was graduated at Gettysburg at an early age, and, after spending some time in scientific researches, entered Hartwick Theological Seminary, where he astonished his teachers by his remarkable aptitude for religious study and investigation. On receiving his degree of D. D. from the Lutheran Church, he was assigned to a mission in Waterloo, N. Y. Removing to New York city [sic] to avail himself of larger facilities for scientific study, he relinquished preaching, and applied himself to the study of medicine and surgery in the office of Dr. Worcester. The natural rapidity with which he acquired information, especially of a complex character, enabled him to complete his course in half the usual time. He was graduated at the New York College of Physicians and Surgeons with a distinguished record, but never practiced this profession to any extent. In 1867 he married Josephine Griswold, who had already won a reputation by her literary work. After this Mr. and Mrs. Fairfield devoted themselves to literature, contributing to newspapers, magazines, and reviews.[612]

After abandoning the pulpit for journalism, Fairfield published two books that received much attention: *The Clubs of New York* (1873) and *Ten Years with Spiritual Mediums* (1875). The work of the wide-ranging Fairfield was not universally praised. In 1883, Eugene Lemoine Didier pronounced *The Clubs of New York* "the greatest piece of downright puffery we have ever had the misfortune to encounter."[613] Didier noted, "Mr. Fairfield cannot write simple, pure,—in a word, good English."[614] An earlier article in the *American Monthly Microscopical Journal* blasted Fairfield's scientific endeavors with the following assessment: "We have read a number of articles from Mr. Fairfield's pen, and we do not hesitate to assert that he is either woefully ignorant of science, or else a consummate humbug."[615] And, if such were possible, Fairfield's interpretation of Edgar Allan Poe was even less well-received.[616]

Like the events in the West Stafford cemetery that Fairfield wrote about, his own story was not without tragedy. The headlines of his obituary, in the form of a letter penned in April 1887 by a longtime associate at the *New York Times*, summarized Fairfield's personal misfortune: "Some of the many tragedies of journalism. Josephine and Francis Gerry Fairfield. Deadly opium leads both to an untimely end. Bright writers whose lights went out unpleasantly." The letter related how Fairfield had graduated from college, studied theology, and then earned an advanced degree in veterinary medicine. But writing was his calling. As his colleague wrote:

In 1865 a bright-eyed, fragile-built, clean-cut young man entered journalism in this city. He was well born, well bred, and his education was along broader lines than those of most of us. He had a peculiar mind, which sought information in unusual channels. . . . No sight more common in Printing House square than Fairfield, with his cigar or pipe, a bundle of books or papers under one arm, and his little wife upon the other.

Then, the brilliant couple began experimenting with opium. His former associate described the dimming of the once-bright lights:

His work was as brilliant as ever, his articles, whether for magazine, weekly or daily, were as readily accepted as ever, but little by little his manner changed, and as he went so went she. . . . Well, there's no use in describing the general decay, suffice it that every day for years the two were seen flitting here and flitting there throughout the newspaper offices that congregate on and about Printing House square. After a little they forsook the realms of daily journalism and continued themselves to the festive magazines and weeklies.

Gradually there were indications of pecuniary pinching.

The natty dress was no longer visible.

Evidence of care and trouble showed upon their faces. . . .

Still They Clung Together,

and the one time in all my quarter of a century's knowledge of them that I ever saw him alone, was a week ago Wednesday, when I saw him standing in front of the office of the New York Times, looking irresolutely up and down the street. . . . That very day, it seems, he took his wife, reduced to a simple skeleton, to a cheap tenement on one of the squares of the city, where she died alone, and he, despairing, broken hearted, with his better half sheered away, sought refuge in a cheap hotel on Third avenue, where he, too, some forty-eight hours thereafter, was found dead alone.[617]

Fairfield was interred in the Old West Stafford Cemetery, perhaps the scene of the tragic events he wrote about a little more than a decade earlier. His obituary in the *Kansas City Times* (April 10, 1887)—entitled "An Opium Eater"—closed the book on him with the following: "Divinity, journalism, literature, spiritualism, medicine, veterinary surgery, opium, Bohemianism, scatter-brained inventions, poverty, degradation, death."[618]

Biographical snippets suggest that Fairfield learned of the exhumation by way of oral tradition or, perhaps—plausible, but less likely, in my opinion—acquired the text from some local printed source, such as a newspaper. Fairfield was born and raised in the community where the exhumation was performed. He knew and valued its oral traditions. His great-grandmother apparently was considered a witch of sorts, so she probably was acutely aware of local supernatural incidents. She was a significant source for the lore that Fairfield absorbed during his early years. Fairfield's biography also raises the specter that the exhumation story is a fabrication. Several contemporary critics accused Fairfield of misunderstanding or deliberately misrepresenting topics he wrote about, which suggests that he was careless with facts or ignored them if they failed to support his preconceptions. What follows does not lead to indisputable conclusions regarding the authenticity of Fairfield's vampire; yet it does illustrate the vampire's malleability in the hands of a determined author.

In 1871, five years before the appearance of Fairfield's article in *Appleton's Journal*, two vampire stories, also set in Stafford, were published anonymously. In "A Yankee Vampire" (*Sacramento Daily Union*, April 29, 1871) a newspaper man arrives in Stafford Springs and learns that Frank Farley has been arrested for exhuming the body of his fiancé's sister. A tale within a tale unfolds, as the jailed Farley relates his story to the newsman. Dialogue advances the narrative, with Farley speaking in a distinctive New England dialect. Farley's beloved Florrie Dunbar was born into a tainted family, doomed for extinction by galloping consumption: "the Dunbars kept dropping off as regular as clock-work"—so regular that every three weeks another family member was buried. After Florrie's remaining sister, Mollie, died, Florrie began to show the dreaded symptoms. In desperation, Farley made his way one night to the old burying ground with lantern and spade. He removed Mollie's still-beating heart, hurried home with it, and burned it. Convincing authorities he was neither insane nor a grave robber, Farley was able to secure his freedom. He married Florrie, and he always swore, "She'd have been dead this very day . . . if the whim hadn't taken me that there might be something in the old superstition."

The image of Mollie's beating heart perhaps betrays some poetic license by the author. While practitioners of the consumption ritual often searched for "fresh" blood in the heart of a corpse, none, to my knowledge, encountered a heart still beating. Amy Lowell, in her 1925 poem, "A Dracula of the Hills" (discussed in chapter 22), stops just short of this example of poetic license:

But her heart was as fresh as a livin' person's,

Father said it glittered like a garnet when they took th' lid off th' coffin.

It was so 'live, it seemed to beat almost.

The *Sacramento Union* was the oldest daily newspaper west of the Mississippi when "A Yankee Vampire" was published. Another writer for the *Sacramento Union* was Bret Harte, who wrote popular tales of the California Gold Rush. Harte was married to the sister of Fairfield's wife, Josephine—

which may explain how Fairfield found a venue for the story of Frank Farley's vampire adventure. Due in large measure to Seba Smith's "Letters of Major Jack Downing" (see chapter 4), the stereotypical Yankeeness of Fairfield's characters already had country-wide recognition.

With a few alterations, "A Yankee Vampire" became, six months later, "Quick Consumption: Every-day story of New England" (*Frank Leslie's Illustrated Newspaper*, November 25, 1871). Although a much shorter narrative (2,656 compared to 4,277 words), it follows the same arc as "Yankee Vampire," including the surname Dunbar, and all dying at three-week intervals. Dr. Bloupil is the attending family physician in both texts. In the newer variant, the anonymous author (surely Fairfield) had the deceased sisters appear to the dying—the classic vampire-as-revenant motif absent in New England's authentic tradition. All three of the Stafford consumption tales have the exhumation taking place at night, conducted by a single individual, which, in the New England tradition, is a remarkably rare performance. In each story, the afflicted wife lives a long, happy life.

Fairfield's writing style (characterized by one critic as "downright puffery") is readily identifiable. His distinctive, recurrent patterns of language preferences—such as word choices, sentence structure, syntax, punctuation, and turns of expression—support the identification of Fairfield as the author of all three of the vampire stories set in Stafford, Connecticut. Their situation in Fairfield's hometown argues against mere coincidence. Other narrative elements point to Fairfield as the author. He favors repetition for effect. Words or short phrases repeated within a span of several sentences act as an echo, reinforcing specific images. When used intermittently throughout the narrative, repetition advances the narrative step-by-step, gradually building a mood or atmosphere. Ancient trees and old architecture, often imbued with ghosts of the past, are another noticeable stylistic element. Fairfield began "A Century Ago" (1876) with drawn-out descriptions of the old houses that connected him to the past. In "Quick Consumption" (1871) the description of the Dunbar family's dwelling suggests its animate, "goblin" nature. In his barely shrouded autobiographical narrative of becoming a writer in New York, "Timothy Tot: A Prose Story with Poetic Passages" (1872), Fairfield, flashing back to his childhood and education, ties the animism of old trees to the imagery of goblins: "There was one gnarled old oak, in particular, west of the house, that always gave me the impression of a thinly disguised goblin, I being in constant anticipation of seeing it take goblin legs and walk off, leaving neither stump nor other vestige of having stood there."[619] Fairfield uses the word "goblin" no less than a dozen more times in this article: there are "goblin moons" and "goblin hills and woods"[620] and many variations on "goblin old house."[621]

Words and phrases used to describe New England's vampire tradition in the three versions show close resemblances. In "Quick Consumption" (1871), Fairfield wrote (emphasis in bold is mine):

> It is one of those weird old superstitions with which the household literature of New England abounds. It is, that the heart of the dead, dying not, **by some strange *rapport*, feeds upon the vitality of the living**—the living being thus actually eaten up of the dead; and weird stories are afloat of the dead having been taken up, and there having been found, **still red and warm, in the midst of ghastly rottenness**, the hearts of some who have died of quick consumption. Whole families have gone of it, one after another, the dead gnawing and feeding upon the vitality of the living, until, as the last dropped into the grave, the red, warm heart in the coffined corpse, having no living relative upon the vitality of

whom to feed, has wasted also and died—died at last in its coffin for want of something upon which to prey.

In "A Century Ago" (1876) it is rendered:

> The old superstition in such cases is that the vital organs of the dead still retain a certain flicker of vitality, and **by some strange process absorb the vital forces of the living**; and they quote in evidence apocryphal instances in which exhumation has revealed a heart and lungs **still fresh and living, incased in rottening and slimy integuments**, and in which, after burning these portions of the defunct, a living relative, else doomed and hastening to the grave, has suddenly and miraculously recovered.

In all the tales, the exhumation takes place at night and is conducted by a single individual. And the afflicted wife (or soon-to-be wife) for whom the ritual was performed, lives a long, happy life.

By way of summary, let us review, in chronological order, the five texts I attribute to Fairfield:

1. April 29, 1871, "A Yankee Vampire," *Sacramento Daily Union*. In Stafford Springs, two grandparents, both parents, and two sons and two daughters—in a single family—die of consumption. Florrie, the remaining daughter and fiancé of the protagonist, is saved when he exhumes and burns the heart of the corpse of the last sister to die.

2. November 25, 1871, "Quick Consumption: An Every-Day Story of New England," *Frank Leslie's Illustrated Magazine*. In Stafford, both parents and six sons in a family die of consumption; then a second family is struck with consumption, and both parents and two of three sisters die. The protagonist saves the third sister, his fiancé, by exhuming the corpse and burning the heart of the last sister to die. Prior to death, in the classic vampire-revenant motif, the dying person is visited by the last deceased.

3. May 30, 1876, "A Century Ago in New England," *Appleton's Journal*. In West Stafford, five of six sisters die of consumption. The last sister is saved when her body is exhumed at night by one person, and her vitals are extracted and burned.

4. 1888, J. R. Cole, *History of Tolland County, Connecticut*. Cole quotes verbatim Fairfield #3 without attribution.

5. March 28, 1915, "Tragic Legend of West Stafford Cemetery," *Hartford Courant*. Almost identical to Fairfield #3, with the addition that it was "the lover of the sick girl" who undertook the exhumation; this change accords to the first two (anonymous) narratives. Did this newspaper's author combine two narratives, or did he retrieve his tale from oral tradition, as he hints when he writes, "a strange tale is told"?

Has forensic linguistics identified the parent of the two orphan texts? It seems beyond doubt that Francis Gerry Fairfield authored the two anonymous narratives. Authorship aside, are these two narratives fiction? Several features argue that the events in the two anonymous texts did not happen as described. First, none of the named individuals can be found in the records for the town of Stafford (or Tolland County). Second, the patterning of the narrative motifs, seamlessly moving the action forward in regular and predictable steps, belies a sense of reality. Each person dying at three-week intervals, for example, is too perfect. In the second text, the vampires appear to those who are nearing death as revenants, a motif noticeably absent in the New England tradition. The happy ending is just icing on the cake.

While these stories deploy varying degrees of artistic license, all have a core element that accurately portrays, not only New England's vampire tradition, but also Stafford's general folklife. It seems likely that Fairfield used a local legend or personal experience story, received from his great-grandmother, as the foundation for the narratives, blending oral history, folklore, and fiction. He created marketable stories by embellishing features of the ritual, animating the characters, and molding an intriguing plot.

***

Francis Gerry Fairfield recalls Joseph Ripley Chandler, who wrote about a vampire exhumation in his hometown of Kingston, Massachusetts, a generation earlier (discussed in chapter 18). Both were born in small New England villages, received good educations, and left home as young men to make a name, and perhaps a fortune, in a large city. Neither forgot his roots, particularly the folklore that, for them, defined a place that had changed and would never be the same again. For both Chandler and Fairfield, their community's vampire narrative was the supernatural lore that moved them to lofty prose and occasional flights of poetry. Fairfield took a turn for the worse, wasting away on opium and dying destitute at a relatively young age, while Chandler died a respected elder statesman, on July 10, 1880.

Chandler and Fairfield also shared a narrative context: a romantic view of the fading lore tinged with an air of superiority for having moved forward into modern times. Fairfield's narrative appeared in *Appleton's Journal of Science, Literature and Art* and Chandler's in Godey's *Lady Book*. The static nature of the printed word belies the fragility of these texts. Yet, they are no more honest, nor less capricious, than the people who created them. Even though texts can tell us much, they cannot stand alone as things in themselves. The wall between history and fiction is like that of a living cell, a porous membrane that leaks in both directions.

21   Thoughts of a Horrid Nature Arose in the Parents

Deep inside, have we humans remained fundamentally the same since the dawn of our existence? Owen Wister (1860-1938) thought so, and he offered a vampire narrative to support his belief. When he inserted the story into his 1915 essay, "Superstition and the Doctor," he was a celebrated author, having penned the best-selling, quintessential Western novel, *The Virginian* (1902). Wister had a talent for finding stories wherever he went, in everything he did, and from everyone he knew. Owen Wister (Dan to family and friends) was born July 14, 1860. His father was a wealthy physician from the Germantown neighborhood in Philadelphia, and his mother, Sarah Butler Wister, was a writer and the daughter of renowned British actress, Fanny Kemble. At Harvard, Wister was a classmate and friend of Theodore Roosevelt (future President), Oliver Wendell Holmes, Jr. (future Supreme Court Justice), and Henry Cabot Lodge (future Senator). Wister showed his eclecticism early: he was editor of the Harvard Lampoon, wrote a light opera, *Dido and Aeneas*, for the Hasty Pudding Club, studied music at the Paris Conservatory, worked briefly in banking (which he loathed), then earned a law degree from Harvard. But he evidently was no more taken with law as a profession than banking.

After several trips out west for health reasons (diagnosed as "neurasthenia"), Wister became enamored of the region, resulting in his most significant literary work. His novel, *The Virginian: A Horseman of the Plains*, defined the western genre. The unnamed protagonist famously responded to an insult from his opponent during a (now) prototypical gunfight facedown:

> The Virginian's pistol came out, and his hand lay on the table, holding it unaimed. And with a voice as gentle as ever, the voice that sounded almost like a caress, but drawling a very little more than usual, so that there was almost a space between each word, he issued his orders to the man Trampas:
>
> "When you call me that, smile!"
>
> And he looked at Trampas across the table. Yes, the voice was gentle. But in my ears it seemed as if somewhere the bell of death was ringing; and silence, like a stroke, fell on the large room.[622]

In addition to novels, Wister also authored short stories, poems, essays, plays, and biographies (including one of his classmate and friend, Theodore Roosevelt) and other nonfiction.[623]

"Superstition and the Doctor," published in the *Boston Medical and Surgical Journal*, was the transcription of a talk that Wister had delivered on April 29, 1912, before the College of Physicians in Philadelphia, and again on December 14, 1914, at the Johns Hopkins Historical Society in Baltimore. Wister explicated his thesis—a theme that runs throughout his writing, fiction and nonfiction—that "we, the human race, have not had time to change much."[624] He elaborated, "We human beings remain so strange, so elusive, so inveterately beyond all logic, that self-contradiction is part of our very harmony."[625] Science, he argued, "has not killed our intuitive affinity for signs, charms, amulets and methods occult."[626] Although Wister, like George R. Stetson, embraced the racially-biased concepts of eugenics, he was convinced that human culture had not evolved to the degree that many anthropologists, including Stetson, had proclaimed. These anthropologists saw superstitions as vestigial "survivals" from lower levels of culture. To Wister, they demonstrated the immutability of human nature, an assertion that he supported by offering several "strange examples of this ingrained ancestral credulity which lies in us so deep, and crops out in such protean aspect."[627]

Wister presented two cases "from personal experience"—his or someone else's, we are not informed. He did not name the state in which these events occurred, but it is certainly Rhode Island. The first instance concerned the "Bonesetter Sweets," a family legendary in southern Rhode Island for having the hereditary gift of setting broken bones: "In a New England State, since Colonial times, has lived a family of natural-born bonesetters. One of these unlicensed hereditary practitioners unquestionably cured Theodosia Burr of hip trouble when the surgeons of New York had failed. Today any jury of that State will take the word of an unlicensed bonesetter against the sworn testimony of an educated, licensed physician."[628] Wister's second narrative dealt with exhuming a vampire:

> In the same State, and within the sound of the whistle of the express trains between New York and Boston, a family who were suffering from anemia dug up their recently buried sister and roasted her liver to stop her from "vampiring" them. Now, at the end of the nineteenth century, for that is when my anecdote occurred, the liver of the New England Yankee woman was roasted because six or seven thousand years ago people believed that the liver was the seat of the soul. Only thus could the vampire be killed. On this liver we fly straight back from New England, through Britain, the Middle Ages, Rome, Greece, Egypt to Babylon, where we have to stop for lack of earlier records. In spite of the telephone, we resemble the mummies—the set fractures of whose bones, by the way, give proof of a clearly developed knowledge of surgery.[629]

Consistent with his general thesis that humanity has maintained a solid core of understanding (at times, misunderstanding), Wister bridged an enormous span of time, across far-flung continents, in a single stroke:

> In the Yankee vampire, Solomon, visiting New England, would have recognized at once an old acquaintance—Lilith—although he might have to visit Borneo or Burma or Uganda, to see divination by the liver in actual practice. In our down-town streets he would very naturally mistake our banks and stock-exchanges for temples of the gods; but wherever he went, he would hum his old maxim, that there was nothing new under the sun. Much that is done by the Hudson and Mississippi they did by Tigris and Euphrates.[630]

I am convinced that Wister's subject was Mercy Brown, even though two details in his narrative do not correspond precisely to the known facts of her exhumation in 1892: consumption, not anemia, was the cause of the Brown family's problem (although anemia is a symptom of tuberculosis); and Mercy's heart, in addition to her liver, was burned. The key to deciphering Wister's vampire narrative is the status of Saunderstown, Rhode Island, as a summer resort.

The Sunday features page of the *Philadelphia Inquirer* (July 7, 1904), under the heading "Gossip from Seashore and Mountain," proclaimed that, "Owen Wister, a Philadelphia novelist, is to become the next thing to a local resident. He has just bought twenty-five acres of land in Saunderstown, four miles distant [from Narragansett Pier] by trolley, and will soon build a large villa. He has leased a cottage there for four years." The Wisters were part of a mélange of affluent socialites who summered along Rhode Island's shoreline, especially near Narragansett Pier. A guidebook published in the 1880s characterized these visitors: "The society at the Pier and in the vicinity is select; merchants, manufacturers, statesmen, men of letters and practical science, and eminent professional characters of every sort, choose this as their favorite summer retreat."[631] A 1906 *Harper's Weekly* article on Narragansett Pier argued that "Narragansett's chief charm is not to be sought in any merely physical combination of land and water and air, but rather in certain of its social aspects." It was not the place, but the people "who shaped its social character." While visitors from Philadelphia and New York mingled with those from Richmond and Louisville, "the tone of the summer's colony at the Pier is rather Southern than Northern, with the warmth and the heartiness of the one and without any of the frigidity and affectation which only too often chills social intercourse in the other."[632] Besides bathing at the mile-long stretch of sandy beach, there was tennis, golf, croquet, polo, horse shows, and, of course, luncheons, dinners, and dances.

Vampire commentators and critics were among these pleasant companions. Moreover, some year-round neighbors certainly had firsthand vampire experiences. You may recall (from chapter 8) that in 1872, William Rose, a resident of Saunderstown, exhumed the bodies of a son and daughter. Just two years later, Benoni Lockwood became Saunderstown's first summer resident, fleeing the intense social life of Narragansett Pier. His family and friends soon followed. His daughter, Frances Willing Wharton, a well-known author, and her husband, Henry T. Wharton, a coal magnate from Philadelphia, began summering at 56 Waterway. Their cousin, novelist Edith Wharton, often visited from her summer home across the bay in Newport.[633] (I discuss her 1926 short story, "Bewitched" —inspired by New England's vampire rituals—in the next chapter.) Just south of the Wharton House was the home of the Grant LaFarge family. LaFarge, son of artist John LaFarge of Newport, was a prominent New York architect. His wife, Florence, was another daughter of Benoni Lockwood; as noted in the Rhode Island Historical Preservation Report on North Kingstown, Florence "entertained the family's many friends and relations at daily croquet games followed by formal tea and her famous conversations, a university in talk for the numerous younger generation of the colony."[634] Up the road, at 25 Waterway, Owen Wister had been renting the Captain Alfonso Gould House for ten years, prior to the construction of his own family compound, completed in 1913, under the supervision of its designer, Wister's friend Grant LaFarge. I would love to have transcriptions of the "famous conversations" held in that close-knit summer neighborhood.

Another summer resident along Rhode Island's shoreline was Reverend Edward Everett Hale, whom I suspect had a hand in authoring an anonymous vampire narrative that appeared in the *Boston Sunday Herald* (August 13, 1899), embedded in an article headlined, "Lingering Belief in 'Hoodooism'." As I noted in discussing this remarkable narrative (chapter 7), unmistakable traces of the Boston Brahmins—the elite culture of New England characterized not only by good breeding and manners, but also by the best-available education and understated wealth—are layered into the *Herald* article, with the at-least implicit presence of not only Hale, but also Ralph Waldo Emerson, James Russell Lowell, and Oliver Wendell Holmes, Sr. (whose son was a friend and former Harvard classmate of Wister). In July 1903, four years after the *Herald* article was published, Hale and Wister attended the same dinner party in Point Judith, Rhode Island, as noted in the *New York Times* (Sunday, July 26, 1903): "At the Point Judith Country Club, Mr. and Mrs. Arthur Hale of Philadelphia gave a luncheon to-day. Their guests were Owen Wister, the novelist; Russell Perkins, Lawrence Perkins, Miss Sibyl Hale, Mrs. Leonard Wood, wife of Gen. Leonard Wood, United States Army, and Dr. and Mrs. Edward Everett Hale of Boston."[635] I must wonder if discussions with Hale (and perhaps other Brahmins) sparked Wister's interest in local vampires.

Yet, another neighbor—a year-round resident with whom the Wisters had a close relationship—may have played the pivotal role in introducing Wister to Mercy Brown. In 1913, after four years of construction, the Wisters finally occupied their new summer compound, which they dubbed Champ de Corbeau, at 1600 Boston Neck Road, about two miles north of their rental cottage in Saunderstown. But, devastatingly, later that summer, Owen's wife (and distant cousin) Mary Channing (known as Molly to family and friends) died, after giving birth to her sixth child, in the new family home. One of their daughters, Fanny Kemble Wister, later recalled the event:

> Molly's life of service to the community and devotion to family was cut tragically short at the young age of 43. On the afternoon of August 24, 1913, Molly called her five children into her room at the family's summer house in Saunderstown, Rhode Island, and told them that she was going to have a baby. According to Fanny, none of Molly's children even knew that she was pregnant. "She told us that she wanted us to grow up to be leaders and always set a good example for our brothers and sisters." That evening, Molly died from complications during the delivery of her sixth child, Sarah Butler Wister. Dan had to break the news to his children, most of whom were too young to fully understand what death means. "We all sat down on a big rock, and Dadda said, 'I have a message for you from your mother. It is goodbye.' In one voice, we all said, 'Goodbye?' 'Yes,' he said, 'she is dead. You have a little sister.' Then he burst into sobs. We all howled, with tears streaming down our faces; it was minutes before we could speak."[636]

Molly Wister had her own connections to Rhode Island, as she was a descendent, through her mother, of William Ellery, a signer of the Declaration of Independence for Rhode Island.[637]

Just three miles up the road from Champ de Corbeau, at 130 Boston Neck Road, was the residence of Dr. Harold Metcalf, the physician who, in his capacity as the Medical Examiner for the towns of North Kingstown and Exeter, was in attendance when the corpses of three women of George Brown's family were exhumed. In an unpublished biography of Metcalf, penned by his son, George, it was disclosed that "he had a very large practice among all walks of life—from the rich to the very poor. The rich came in the summer, for Wickford and vicinity in those days was a fairly popular summer resort. Among his well-to-do patients were the families of Owen Wister, the novelist . . . . While father enjoyed his contacts with his more prosperous patients and became good friends with some of them he never catered to them at the expense of his poorer patients."[638] Was he good friends with the Wisters? We do not know, but it seems more than likely that Metcalf, their local physician, was attending to Molly Wister when she gave birth and subsequently died from heart failure due to post-natal complications.[639] According to his son, Metcalf had delivered many children and, perhaps, even had experience in Caesarean deliveries: "I didn't get this from father but from mother or from sister Mary, who was much better versed in family history than the rest of us children. I can't go into details, but the gist of the story is that father once performed an operation by lantern light on a kitchen table in the backwoods of Exeter and my impression is that it was a Caesarean."[640] It is almost certain that the unnamed vampire in Wister's narrative was Mercy Brown and that his key vampire informant was Harold Metcalf. When Wister refers to the two cases "from personal experience," might he be referring to Metcalf's own experience with the Brown exhumations and, perhaps, his contact with some of the Bonesetter Sweets? I can imagine Metcalf urging Wister to create an essay built around these two narratives: "Owen (or Dan, if they were good friends), these are fascinating medical stories, and you're a writer. I think you could produce a wonderful piece built around them."

Perhaps it was fitting that Owen Wister died at Champ de Corbeau, in Saunderstown. On July 21, 1938, just one week after celebrating his seventy-eighth birthday, he succumbed to a cerebral hemorrhage.[641] A noteworthy passage in *The Virginian* occurs when, at a dramatic point, Wister abandons the narrator's voice. Addressing readers directly, he ponders, in effect, whether good ends justify evil means. He states that he does not believe so, and, furthermore, those who believe they do deceive themselves. "But this I can say: to call an act evil instantly begs the question," he asserted. "Many an act that man does is right or wrong according to the time and place which form, so to speak, its context; strip it of its particular circumstances, and you tear away its meaning."[642] Engaging readers through a cooperative argument dialogue, a rhetorical method used by philosophers from Socrates to Wittgenstein that is based on asking and answering questions to stimulate critical thinking and to draw out ideas and underlying presuppositions, Wister asks them to consider the morality of several hypothetical situations. His conclusion that, essentially, context *is* meaning underlies my insistence on interpreting New England's vampire tradition in the fullest context whenever possible.

Coursing through Wister's writing is a belief that there is a wellspring of human nature—grounded in the species, stretching back before recorded time—that is manifest in every individual. Time and circumstance change, but this essence, judged better or worse depending on the context, remains stable and unchanging. Years before he penned his iconic novel or brief vampire narrative, Wister had arrived at his thesis regarding the human condition. In a journal entry of August 4, 1885, while in his beloved Wyoming, preparing to attend a roundup in "beastly hot" temperatures, Wister asserted: "It is a confusion of language to speak of one spirit as ousting another or to describe progress in any way that implies more than one agent. Change in matters of thought is not wrought by the action of a thing upon another thing. There is but one thing. . .. The chameleon simile is nearest the truth. Whatever the momentary colour may change to, the same beast is beneath all the while."[643] In *The Virginian*, one of Wister's characters says, "I expect in every man you'd call sensible there's a little boy sleepin'—the little kid that once was—that still keeps his fear of the dark." In that novel's introduction, Wister wrote, "His wild kind has been among us always, since the beginning: a young man with his temptations, a hero without wings."[644] Scratch the surface of the rudest savage and you will find the soul of a nobleman; look deep inside the civilized nobleman and you will see the heart of a savage.

Consumption was an untamed menace that threatened to extinguish families and communities. Its old remedy revealed human culture's persistent fearful, but hopeful, childhood. Yes, Wister called her a vampire. Did he smile when he called her that name?

∗∗∗

Walter Prichard Eaton (1878-1957) may have exceeded even the bounds of poetic license when he wrote "In Old South County" (*Outing* magazine 1912). The subtitle of the article hints at how Eaton approached his topic: "A favored land where the leisurely spirit of the Old South still lingers in bustling New England." Midway through the article, Eaton introduces a vampire narrative. In the prior scene, he quoted anecdotes and quips offered by his stagecoach driver, a local character steeped in South County lore (which recalls the coach driver who took Jonathan Harker across the Borgo Pass to Castle Dracula, in Bram Stoker's 1896 novel). The authorship of the following narrative, however, is not acknowledged nor is it set off in quotation marks:

> But certain of the ancient traits persist of a grimmer nature—tendencies to strange superstitions, relics, it may well be, of the primitive beliefs of the Indians and negroes [sic], grafted upon the credulous countryside back in the days when New England witchcraft was still a fresh memory, and not yet wiped out by modern enlightenment.
>
> In the first year of the twentieth century, in a small hamlet of South County, an old woman died and was buried. Scarcely had her body been put in the ground when a series of minor misfortunes befell. A cow cast her calf; a child had the measles; a fisherman's power dory slipped her moorings and was carried down the bay; a farmer's haystack mysteriously took fire; and so on through an easily imagined list. The people of the hamlet met in council and decided that the dead woman was a "vampire" and her soul was haunting the place and making the trouble. Just how the term vampire, which they freely used, applied, it is difficult to see. Perhaps investigation could have traced its use back to some negro superstition of slavery days. At any rate, such was the term the villagers employed.
>
> A committee was appointed who dug up the woman's body, cut out her heart and burned it, which was supposed to prevent further possibilities for mischief. This, of course, takes us directly back to the most primitive beliefs of the most primitive peoples. Probably, to many readers, the story will seem utterly incredible. Yet it is true, and its scene was Rhode Island in the twentieth century, and its actors were not savages but Anglo-Saxon farmer and fisher folk, living within easy range of Narragansett Pier and Newport. A famous scientist has recently asserted that a return of the belief in witchcraft is always possible, in the most enlightened nations. Here, surely, is data for his argument.[645]

Several elements in Eaton's narrative contradict what we know about New England's vampire tradition:

- The series of events that followed the old woman's death—stillborn calves, measles, and other mysterious phenomena—may be typical signs of witchcraft, but consumption was the sole symptom associated with New England's version of vampirism; these motifs recall the *Boston Globe* (January 27, 1896) parody discussed in chapter 17.
- The people in the hamlet used the term "vampire," which we know was not the case in Exeter or, indeed, among insiders throughout the region.
- Eaton appeared eager to connect the practice to African American culture, contrary to the best available evidence.

The closest match to Eaton's story is the Brown family incident, which occurred in Exeter, in 1892. In both narratives, a council of decision-makers was convened to evaluate the situation, and a woman's corpse was exhumed, and her heart (and liver) extracted and burned. If the Brown family event *was* Eaton's source, then either the narrative he received was incorrect in relation to the known facts or Eaton changed it. In the Brown family exhumations, three corpses were exhumed but only one was mutilated; in Eaton's narrative, the corpse of "an old woman" was exhumed and mutilated. While it is possible that Eaton's narrative is based on another exhumation case, it seems unlikely that such an event would have occurred at the turn of the century and failed to have been recorded in the popular press or documented in local oral tradition. Ultimately, Eaton's conflation of witchcraft and vampirism undermines his narrative's credibility. Perhaps Eaton paired the vampire incident with New England's witchcraft history to provide a supernatural context that was more familiar to his readers (and maybe himself, as well).

Eaton was an outsider whose experience in the local community did not extend beyond that of a visitor gathering material for a travel article featuring local color—which is to say, not sufficient to establish his credentials as a knowledgeable insider or even, as in the case of Owen Wister, a longtime summer resident. Eaton acknowledged his debt to the writings of one of South County's most enthusiastic, tradition-steeped narrators, Thomas Robinson Hazard. Eaton opened his essay with a discussion of jonnycakes, drawn from "Shepherd Tom" Hazard's book, *The Jonny-Cake Papers*. Eaton wrote, "for those who know Rhode Island intimately her true claim to glory is her Jonny Cake."[646] Tellingly, Hazard, in his inclusive and entertaining essays on the traditions and eccentricities of South County, scrupulously avoided mentioning the vampire tradition (which, given his intimate knowledge of the community, was undoubtedly deliberate). Closing the circle, Eaton returned to South County's jonnycake worship, echoing Hazard's style: "You are in old South County, and your cup of summer happiness is full, for even as the sun stands in the meridian, warning you it is noon, and you put your tiller hard over, there comes to your nostrils the scent of Jonny Cake cooking, three miles against the wind, and as your little craft takes the bone in her teeth and rips up the Pond your mouth begins to water for the ambrosia of Olympus."[647]

Like Owen Wister and George Stetson, Eaton juxtaposed South County with Newport and Narragansett Pier, using the latter as paradigms of civilized behavior: within a few miles of wealth, culture, and education, there is a place where the superstitions of a "lower level of culture" survive. The quaint-but-quirky narrative approach employed by Eaton accords with articles written for periodicals, such as *Outing*, that aimed to provide readers with a sense of not only the beautifully unique features of a place (illustrations by Walter King Stone assisted in this regard) but also the strange aspects that set it apart. Eaton's primary literary genre, however, was neither travel nor adventure, but the theater, about which he wrote several books. Walter Prichard Eaton was born in Malden, Massachusetts, and graduated from Harvard. He was employed as a drama critic for several popular periodicals before accepting a position as Associate Professor of Playwriting at Yale University. It would not be remarkable if the residents of South County perceived his vampire narrative to be aloof and out of kilter, if not an offensive caricature.

∗∗∗

An early piece of romantic fiction illustrates the jarring disjunction of a consumption narrative when its core ritual disregards traditional boundaries. In the New England tradition, a troublesome corpse is neutralized by burning its vital organs or turning it face down. In a few cases, vital organs may be removed after death but prior to interment, often at the request of a dying sibling. "Consumption," a story published in the *Pawtucket* [RI] *Chronicle* in 1829 (October 3), takes preemption to a ghastly extreme: the consumptive sibling must die (or be killed) before her sister shows signs of the disease.

The anonymous author's introduction to the story proceeds along familiar lines, describing the awful arc of consumption's devastation upon families, particularly those who should be in the prime of life but instead are declining. The story tracks the small family of James and Mary Lyndon, recent immigrants to the Colony of Rhode Island from England. Their older daughter, Elisabeth, was sixteen when they embarked on their journey. But after four harsh winters, as her beautiful countenance began to fade, a horrible, but familiar trajectory took shape:

> When it had once laid its finger upon one of a family, the whole prepared to die—for their experience and their superstition, both made it hereditary. They had a tradition which went to say that, when one of a family circle was laid in the grave, by consumption, decay commenced not upon the body, until the disease had selected its next victim. It was a simple superstition, like many other ideas of our puritanical fathers; but not the less strongly adhered to, for being simple. We had an illustration of this belief, from a son of one of the first settlers of the Island of Rhode-Island [sic]. To believe the story to be a true one, is no harder task than to receive as truth, many historical facts connected with the first settlement of New-England [sic].[648]

While the more beloved younger daughter, Hannah, was yet free of symptoms, the family became fixated on preventing what seemed inevitable. Elisabeth began praying that she might meet death before consumption completed its gruesome work, for she shared her parents' belief. "A residence of four years in this country, had made the Lyndons partakers of all the superstitions of its earlier settlers," the author wrote. "Among the rest, they had become impressed with the belief, that if one of their daughters died of Consumption, the other must follow. But if, by any means, the invalid was taken out of the world, before the heart-appaling [sic] disease could effect its object, the younger child would be saved."

At this point, "a horrid, fiendish idea . . . came across the brain" of James Lyndon. To save Hannah and ensure that he should not "be left childless in his old age," Elisabeth "should *die by his own hand!*" This notion, which arrived in a "frenzy," took hold and "could not be banished." One evening, at sunset, Elisabeth walked to her favorite place, a cliff overlooking the sea, to find solace and seek guidance. Startled by her father's footsteps, she looked up into his frightening gaze. As "she shrank back," he kissed her and said, "you must die—you *must* die, and not by consumption!" As "the horrid truth" took hold, she asked her father to pray with her. "Her voice faltered, she fell again into the arms of her father—the life-blood gushed from her mouth, for her heart was broken. . . . and she died." Steering between the Scylla of consumption and the Charybdis of death by a father's own hand, the author lands readers safely on the shore of a broken heart.

The epigraph for this story— "Start not— 'tis but Fancy's Sketch"—is taken from the lyrics of an 1817 song, which begins: "Here mark the poor desolate maid, by a parent's ambition betrayed. . .. And here stands the murderous wretch, but mark me . . . 'tis but fancy's sketch." This story is, indeed, a fancy, but what of the core belief that a consumptive's corpse will not begin to decay until the family's next victim has been selected? Like the narrative, this belief is a distorted echo of the New England tradition. I found no documentation for the superstition that a sibling would be spared if her consumptive sister died from a cause other than consumption. However, there is a widespread belief, found in both Britain and the United States, that a when a corpse fails to stiffen, another member of the family is going to die.[649] The lack of any Rhode Island record for John Lyndon and his family strengthens a conclusion that the story is imaginary.

22   They Drove a Stake through Her Breast

Mrs. S. E. Farley, "Popular Prejudices," 1842

Did Mrs. S. E. Farley write the first fictionalized treatment of New England's *authentic* vampire tradition? While the superstitions that Farley places in her 1842 essay, "Popular Prejudices," are well documented in American folklore, the people and events in her narratives cannot be verified. *Where* she places them subverts their authenticity: "I will confine my observations to my own immediate neighbourhood," she wrote. "Who would believe there are individuals in some of our largest cities, whose doors are daily thronged, not with the ignorant and degraded classes only, but with many comparatively cultivated and enlightened persons" who, nonetheless, seek what "wisely and graciously" is "hidden from the view of mortals."[650]

In her own city, Farley seems to have been surrounded by credulous indulgers in superstitions of all sorts, for her neighbors would: use the skin of a black cat to cure a sore throat, and the blood of a black hen to cure shingles or rheumatism; never commence a journey or start a business venture on a Friday; interpret the howling of a dog before the door or window of a sick person as an omen of "dissolution," and a ringing sound in the head as a "death-bell for somebody"; carry the "lucky bone" from a calf's joint to bring money your way; slaughter swine and cattle on the waxing moon" so the meat will not "shrink in the pot"; plant root crops, such as potatoes, turnips, and onions, on the waning moon so that they do not "all grow to tops"; and, stir melting butter "against the sun" to prevent it from turning to oil. In addition to such well documented folk beliefs, Farley's neighbors also told her of dreams that came true, wounds that had been charmed, warts that were magically cured, and "undiscovered mines of gold or silver" that were detected by "rods of the witch-hazel."

Farley sharpened her focus. The neighbor on her right "is a firm believer in 'apparitions,' as she calls them, having had guests of that description on several occasions, of whose supernatural powers she gives a marvellous [sic] account."[651] Her neighbor on the left had the most amazing—yet to us, familiar—story (italics in the original):

> In the house on my left lives a stout athletic man, of little more than middle age, of good natural abilities, and by no means destitute of cultivation, whose ruling error is of a graver cast. I had noticed his wearing a small box suspended to his neck by a cord, and, having once alluded to the circumstance, my neighbor gave me the following relation.
>
> "My parents had twelve children, each of whom as they arrived at the age of maturity, sickened and died of consumption. Just twelve months elapsed between the different burials, until eleven sons and daughters were laid in the grave. When my last brother died I had just attained majority. The sympathies of the people around us were strongly excited. When the grave was digging for Joel, some of our friends opened the coffin of the next older child, and found the body as fresh and fair as if the soul had just departed. *As each brother or sister died, they fed upon the life of the next in age.* Our friends then urged upon

my parents and myself the necessity of *burning the heart* of my brother and wearing the ashes about my person, as the only means of saving me from a like fate. But I could not consent to such a course, and he was buried. Two months passed away, and I could no longer conceal from my anxious parents the ravages of disease. They again urged burning the heart of my brother as being their only hope, and as I continued to resist, my mother called the aid of a still more powerful advocate. My wife, to whom I was then engaged, entreated me to consent. 'If I am restored to health by such means, Abby,' I replied, 'I cannot live here. You must leave home and friends, and go with me to some uninhabited spot.' 'I will, to any part of the world;' was her firm response. So the deed was done. I came here with Abby when there were no inhabitants but bears, catamounts and loupcerviers. I hunted, felled trees, tilled the soil, and built this house with my own hands; yet I am, as you perceive, still strong and hearty."

"But may not your recovery be attributed to air, exercise, change of climate and different mode of living?" I inquired. He shook his head, touched the box, (*the charm*) and turned away.[652]

Although this neighbor's story is presented as a first-person narrative, its reality is dubious. The apotropaic (warding off evil) ash pendant is problematic. One might argue a larger context for ash pendants, but the Victorian mortuary art that was made from hair, for example, has no connection to the vampire ritual. Nor does the Paracelsian amulet that contains *spiritus*. The Paracelsian "doctrine of amulets," Karen Gordon-Grube wrote, "holds that spirits are corporeal and can be caused to prevent or to cure some particular disease by hanging the amulet in which the spirits are contained about the neck or other part of the body."[653] The ancient healing magic of preserving the cremated remains of a loved one is a powerful motif. But in New England folk medicine, the neck amulet as described by Farey seems more akin to wearing an asafetida bag (filled with various herbs and concoctions) to cure a cold. In the vampire tradition, burning the heart and then, sometimes, administering the ashes to the sick, terminates an asymmetrical, harmful relationship. Keeping the deceased's ashes in close contact extends that fatal familial connection. Yet the small box at the neighbor's neck served Farley's narrative purpose, leading him to tell its back story.

There are elements in Farley's sparse biography that suggest she may have had firsthand or community-based knowledge of authentic vampire incidents. Mrs. Farley was born Sarah E. Foster (ca. 1808-1880), the fourth of nine children of Maria Emerson and Major Robert Foster. During the War of 1812, her family relocated to South Union, Maine, from Newburyport, New Hampshire. Major Foster purchased a farm and mills and at one time kept the only store in Union. In his history of the town of Union, John Langdon Sibley (see chapter 11) recorded that "in 1832 and 1833, a few persons put in practice the proverb, that the burning of the lungs of relatives who died of consumption would cure that disease in the living." He wrote that "one body was exhumed several months after death, and the vital parts were burned near the grave, which was in the Old Burying Ground."[654]

Sarah married a widower, William Jewett Farley, in 1829. He was born in Waldboro, Maine (about 1802) and graduated from Bowdoin College in 1820. Five years later, he moved to Thomaston, Maine. His first wife, Alice—who died in 1827 at age twenty-six—was the daughter of Joseph MacKeen, the first president of Bowdoin College. William Farley and Ephraim Wilder Farley (1817-1880), a kinsman, practiced law together in Thomaston. At the Bowdoin College commencement of September 13, 1836, E. Wilder Farley (as he was known) gave a "dissertation" on "The Character and Influence of Popular Prejudices." Is it just coincidence that, six years later, Mrs. Farley published her article, "Popular Prejudices," which contained an exhumation narrative? While I believe there is a connection between the two, I've not found a copy of E. Wilder's dissertation, or even a summary of its contents.

Sarah Farley lived in a time and place where existence itself was consumptive: the harsh environment and rigorous demands of daily subsistence drained many people of more than they could replenish. Maine's local town histories are liberally sprinkled with disheartening accounts of promising young inhabitants who died in their prime, unable to overcome some minor ailment, such as a cold. Farley's mother died in 1831, at the age of fifty-four; her younger brother, Edward, also died in 1831; and her older sister, Martha, died the next year. In 1839, Mrs. Farley's husband, William, died; he was just thirty-six years of age (and the first member of Bowdoin's class of 1820 to die).[655] The young Mrs. Farley had to face life as a single mother with a two-year old daughter, having buried her husband and their three children who died in infancy.

Small wonder that Farley became a "devotional writer" whose inspirational essays focused on overcoming life's obstacles through piety and faith. Yet, despite her eventful life—and writings that reflected on the sorts of hardships she endured—contemporary scholars have distilled the import of her entire existence to the opening sentence in her essay, "Domestic and Social Claims on Woman" (*Godey's Magazine*,1842): "As society is constituted, the true dignity and beauty of the female character seem to consist in a right understanding and faithful and cheerful performance of social and family duties."[656]

Farley's condemnation of the vampire practice was the norm: virtually everyone stood against superstition. But the opening paragraphs of "Popular Prejudices" imply that virtually everyone, including her near neighbors, were irredeemably superstitious—notwithstanding that they also may have been intelligent, polite, and neighborly. Farley concluded the exhumation story with an admonishment that contains, as far as I know, the first explicit connection between a New England consumption ritual and vampires: "Surely when such things are believed and practised in our own country, we need not wonder at the Vampyre tales of other lands."[657]

Mary Andrews Denison, "Old Superstition," *Home Pictures*, 1853

In 1853, Mary Andrews Denison was on the road to a promising career as a writer, with the publication of her third novel, *Home Pictures*. In the preface, Denison's husband, Reverend Charles Wheeler Denison, a Baptist minister, wrote, "The title of the book sufficiently indicates its character. It is a series of Pictures of Home, as supposed to be viewed by a country girl, who becomes the *wife of a merchant*, and sketched by her own hand from real life." Many of the chapters first appeared in *The Olive Branch*, a periodical "devoted to Christianity" and "polite literature," which, at that time, was under the assistant editorship of Reverend Denison. This weekly religious newspaper, which began publication in Boston in 1836, was one of the first papers to publish continued stories. The following exhumation narrative—a story within a story that comprises the chapter entitled "Old Superstition"—may have appeared first in *The Olive Branch*:

> One learns many a curious little thing in a village like this. I listened to the narration of a most singular incident yesterday at the house of a neighbor.
>
> It seems that there is an old superstition, strongly believed by the credulous even at this day, that if the heart of the last deceased member of a consumptive family is taken from the body and burned, and the ashes reserved as a medicine to be given to the rest in small doses, no other person of that family will die of this terrible scourge. Various reasons are assigned as causes for belief in the efficacy of this curious experiment. Among them, one that in that dead heart there is a drop of blood which retains its color and freshness, by preying upon the vitality of those connected to it when living, by natural ties.
>
> Several members of a large and respectable family had been early taken from earth by consumption; and, after following the body of an amiable sister to its final resting-place, the survivors met to talk over past events, and to mourn together for their loss. Each brother and sister felt the hectic glow, with its fitful fever feeding on their cheeks—each knew that the seeds of an insidious disorder were deeply sown in their feeble constitutions. They painfully realized how hopelessly doomed they were to certain and early death.
>
> Among the matters discussed was a proposition, made by a friend of the family some time previous, to test the efficacy of this strange remedy—the roasted heart of the buried sister. No wonder they shuddered as they thought of it, standing sorrowfully together, a little remnant, soon to be uselessly laid by—nor that they each and all shrunk back from the idea of eating their own flesh and blood. But one after another they submitted to the alternative. The physician was consulted, and requested to apply the knife to the corpse after it should be taken out of the tomb. He hesitated, and persuaded them to relinquish the idea, at once senseless and heathenish, and they desisted. But another fell a victim to the disorder, and they determined, at all events, to perform what they considered their duty.
>
> Again the doctor was summoned, and this time he complied with their strange request.
>
> Accordingly, at midnight he repaired, with a few of the family, to the old burial-ground, and, with a dark lantern, they all stood beside the grave in the stillness of the ghostly hour, while the aged sexton threw up the damp clods, and finally lifted the door that led into the tomb. The heart was carefully separated from the body by the surgeon's knife, and placed in charge of one of the brothers. As if to verify the truth of the assertion, there was, truly enough, a drop of fresh, red blood in its centre; and shocking as was the ordeal in prospect, they almost exulted as they fancied that the true and only successful remedy had been at last discovered.
>
> They burned the heart to ashes, and used it as a medicine. But, alas for human hopes! the hand of the destroyer was not stayed. Long since, every soul of that family had gone to its last account. So much for old superstitions.[658]

Whether fictional or based on an actual event, I cannot say, but every motif in this story accords with the New England vampire tradition. Denison undoubtedly had heard of or read about at least one of these exhumations. The details of her life, though sparse and sometimes conflicting, argue against her firsthand encounter with the tradition. Mary Ann (née Andrews) Denison (1826-1911) was born in Cambridge, Massachusetts, and married Reverend Denison (seventeen years her senior), a Baptist minister and editor of the *Emancipator*, New York's first antislavery journal.

Mary Denison wrote some eighty popular novels that sold more than a million copies, including *Edna Etheril, the Boston Seamstress* (1847), *Gertrude Russell* (1849), *Home Pictures* (1853), *Opposite the Jail* (1859), *The Mill Agent* (1864), and her best-selling, *That Husband of Mine* (1877). Her success was founded on an ability to weave together various elements of popular forms, such as murder mystery, adventure, and suspense, within the framework of domestic life. Her biographer observed that "Mary Denison's strength lay in her ability to gauge and satisfy the lowest common denominator of current popular taste. Her forte was in combining romance and reality, and in technique and popularity she matched better known women novelists of the day. To thrilling and suspenseful incidents of temptation, jealousy, and revenge, she joined appeals to domesticity, connubial faithfulness, and the homely virtues."[659] Her work touched upon the diverse issues of the day, including slavery, temperance, and class antagonism, with unexpected twists and turns. In *Old Hepsy* (1858), the heroine believes she is white but discovers that her father is Black and that her parents were half-siblings. Despite such astonishing contrivances, however, Denison's writings fundamentally reinforced conventional morality.

Denison likely learned of New England's vampire practice through the popular press. Two years before the appearance of *Home Pictures*, several newspapers published two conjoined exhumation narratives, one from Cayuga County, New York, and the other from Litchfield County, Connecticut (see chapter 6). Denison could have used them to mold her narrative, as their particulars are a close match.

Amy Lowell, "A Dracula of The Hills," 1925

MacEdward Leach scolded writers who misrepresent a community's genuine folklife. But he acknowledged those who succeeded, writing that "much genuine folklore is used in popular regional literature accurately and artistically."[660] The regional writings of Amy Lowell and Edith Wharton lean to the latter.

Despite the title of her lengthy narrative poem, "A Dracula of the Hills" (first published in *Century Magazine*, in 1925), Amy Lowell steered away from European vampire lore, pursuing, instead, the New England tradition. Lowell's vampire, Florella, kills silently from the grave, beginning with her dearest family member, her husband. Written in blank verse, in iambic pentameter, like the other free-verse poems in her posthumous collection *East Wind* (1926), Lowell's vampire narrative unfolds in the style of a traditional ballad: through dialogue, rendered in dialect, the narrative, written as a memorate (supernatural personal-experience story), leaps from scene to scene, lingering on the crucial episodes that move the plot toward its dramatic climax. Echoing the montage technique of the nascent film industry of the time, she builds terror as her impressionistic sequence of images unfolds the New England vampire theme. This poem, like a ballad or memorate, should be performed; it is aural art transcribed.

Lowell's dark, atmospheric tale is set in western Massachusetts during the late 1800s. In this story within a story, an aging Becky Wales, now living during a time of telephones and automobiles, looks back to her childhood— "'Twas all of forty year ago," she says—when she lived "t'other side o' Bear Mountain to Penowasset." Foreshadowing imminent horror, Becky tells us there were "some fearful strange things I can't never lose a mite of, no matter how I try." Unsure how to characterize the tradition— "I don't know as you'd call it a custom"—she recognized that it was recurrent— "'Twarn't th' first time th' like had happened, I know."

Becky's dwindling neighbor, Florella, becomes alarmingly animated as she foresees a supernatural pathway for cheating death:

"'Tain't no use lyin' to me, Becky Wales, I know I'm dyin'.

But I won't die. You'll see.

I'll find some way o' livin'.

Even ef they bury me, I'll live.

You can't kill me, I ain't th' kind to kill.

I'll live! I'll live, I tell you,

Ef there's a Devil to help me do it!"

She screamed this out at me, settin' up in bed

An' p'intin' with her finger.

The town's doctor confirms Florella's dire prognosis— "Dr. Smilie said ther' warn't nothin' to do for her"—prompting her to seek out the local sorceress: Florella "took a notion to see Anabel Flesche . . . a queer sort of woman . . . she lived in a little shed of a place over Chester way." Despite the witch's knowledge of "herbs and semples . . . an' things like that . . . Florella didn't change none" but kept "sinkin' an' sinkin'" until "ther' warn't nothin' lef' of her but eyes an' bones." We learn what was killing Florella: "'Worryin' 'bout th' life was leavin' her, an' all eat up with consumption.' It was not galloping consumption, for she took two years to die."

Staring into the open coffin at Florella's funeral, Becky notes, "Ther' was a queer, awful smile 'bout her mouth. It made her look jeery, not a bit th' way Florella used to look." Soon, Florella's husband began to show the dreaded symptoms. Near death, Joe acknowledges her grasp:

"Florella had a mighty strong will," says Joe agin.

"She owned me body an' soul, an' that was a rare pride to me."

. . . "I guess she owns me still," he says.

Joe dies, and Becky relates the evident implications:

About a week after th' funeral, Father met Anabel Flesche.

"So Joe Perry's dead," whined Anabel, an' Father was sure th' old hag looked pleased.

He only said, "Yes, he's dead," an' was pushin' on when Anabel stopped him.

"Florella's a determined woman," she cackled, "ain't you afeerd she'll try somebody else?"

"What th' Hell do you mean?" cried out Father.

"She loved life," said Anabel, in a queer, sly way,

"Joe's gone, but ther's others."

In a scene played out in many consumption rituals, the town's decision-makers gather to interpret the uncanny events, summoning past experiences:

Father an' Jared Pierce went straight to th' Selectmen,

An' told 'em what Anabel was hintin'.

Then some old people rec'llected things which had happened years ago,

An', puttin' two an' two together, they decided to see for themselves.

At night—"so's not to scare folks"—with lanterns, pickaxes, and spades they raise the coffin and open it.

Florella's body was all gone to dust,

Though 'twarnt' much more'n a year she'd be'n buried,

But her heart was as fresh as a livin' person's,

Father said it glittered like a garnet when they took th' lid off th' coffin.

It was so 'live, it seemed to beat almost.

Father said a light come from it so strong it made shadows

Much heavier than th' lantern shadows an' runnin' in a diff'rent direction.

Oh, they burnt it; they al'ays do in such cases,

Nobody's safe till it's burnt.

Now, sir, will you tell me how such things used to be?

They don't happen now, seemingly, but this happened.

You can see Joe's grave over to Penowasset buryin'-ground

Ef you go that way.

The church-members wouldn't let Florella's ashes be put back in hers,

So you won't find that.

Only an open space with a maple in th' middle of it;

They planted th' tree so's no one wouldn't ever be buried in that spot agin.

Born into a wealthy, aristocratic, and prominent family in Brookline, Massachusetts (near Boston) Amy Lowell (1874-1925) was educated at home and in exclusive schools. Her knowledge of the vampire tradition surely was secondhand. She wrote that her upbringing was "very cosmopolitan" and that "the decaying New England" had been "no part of my immediate surroundings." Regarding the native Yankee, she admitted to being "a complete alien."[661] Where, then, might she have learned of the vampire tradition? Lowell answered the question in a letter (in 1921) to Glenn Frank, editor of *Century Magazine*: "The last case of digging up a woman to prevent her dead self from killing the other members of her family occurred in a small village in Vermont in the '80's. Doesn't it seem extraordinary?" She said her source was the "American Folk-Lore Journal."[662]

The article in the *Journal of American Folklore* (JAF) that matches Lowell's brief characterization is Jeremiah Curtin's "European Folk-lore in the United States," published in 1889. But the details of this case (see Chapter 15) do not match the brief description Lowell wrote in her letter. The Woodstock vampire was a man, not a woman (although a woman was Curtin's informant) and the incident, *reported* in the 1880s, occurred about 1829. It may be that Lowell first encountered the New England vampire tradition in Curtin's article, but since her poem included several salient motifs of the consumption tradition absent in Curtin's article, she surely used other sources.

Lowell's poem bears a striking resemblance to the two vampire narratives undoubtedly written by Francis Gerry Fairfield, published in 1871 (see Chapter 20). Selected excerpts highlight their shared motifs.

"A Yankee Vampire," Sacramento *Daily Union*, April 29, 1871 (emphasis mine):

- *Florrie* Dunbar is a main character.
- "The doctors called it *galloping consumption*."
- "They told strange stories, too, about old Gramfer Dunbar having *threatened to come after the rest of the family when he died*; and, the fact is, there was a sort of vampirism in the very appearance of the old gentleman."
- ". . . and the neighbors had started the idea of taking up old Dunbar's body and bunring it."
- "when I removed the heart I aver, sir—I am prepared to swear—that it beat just as regularly as my own at this very instant. *It was a fresh, warm, faintly-beating heart* that I held in my hand—not a dead heart at all."

"Quick Consumption: Every-day Story of New England," *Frank Leslie's Illustrated Newspaper*, November 25, 1871 (emphasis mine):

- *Florry* Calhoun is a main character.
- "Hopeless case, I'm afraid," grumbled the oracular Bloupil." "No use," muttered he, monosyllabically. "Beyond human skill already."
- "I saw her in her coffin. She was absolutely wasted to skin and bones, as if some horrible something had eaten away, buzzard-like, every ounce of flesh from that *grinning anatomy*—as if, in fact, some horrible vampire had sucked the arteries of vitality dry, leaving of her nothing but a mere withered anatomical framework."
- "The corpse crumbled—crumbled as I began to remove the white grave-clothes, reeking with horrible mildew. But I found it at last. It was—or so I fancied—*a red, warm, human heart*, lounging prone upon the bare, fleshless spinal column in the middle of the coffin."

Had Lowell read Fairfield's narrative(s)? The vampires of both, while explicitly based on the New England model, trace their ancestry to pre-industrial Europe. In a discussion of vampires as creatures of the imagination, Koen Vermeir noted that some "magically oriented writers" of

Europe's Early Modern period (ca.1650 to 1750) believed that the human imagination was "so powerful that it could act after the death of a person, and it was this strong imagination that turned him or her into a vampire."[663] These writers supposed that "the imagination of the corpse, still active because of the continuing operation of the vital powers, sent out noxious vapours or even a semi-corporeal avatar that could kill specific surviving relatives." [664] Lowell's Florella vows to find a way to live, even if buried. Her dying husband acknowledges her "mighty strong will," admitting, "She owned me body an' soul" and "I guess she owns me still." It was the powerful imagination of Fairfield's old Gramfer Dunbar that animated his threat to "come after the rest of the family when he died." The willful and vocal assertion to continue living after death, at the expense of their living kin—a dramatic foreshadowing device—separates these fictional creations from their documented New England counterparts, as does Florella's willing submission to the Devil.

Lowell's choice of "Dracula" instead of the generic "vampire" for her poem's title is telling. The term "vampire" did not appear in her letter to Glenn Frank in which she commented on the "extraordinary" New England custom. Lowell's selection of the literary Dracula suggests that she assumed her readers would know the novel or its stage productions and be able to link Florella with the Count. By the early 1920s, when Lowell had completed her poem, Dracula was well on the road to dominating the vampire genre. *Dracula* and *vampire* became synonymous.

Edith Wharton, "Bewitched," 1926

In "Bewitched," Edith Wharton's visual portrayal of the landscape (natural and built) and her characters—and how these fading New England images reflect each other—is vivid. Her subtle hints build tension as the short story progresses, sowing more doubt than resolution. I'm not surprised that Wharton and Henry James ("The Turn of the Screw") were close friends: both allow the artful psychology of doubt to worm its way into the reader's psyche, where it lingers. Wharton's footprints in the snow is a hoary motif in folk literature, yet performs nonetheless, as does this image early in the narrative: "The snow continued to fall in a steady unwavering sheet against the window, and Bosworth felt as if a winding-sheet were descending from the sky to envelop them all in a common grave." Near the story's climax, an afflicted family member confirms this foreshadowing, "Don't it sometimes seem zif we was all walking right in the Shadow of Death?"[665]

New England's history of exhuming vampires also lurks in the shadows. Ironically, its most graphic portrayal, a stake through the breast, is also its least authentic:

> "Prayer ain't any good. In this kind of thing it ain't no manner of use; you know it ain't. I called you here, Deacon, because you remember the last case in this parish. Thirty years ago it was, I guess; but you remember. Lefferts Nash — did praying help him? I was a little girl then, but I used to hear my folks talk of it winter nights. Lefferts Nash and Hannah Cory. They drove a stake through her breast. That's what cured him."
>
> Sylvester Brand raised his head. "You're speaking of that old story as if this was the same sort of thing?"
>
> "Ain't it? Ain't my husband pining away the same as Lefferts Nash did?" The Deacon here knows—"

>The Deacon shook his head. "The man's a sick man — that's sure. Something's sucking the life clean out of him."
>
>"Oh—" Orrin Bosworth exclaimed.
>
>The Deacon stirred anxiously in his chair. "These are forbidden things," he repeated. "Supposing your husband is quite sincere in thinking himself haunted, as you might say. Well, even then, what proof have we that the...the dead woman...is the spectre of that poor girl?"
>
>"Proof? Don't he say so? Didn't she tell him? Ain't I seen 'em?" Mrs. Rutledge almost screamed.
>
>The three men sat silent, and suddenly the wife burst out: "A stake through the breast. That's the old way; and it's the only way. The Deacon knows it!"

Later, the mysterious, unnamed wasting ailment is linked to certain families:

>"They say her lungs filled right up...Seems she'd had bronchial troubles before...I always said both them girls was frail...Look at Ora, how she took and wasted away!... Their mother, too, she pined away just the same. They don't ever make old bones on the mother's side of the family."

When Wharton alludes to New England's vampire tradition without mentioning vampires (or even consumption), she reflects the ground-level reality of its participants. In Everett Peck's family story of Mercy Brown (chapter 9), for example, neither word—*vampire* nor *consumption*—was mentioned; it was "this awful thing" caused by "some mysterious disease."

Edith Wharton (1862-1937) had ample opportunity to learn of this tradition from both published and oral sources. She was raised in New York City, with access to periodicals that regularly published such accounts. As a young child, she summered with her family in Newport, Rhode Island, a few miles from where Mercy Brown was exhumed in 1892; after marriage, in 1885, she and her husband bought a house in Newport ("Land's End") where they lived each year from June to January, until about 1904. The Whartons often visited Saunderstown, in North Kingstown, a ferry ride from Newport. Edith's cousins, Henry Wharton and his wife, Frances Willing Wharton, were among Saunderstown's summer residents, many of whom had national reputations in the literary, artistic, and political realms. Frances, the daughter of Benoni Lockwood (the son of a Providence trader who operated in the West Indies, also named Benoni), was, like her cousin Edith, a well-known writer.[666] Saunderstown was the home of William R. Rose, who performed the consumption ritual in 1872 (see chapter 8). Tiring of Newport, Edith and her husband bought a summer house in the Berkshire Hills of western Massachusetts, fertile ground for therapeutic exhumations, as well as the setting for Amy Lowell's "A Dracula of the Hills," published just one year before Wharton's short story.[667]

Manly Wade Wellman, "Chastel," 1979

Manly Wade Wellman (1903-1986) was well positioned to write a story about New England vampires. His ancestry, upbringing, education, and work experience all pointed him toward his favored topics, the supernatural and the South, both rooted in tradition and conflict. Wellman was born in Angola, Africa, where his father, a doctor, purportedly saved a witch doctor's eyesight. As a boy, Wellman learned African stories and magical traditions. At various times, he worked as a harvest hand, cowboy, and roadhouse bouncer.[668] After graduating with a degree in English from Wichita State, he earned a law degree from Columbia, then began a career as a newsman in Kansas.

Wellman became friends with Ozark folklorist Vance Randolph, accompanying him on collecting trips. On one of these forays, Wellman met folk musician Obray Ramsey, who became a lifelong friend and inspiration. Wellman began writing tales about the Ozarks and, in the 1930s, widened his scope and began publishing stories in pulp magazines, including *Weird Tales*, *Wonder Stories*, and *Astounding Stories*. When his income from writing began to decline during the Depression, he took the position of Assistant Director of the Work Project Administration's New York Folklore Project. He later relocated to rural North Carolina, where he became absorbed with local folk traditions, especially legends and music. Descended from Confederate general Wade Hampton (his namesake), he researched and wrote about Southern history, especially the Civil War. One critic wrote that Wellman had "the eye of an anthropologist and the storytelling gift of a likable teacher."[669]

Wellman's short story, "Chastel," first published in 1979, blends New England's vampire tradition with the familiar fictional vampire. It is June in a village near Jewett City, Connecticut, and a small group is watching rehearsals for a musical adaptation of Dracula, entitled "The Land Beyond the Forest." Among them is the occult detective and author of Vampiricon, elderly Judge Keith Hilary Pursuivant, essentially Bram Stoker's Van Helsing in a "tailored blue leisure suit." Lee Corbett, also an authority on the supernatural,

> remembered the story in Pursuivant's book about vampires at Jewett City, as reported in the Norwich Courier for 1854. Horace Ray, from the now vanished town of Griswold, had died of a "wasting disease." Thereafter his oldest son, then his second son had also gone to their graves. When a third son sickened, friends and relatives dug up Horace Ray and the two dead brothers and burned the bodies in a roaring fire. The surviving son got well. And something like that had happened in Exeter, near Providence in Rhode Island.

In an earlier conversation, when Pursuivant mentioned the "lively vampire folklore" in Rhode Island, Corbett had suggested that they "leave Rhode Island to H. P. Lovecraft's imitators."

As Wellman unfolds his tale, we learn that the play's leading lady, Gonda Chastel, is a descendent of the Ray family and that her mother, whom Pursuivant had fallen in love with some sixty years earlier, is entombed in the local cemetery. Pursuivant naturally had the foresight to prepare for the impending assault from the past: "He looked at jottings from the works of Montague Summers. These offered the proposition that a plague of vampires usually stemmed from a single source of infection, a king or queen vampire whose feasts of blood drove victims to their graves, to rise in their turn. If the original vampires were found and destroyed, the others relaxed to rest as normally dead bodies." Summers's book, *The Vampire in Europe* (1929), obviously was a source for Wellman's (and, thus, Pursuivant's) vampire history and folklore. Both the Ray case and that of William Rose, of Rhode Island, appear on the same page in Summers's book (see chapter 8).

Charles M. Skinner's book, *Myths and Legends of Our Own Land* (1896), also makes an appearance when Pursuivant reads an excerpt of a vampire legend from Schenectady, New York;

"The Green Picture" is the tale that I am certain inspired Lovecraft's short story "The Shunned House." Wellman's description of the book is indisputable evidence that his knowledge of it was firsthand: "Judge Pursuivant sat in his cubicle, his jacket off, studying a worn little brown book. Skinner, said letters on the spine, and Myths and Legends of Our Own Land. He had read the passage so often that he could almost repeat it from memory: 'To lay this monster he must be taken up and burned; at least his heart must be; and he must be disinterred in the daytime when he is asleep and unaware.'" In "Another Look at Charles Skinner," John Bealle noted, in 1994, that Skinner "perceives himself within a literary tradition, as providing source material for the Hawthornes and Irvings to come."[670] Lovecraft and Wellman were two such beneficiaries of Skinner's narrative stock.

Like Edith Wharton in "Bewitched," however, Wellman cannot resist the stake through the breast motif, even though Pursuivant had just reviewed Skinner's tale and previously informed his companions that the Ray vampires were dispatched by burning their corpses. But Wellman needed the venerable occult detective to employ an even more venerable weapon once again. So, when the time for action arrives, Pursuivant

> put aside the notes . . . and picked up his spotted walking stick. Clamping the balance of it firmly in his left hand, he twisted the handle with his right and pulled. Out of the hollow shank slid a pale, bright blade, keen and lean and edged on both front and back. Pursuivant permitted himself a smile above it. This was one of his most cherished possessions, this silver weapon said to have been forged a thousand years ago by St Dunstan. Bending, he spelled out the runic writing upon it: Sic pereant omnes inimici tui, Domine. That was the end of the fiercely triumphant song of Deborah in the Book of Judges: So perish all thine enemies, O Lord. Whether the work of St Dunstan or not, the metal was silver, the writing was a warrior's prayer. Silver and writing had proved their strength against evil in the past.

To learn if "silver and writing" prevail again, read Wellman's tale, which is indebted to the vampire folklore and the history of New England.

In both Wellman's "Chastel" and his nonfiction essay, "The Vampire in America," we learn his sources of information about America's vampires: Montague Summers's *The Vampire in Europe* (1929), which included the accounts of Horace Ray of Connecticut, William Rose of Rhode Island, and the unnamed woman from Chicago; and Charles Skinner's "The Green Picture," in *Myths and Legends of Our Own Land* (1896), which located the vampire belief in Rhode Island. Wellman also arrived at the conclusion I made independently, some years later, regarding the probable relationship between Skinner's tale, "The Green Picture," and Lovecraft's "The Shunned House." Wellman wrote: "This intriguing story returned eastward to Providence, home of Howard Phillips Lovecraft, six years old in 1896 and already beginning to read and treasure such things. He was to use Skinner's account for the plot of his own creepy tale, 'The Shunned House,' with a climax he successfully hoped would make the hair stand up on the reader's head."[671]

***

The century-and-a-half fictional line from "Consumption" (1829) to "Chastel" (1979) followed interpretive trends in the broader contexts of New England's consumption ritual, with one notable exception: Mrs. Farley's introduction of "vampyres," in 1842, preceded by more than a decade

Franklin Hough's mention of "searching for vampires," in his 1853 *History of St. Lawrence and Franklin Counties, New York*. During the first half of the nineteenth century, the European vampire was perhaps more familiar to literary authors, such as Farley, than it was to local historians, medical practitioners, and journalists. Francis Gerry Fairfield's Yankee vampire stories of the early 1870s bridged a wide lexical gap, for it was not until the 1880s and '90s that the term *vampire* was applied liberally to the New England tradition by narrators of all sorts. But the word immigrated with connotations that engulfed narrators of the local tradition. The notoriety of Rhode Island's vampires, launched by the Mercy Brown event of 1892, received a boost from George Stetson's 1896 article. That same year, when the *Boston Globe* transformed Rhode Island vampires from hard news to infotainment, they began traveling the bifurcated path of object lesson and caricature. Europe's classic vampire, represented by Dracula after 1897, haunted both. Amy Lowell's rustic Florella *was* a Dracula. A stake through the breast destroyed Edith Wharton's thinly veiled vampire. Despite Manly Wellman's undeniable knowledge of New England's authentic tradition, he would not abandon the Dracula model, as his Connecticut vampire left the grave, traveled through decades to infect others, and finally was dispatched with a magical silver spike through its heart.

Nancy Kinder's "The Vampires of Rhode Island," which appeared in the October 1970 edition of *Yankee* magazine, was a twentieth-century watershed event, reigniting embers that had been dormant for decades. She wrote that "the family and friends" of Edwin Brown "unanimously agreed that it must be a vampire that was sucking his blood and causing his loss of strength."[672] The state's vampiric reputation crested in 1975 when Raymond McNally, who linked Dracula to the historical Vlad the Impaler, dubbed Rhode Island the "Transylvania of the Western World."[673] Newspaper articles and television documentaries fed back into a thriving oral tradition that had informed their production, transforming unfortunate victims who died of pulmonary tuberculosis into fanged fiends still roaming the countryside. In the wider vampire context, films, many based explicitly on Dracula, were omnipresent; novels, notably Anne Rice's vampire chronicles, initiated in 1976, were published in increasing numbers; and television shows, such as *Buffy the Vampire Slayer* (1996–2003), featured vampires as both protagonists and antagonists. Internet users eagerly embraced vampires. Heightened popularity, combined with an expanding awareness of New England's tradition, induced authors to write novels featuring American vampires. Mercy Brown, who died in relative obscurity at the age of nineteen in 1892, is now an international celebrity. She has achieved immortality and, in that sense, *is* a vampire.

Conclusion: Disease and Death, Fear and Hope

When the consumption ritual is viewed collectively—as in this book—its strangeness fades. It was a reasonable (if not rational) response to an intractable disease. But this vantage was not available to most New Englanders who encountered the ritual. For them, it was a singularity, an extraordinarily barbaric ceremony, out of place in civilized society. Those afflicted, however, could relate to Voltaire's description of medical practitioners, allegedly penned in 1760: "Doctors are men who prescribe medicines of which they know little, to cure diseases of which they know less, in human beings of whom they know nothing." The consumption ritual was a crossroad where hope intersected with fear of disease and death.

Eyewitness accounts record these poignant events, providing a perspective unmediated by intervening voices. No first-person narrative is more compelling than that of Reverend Justice Forward, of Belchertown, Massachusetts. His decision to exhume the bodies of his mother-in-law and daughter, in 1788, was made in consultation with others in the family and community, and in relation to various cultural systems, including kinship, folklore, medicine, and religion, each with its own premises and operational rules for identifying choices. My examination of another first-person recollection, by a man who had directed the exhumations of four family members in the mid-1800s, revealed the prominent role kinship played in disseminating the ritual across generations and from one place to another. New Englanders seemed constantly on the move, looking for better opportunities. Family and community networks were vital in creating and sustaining stories that introduced the ritual and affirmed its healing powers. Outside the bounds of kinship and community, traveling healers, some of them foreigners, also played a role in introducing and spreading the tradition.

Family history stories confirm the variability of oral tradition. Those of Everett Peck and Reuben Brown, from Exeter, Rhode Island, show the significance of role relationships and feedback in the storytelling process. The insider-outsider views (the esoteric-exoteric factor) that shape consumption narratives are exemplified by newspaper interpretations that appeared immediately following the Brown family exhumations in 1892. Current mass media stories of Mercy Brown are in stark contrast to the family stories. Peck's older relatives, warning youngsters away from the site of "this awful thing," narrated a connecting web of people, places, and events that stretched into the past. Mercy's story played a central role in this family's initiation ritual that was performed periodically in the Chestnut Hill Cemetery. Contemporary media package the narrative of Mercy Brown with other spooky Halloween stories and activities to entertain a mass audience. Present-day oral tradition reinforces and, in turn, is reinforced by mass media treatments. The creation and dissemination of the Nellie-Vaughn-is-a-vampire legend, likely a result of confusing her with Mercy, is a textbook example of the symbiotic relationship between oral tradition and mass media.

Newspapers relegated consumption rituals to "the outer circle of deviance," where issues that "mainstream culture treats . . . with derision and contempt" reside. Sensationalized vampire stories attract readers and require little in the way of context development. Journalists understood that such topics can be "ridiculed, marginalized, or trivialized."[674] Vampire narratives allowed writers and readers to congratulate themselves for having traveled so far along the road to civilization. New England's local histories, unlike newspapers, typically ignored consumption rituals. Many of these town histories were written during the latter half of the nineteenth century, when the "old New England" was a world that appeared to be slipping away, if not already gone. But it was an idealized place, a "constructed historical region" that "is still the pastoral version of New England—post-Puritan, premodern, rural, and Anglo-Saxon—that many people carry in their heads today."[675] Local chroniclers skirted topics that detracted from cherished images of their idyllic towns. In his county histories of upstate New York, Benjamin Hough wrote of vampire incidents: "Our space forbids the details, revolting to humanity, and regard for the living, leads us to pass unnoticed these heathenish mutilations of the dead."[676]

Beneath the surface of the romanticized folk lurked the uncouth savage who practiced cannibalism. America's nineteenth-century medical establishment, pushing to consolidate its authority, used such practices to marginalize competing medical systems. In the biomedical paradigm's emerging sovereignty, both strands of cannibalism—the fading Paracelsian doctrine that prescribed mummy as a curative and the family-based folk tradition—were off limits. Physicians joined historians, newspaper publishers, and other members of the establishment in condemning the consumption ritual. Their condemnation found a scientific rationale in the theory of unilineal cultural evolution advanced by comparativists, including folklorists and anthropologists. As human culture advanced from savagery through barbarism to civilization, the theory went, beliefs and practices carried over from lower stages hindered civilization's consummation. These "survivals" needed to be identified so that they could be eradicated. In 1896, anthropologist George Stetson introduced New England's vampire tradition to a wide audience by asserting that the "belief in the vampire" arose because "the barbarian" was "unable to distinguish the objective from the subjective" and, therefore, ascribed "all natural phenomena to good and evil spirits."[677] Referring to southern Rhode Island, Stetson wrote, "Naturally, in such isolated conditions the superstitions of a much lower culture have maintained their place and are likely to keep it and perpetuate it, despite the church, the public school, and the weekly newspaper."[678]

Some authors who wrote about New England's vampire tradition straddled the line between condemnation and romanticism. Vampire rituals provided plot, character, and local color for community-based authors, whose narratives wavered between fiction and nonfiction. Where sticking to facts might narrow the appeal of a good story, writers often chose aesthetics over authenticity. Three versions of a vampire narrative from Stafford, Connecticut, penned by Francis Gerry Fairfield, exemplify history's malleability in the hands of a determined author. I suspect that Fairfield's "Quick Consumption: Every-day Story of New England" (*Frank Leslie's Illustrated Newspaper*, November 25, 1871) was the basis for an article that appeared in newspapers a few months later. Stripped of narrative, the following text was embedded in "New England Superstition," which was published in both the *Elk County* [PA] *Advocate* (March 14, 1872) and the *Atchison* [KS] *Daily Patriot* (March 2, 1872):

> There is a strange vampire superstition associated with that scourge of New England, "quick consumption," which differs materially from the Tartar and Semitic legends of the vampire. . . . Some analogy there is between this and the New England superstition of quick

consumption; but in the New England version, an actual demon is presupposed. By some strange aberration of the ordinary laws of dissolution, the heart is supposed to retain its vitality after death, and live on in its coffin, drawing by some weird, sympathetic influence, its support from the vitality of some living relative, who wastes into the grave in consequence of having to sustain two vitalities instead of one, and thus whole families drop off one after another, with singular regularity of interval. The remedy is to take up the body of the dead and burn the organ supposed to exercise the deadly vampire function; and odd legends are afloat of instances in which the remedy has been successfully tried.

Unlike Fairfield, this article linked the New England tradition to the Old World.

※※※

The vampire is a prime image for metaphorical extension. Philosophers and writers have applied it to unnatural predators, from politicians and businessmen to economic systems, including slavery. In 1836, the year before Elijah Lovejoy was murdered by a mob, he wrote of slavery in the abolitionist newspaper he edited, "In every community where it exists, it presses like a night-mare on the body politic. Or, like the vampire, it slowly and imperceptibly sucks away the life-blood of society, leaving it faint and disheartened to stagger along the road to improvement."[679] Another abolitionist, Joseph Horace Kimball, also died young. But his killer was consumption. A lengthy poem commemorating his death was composed by young Quaker abolitionist, Eliza Earle. It contains the following stanzas:

> Thus, though Consumption's vampire grasp
>
> had seized thy mortal frame,
>
> Thy ardent and inspiring mind,
>
> untouched, remained the same.

Which returns us to the inscription on the gravestone of Simon Whipple Aldrich (1814-1841), located in a North Smithfield, Rhode Island, cemetery (see Introduction). Only the first two lines are visible because the stone has been set in cement. The lightly incised graffiti on it reveals that it was known to legend trippers (people who visit sites believed to be haunted) at least as early as the "1961" scratched onto its slate surface. Its broken top and cemented base suggest that someone broke the stone, perhaps while attempting to steal it. Since the poet's mother was born in the town of North Smithfield, I wonder if the two families were acquainted. Because the gradual wasting-away of consumptives mirrors the reported symptoms of vampire attacks, the vampire is an apt metaphor for consumption.

I believe that the vampire tradition was introduced to New Englanders as a medical practice by immigrants from various regions of Eastern and Northern Europe during late eighteenth and early nineteenth centuries.[680] Both European and American versions of the tradition share the belief that deceased kin can become possessed by evil spirits (or transform into evil beings) and then begin attacking their living relatives. Both also include measures to identify and destroy the evil spirits (beings). The following three European examples could have been recorded in New England:

- Korbesz, Hungary: "At night they went to the cemetery, opened the grave, cut out the dead man's heart, stuck it on a pitchfork and burned it."[681]

- Amarasesti, Romania: "Fifteen years ago, in the village of Amarasesti, a very old woman, the mother of Dinu Gheorghita and his brother, died. After some time, the children of both her sons started to die, one after another. The sons were convinced that their mother had become a vampire. So one night, they opened her grave, cut her in two, and buried her again. But the deaths continued. So they dug her up a second time. They were shocked to find that the two halves had grown together again. They took the corpse to a forest where they removed her heart, from which fresh blood was flowing. They cut the heart in four pieces and burnt it. Then they also cremated the corpse. The ashes were gathered. The ashes of the corpse were reburied in the grave. And the ashes of the heart were taken home and mixed with water, so that they could be drunk by the victims. From that moment the deaths ended."[682]
- Cusmir, Romania: "This time they succeeded to cut out a part of his heart and his liver, which they burnt. The ashes were gathered, mixed with water, and given as medication to those who were ill. This is the only way in which those victims can be cured."[683]

Unlike in New England, however, these European traditions are enmeshed in elaborate cultural systems that explain why and how a corpse becomes a vampire and prescribe measures to prevent that from happening. There are also rituals to ensure that the vampire remains in its grave, to ward off its attacks, and to locate its grave.[684]

**

Decapitation as a means of destroying a vampire does not appear in New England vampire narratives. But if a decapitated corpse could talk, JB, who was disinterred (twice!) in Griswold, Connecticut, probably could tell us a familiar story. As I discussed in *Food for the Dead*, in 1990, Connecticut State Archaeologist Nicholas Bellantoni directed the excavation of a newly discovered, unmarked cemetery, threatened by erosion.[685] Research showed that this family cemetery was in use from 1757 until about 1830. Twenty-eight of the twenty-nine burials uncovered were interred in a manner typical for that place and time. The exception (Burial #4) was found inside a stone-lined crypt, in a wooden coffin whose lid contained an inscription spelled-out in brass tacks: JB AE 55. Peering into the coffin, Bellantoni noticed that JB's skull and thigh bones were arranged in a "skull and crossbones" pattern on top of his ribs and vertebrae, which were also rearranged. Later analysis revealed that JB's ribs contained lesions that indicated he had a chronic pulmonary infection, likely tuberculosis.[686] Interred next to JB were, also spelled-out in brass tacks on their coffin lids, IB 45 (female) and NB 13 (subadult), likely JB's wife and child.

There is no doubt that JB's corpse was deliberately disinterred, his head and femurs were disarticulated within the coffin, and then his remains were reburied. While JB cannot tell us directly what happened to him and why, viewing his plight against the backdrop of the consumption epidemic in nineteenth-century New England and its associated vampire tradition yields a plausible narrative. As a last resort—to save the lives of the family and stop consumption from spreading into the community—JB's body was exhumed so that his vital organs (or entire corpse) could be burned. But JB was in an advanced stage of decomposition. Finding no vital organs or flesh to burn, the ritual's performers devised an alternative. They repositioned JB's skull and thigh bones, thus confusing and disabling the resident evil spirit. Both procedures have precedent in Europe, including Great Britain—and among some immigrant communities in America.[687]

Initial DNA analysis of JB's remains revealed only a generalized European lineage. Newer technology has identified a matching surname: Barber. Research disclosed an obituary in the *Norwich [CT] Courier* (August 16, 1826) for a 12-year-old boy named Nathan Barber, son of John Barber, who died in Griswold, on July 27, 1826. The grave near JB's, containing a coffin with the notation "NB 13" tacked on the lid, might well be that of this Nathan Barber, who died during his thirteenth year. The New London County Map of 1854 shows a house site for "J. Barber" near the Walton Cemetery. As new evidence comes to light, the extraordinary story of JB will continue to evolve.

***

The apprehensions once visible on the living faces of JB's family were made tangible in JB's repositioned bones. In JB's twice-exhumed body we read disease, death, fear, and hope. These primary elements of the human condition are braided into a single strand in the consumption ritual. We can separate them conceptually, but following one leads to the others. Fear of death saturates the tense dialogue between disease and hope. Fear is a universal response involving the entire organism—physiologically and emotionally. It has been addressed by seemingly everyone, from poets, philosophers, and politicians to actresses, comedians, and war heroes. One common thread is that uncertainty creates fear. The palpable anxiety that attends fear is a survival tool, steering us away from possible harm. Paradoxically, though, while we instinctively shrink from perceived danger, confronting our demons sometimes may be the best (or only) way forward. Freedom, it is often said, lies on the other side of fear. The consumption ritual was an explicit confronting of disease and death, a bridge from fear to hope.

The "snobbery of chronology" (as C. S. Lewis phrased it) may entice us twenty-first century humans to pat ourselves on the back to the point of breaking an arm over how far we have traveled on the path of knowledge. We have more facts at our disposal than did our ancestors. We better understand some formerly mysterious diseases, such as tuberculosis. For many people, a fear of God has been replaced by a fear of microbes, a superior fear, some might argue, because it is grounded in science instead of faith. Yet many of the challenges, and associated fears, that our predecessors struggled with, including disease and death, have not vanished, even if our views of them have changed.

Just as the cause of consumption had finally been established, pandemics began their global assault and have continued to the present. Influenza viruses have been especially devastating. Among the worst was the Spanish Flu pandemic (a misnomer, as it did not originate in Spain), which killed an estimated 500,000 people worldwide, between 1918 and 1920. More recent flu virus epidemics include the Asian Flu pandemic (1957-1958), Swine Flu pandemic (2009-2010), AIDS (acquired immune deficiency syndrome), caused by the human immunodeficiency virus (HIV), is both pandemic and epidemic (1981-present day). There is no known cure, but it can be successfully treated with medication.[688]

These devastating infectious outbreaks have shown the adaptability of our pathogenic adversaries. With generations measured in hours, bacteria and viruses mutate rapidly. New strains of known viruses appear faster than scientists can create effective vaccines, a fact that demonstrates the vulnerability even of modern medicine. We can now add novel (new) coronavirus to this list. COVID-19 (the abbreviation for "coronavirus disease 2019") is a new strain of coronaviruses, which range from common cold viruses to those that cause much more serious diseases, including severe acute respiratory syndrome (SARS) and Middle East respiratory syndrome (MERS). The Centers for Disease Control and Prevention (CDC) declared that COVID-19 "is the first pandemic known to be caused by a new coronavirus."[689] And new strains of Covid continue to mutate.

Uvistra Naidoo, a pediatrician and research scientist, recently discussed his personal battle with tuberculosis, arguing that this scourge could be eliminated if "research priorities and overlying policies" were reframed ("Eliminate the TB Scourge," *New York Times*, May 19, 2016). To bolster his argument, he offered some startling statistics: "Today we have an ineffective TB vaccine, insufficient diagnostic tools, TB drugs with pernicious side effects, a growing problem of bacterial resistance to current treatments, and an inexplicable lack of urgency, even though one-third of the planet's population is infected by TB, according to the Centers for Disease Control and Prevention. Some 9.6 million people worldwide became ill with the disease in 2014, and 1.5 million people died." His summary of the historical context amplifies the incredulity of inattention: "Since the era of Hippocrates, 'consumption' has plagued the human race, taking an estimated one billion lives. But only two new drugs have been approved for treatment of multidrug-resistant TB in more than 40 years. This is in sharp contrast to other devastating diseases." This physician/TB patient closed his plea for action with a personal note: "As air fills my now-scarred lungs, I remain hopeful for the day when TB will be a disease of the past. With a global commitment, we can make that happen."

When New Englanders accepted the consumption ritual as a folk medical practice, they avoided blaming one another, or themselves. Their scapegoat was a discarded human vessel inhabited by an evil spirit: the corpse that was once a person had simply become a habitat, leaving its former tenant guiltless. Stressing the question of "How can I get well?" instead of "Why am I sick?" directed these New Englanders to focus on a cure rather than a cause.

***

Fixing and maintaining the boundary between life and death is a crucial need in every culture.[690] New England's consumption ritual rested on the belief that the conclusive sign of death is putrefaction.[691] A decomposed corpse is dry, indicating that it is inert, and death is complete. As folklorist Alan Dundes pointed out, the widespread folk principle that "liquid is life" underlies this and many other customary beliefs and practices.[692] In the consumption ritual, the drying (and, thus, dying) process characteristically was hastened by burning, either the entire corpse or just the heart or other vital organs. Blood, the fountain of life, is the body's essential liquid. Death was viewed as incomplete if there was liquid (interpreted as "fresh") blood in a corpse. Blood was the vital element in which the soul resided ("the blood is the life"). Although people understood, through ordinary observation, that blood coagulates following death, probably few, if any, were aware that blood can reliquefy naturally.[693]

Montague Summers pointed out the vital connection between a corpse's decomposition and its soul, and, therefore, the vampire practice, writing, "So long as the body remains the soul might be in some way tied and painfully linked with it . . . . But the dissolution of the body meant that the soul was no longer detained in this world where it had no appointed place, but that it was able to pass without let or hindrance to its own mansion prepared for it and for which it was prepared."[694]

Proper funeral rites insured that the soul's earthly vessel could dissolve peacefully and, thus, the soul itself could be released to journey to its rightful place. "It was in later years, especially under influence of Slavonic tradition," Summers continued, "that not only love but fear compelled them to perform this duty to the dead, since it was generally thought that those whose bodies were not dissolved might return, reanimated corpses, the vampire eager to satisfy his vengeance upon the living, his lust for sucking hot reeking blood, and therefore the fulfillment of these funeral duties was a protection for themselves as well as a benefit to the departed."[695]

Anthropologist Bilinda Straight summarized the relationship between body and soul that "led some 18th- and 19th-century New England families to mutilate and partially consume the corpses of their loved ones because they thought that those corpses might quite literally be consuming the living from the grave—causing their tuberculosis. These New Englanders did not aim to harm the souls of their loved ones; ending a demonic form of corpse spirit possession that consumed the living would be closer to the mark." Ambiguous corpses called for ambivalent actions. "Although death is nearly always an unwelcome, heartrending disruption of the social order and the lives of the bereaved," Straight wrote, "epidemics like tuberculosis are appropriately viewed as grievous threats to the social fabric, and the usual mortuary rules may not fully apply."[696]

Uncertainty over the demarcation of death was a troubling issue for both traditional and official cultural systems. In the early nineteenth century, the medical definition of death as the point at which the heart and lungs cease to function—that is, the cardiopulmonary standard—began to displace the older criterion of putrefaction. Yet, as late as 1885, the medical establishment remained wary of the change, as remarks in *The British Medical Journal* attest: "It is true that hardly any one sign of death, short of putrefaction, can be relied upon as infallible."[697] This misgiving corresponded to the continuing folk conception of decomposition as the boundary between life and death. The current concept of "brain death," generated in response to medical technology such as the artificial ventilator, which keeps oxygen circulating when the heart and lungs cannot, has reintroduced the figure of the living dead and made old questions once more relevant. Now, to many, death is not a biological event, but a process whose determination requires social consensus. Because of the new ambivalence, some physicians have advocated a return to the traditional cardiopulmonary standard. One neurologist, for example, has argued that the "proper" biological definition of death is "the event that separates the process of dying from the process of disintegration" and that, therefore, the "proper criterion" of death is the "permanent cessation of the circulation of blood." Old questions have been newly framed. Is the "vital principle" of life located in, or produced by, blood, or a single organ or part of a single organ? Is the "soul" represented throughout all organs, tissues, or cells? Does death occur and unique "personhood" end when a small number of organs, or perhaps only one, permanently cease to function? Or must the entire organism go through such a process before death is defined?

A by-product of the medical profession's inability to successfully treat tuberculosis was the perpetuation of the ambiguous vampire figure. Now, the biomedical paradigm's failure to medicalize death has reintroduced the ambiguity, and fear, of the living dead. As one medical scholar remarked, "The new death, with its ambiguous figure of the living cadaver, has rekindled doubts about error and premature declarations of death. . . .. Technologically orchestrated deaths appear intuitively to many people to be unnatural. . . .. We worry that individuals who die bad deaths suffer unduly, and, even though most of us consider such thoughts irrational, even some health-care practitioners may be harrowed by the idea that this suffering will come back to haunt the living."[698] The past is prologue, for we continue to be haunted by serious questions concerning the determination of death. The undead are still with us.

***

Living in a random world of haphazard events is hell on earth. For many New Englanders, routine existence must have seemed like involuntarily participating in a life-or-death lottery whose mode of selection was unfathomable. Consumption-ritual participants were, culturally at least, Protestant Christians. Grafted on the rootstock of folk traditions, New England's Protestantism encompassed a broad range of beliefs and practices. In the introduction to his collection of essays on magical medicine, Wayland D. Hand addressed these apparently unlikely companions:

> Christian miracle is … often invoked to insure the success of a ritual that may otherwise be for the most part profane in character. … These strange juxtapositions are countenanced, apparently, by the need and desperation of the moment. In sickness, more so than in any other vicissitude of life, people will throw all caution to the wind, as it were, and resort to trials and actions that they would not even consider under ordinary circumstances. … In ultimate ordeals where life itself hangs in the balance, resort to magic as well as to religion is commonplace, if not universal.[699]

Hand affirmed that, while it may be "baffling," it is "not hard to understand, that in times of crisis these almost antithetical belief systems are shared and invoked as much by the next of kin and by friends as they are by the sick and the dying. Together, at any rate, sacred rite and magic ritual conspire with faith and devotion to the sick to bring about the miracle of healing."[700]

I believe that Hand aptly summarized the interplay between folk beliefs and official religious beliefs. Certainly, the resort to "supernatural" healing, officially sanctioned or not, is a common response to life-threatening predicaments, such as disease. Folklorists have noted that folk beliefs of the kind labeled "superstitions" are brought into play during circumstances of uncertainty, where people perceive they are at risk yet have limited control of the outcome.[701] But the New England consumption ritual was not incorporated into official religious elements, even when employed by churchmen. Evidence indicates that the ritual remained wholly secular in its performance. Official cultural systems judged these beliefs and practices as foolish superstitions from a by-gone era. But the ritual, because of its inclusion of corpse mutilation and, sometimes, cannibalism, was sternly condemned. Ultimately, however, it was implicitly rooted in the prevailing Protestantism of the region, which envisioned an afterlife where a soul grows as a body decays; death is transfiguring and redemptive. For people suffering in the here and now, knowledge without power amplifies their malaise. Agency—in this case, a theory of consumption's cure—opened the door to intervention. The consumption ritual empowered a community to literally look death in the face, offering a portal through disease and death, from fear to hope.[702]

Notes

Introduction: Medical Mysteries

1. History Collection, "20 of History's Most Devastating Plagues and Epidemics," accessed March 30, 2020, https://historycollection.co/20-of-historys-most-devastating-plagues-and- epidemics/.

2. Owen Jarus, "20 of the Worst Epidemics and Pandemics in History," Live Science, posted March 20, 2020, accessed April 1, 2020, https://www.livescience.com/worst-epidemics-and- pandemics-in-history.html.

3. John Duffy, *Epidemics in Colonial America* (Baton Rouge: Louisiana State University Press, 1953), 4.

4. James Deutsch, "The Values—and Dangers—of Folklore during a Global Pandemic," Folklife, Smithsonian Center for Folklife & Cultural Heritage, posted March 25, 2020, accessed April 2, 2020, https://folklife.si.edu/magazine/values-dangers-of-folklore-global-pandemic?fbclid=IwAR0yvd0YOVlAGF8xIHPPYxw6lb-7aVA1cegEXT7GEb-6gwU2Ecz5SM1NLkE.

5. The Coronavirus Collection: Prevention and Treatments II, Snopes Investigates Rumors and Misinformation Surrounding the Prevention and Treatment of COVID-19, published March 24, 2020, accessed April 2, 2020, https://www.snopes.com/collections/coronavirus-collection- prevention-treatments-2/.

6. Alan C. Swedlund, *Shadows in the Valley: A Cultural History of Illness, Death, and Loss in New England*, 1840-1916 (Amherst & Boston: University of Massachusetts Press, 2010), 84.

7. Duffy, 4.

8. Duffy, 101.

9. Duffy, 102.

10. An example of the belief in a "hereditary predisposition" to consumption can be found in John Warner Barber, *Connecticut Historical Collections: Containing a General Collection of Interesting Facts, Traditions, Biographical Sketches, Anecdotes, &c., Relating to the History and Antiquities of Every Town in Connecticut, with Geographical Descriptions* (New Haven: J. W. Barber: Hartford, A. Willard, 1836), accessed 4/15/20, https://babel.hathitrust.org/cgi/pt? id=nyp.33433081883765&view=1up&seq=15. Barber writes of Dr. Lemuel Hopkins (1750-1801), a physician and poet from Waterbury, Connecticut: "Dr. Hopkins is said to have been determined to the study of physic when young, by observing the gradual decline of some of his connexions, who were sinking under a consumption. This inclination to medical pursuits were strengthened by the circumstance of an hereditary predisposition to the same disorder which existed in the family. ... He fell a victim, we are told, to the exercise of an improper remedy in his own case, occasioned by his dread of a pulmonary complaint."

11. See Stith Thompson, *Motif-Index of Folk Literature: A Classification of Narrative Elements in Folktales, Ballads, Myths, Fables, Mediaeval Romances, Exempla, Fabliaux, Jest-Books and Local Legends* (Bloomington, Indiana: Indiana University Press, 1955-58): Motif H1010 Impossible tasks; see also H1371 Impossible quests and Q512 Punishment: performing impossible task.

Chapter 1: From Consumption Rituals to Vampire Narratives

12. Murray J. Leaf, "Ritual and Social Organization: Sikh Marriage Rituals," in *Frontiers of Anthropology: An Introduction to Anthropological Thinking*, ed. Murray J. Leaf (New York: D. Van Nostrand Company, 1974), 128, 134.

13. Lauri Honko, "Memorates and the Study of Folk Beliefs," *Journal of the Folklore Institute* 1 (1964): 17–18.

14. Honko 1964, 10.

15. Arnold van Gennep, *The Rites of Passage*, trans. Monika B. Vizedom and Gabrielle L. Caffee, introd. by Solon T. Kimball (Chicago: University of Chicago Press, 1960).

16. In *Symbolic Immortality: The Tlingit Potlatch of the Nineteenth Century* (Washington D. C.: Smithsonian Institution Press, 1989), 15, anthropologist Sergei Kan proposed the central position of symbolic immortality, "a continuous symbolic relationship between our finite individual lives and what has gone before us and what will come after."

17. Kan 1989, 13.

18. Taking this interpretive path further, Maurice Bloch argued that "throughout the world, many religious and political rituals revolve around the theme of transforming individuals from prey (victims) to predators (hunters/killers). Bloch sees the ritual movement from prey to predator as an affirmation of human transcendence over the forces of death, a denial of the transience of life and social institutions." Quoted in Beth A. Conklin, *Consuming Grief: Compassionate Cannibalism in an Amazonian Society* (Austin: University of Texas Press, 2001), 233–34. Ingesting the vampire's ashes transforms predator into prey. Living family members avoid being consumed by consuming their predatory (un)dead.

19. Keith Thomas, *Religion and the Decline of Magic: Studies in Popular Beliefs in Sixteenth and Seventeenth Century England* (London: Weidenfeld and Nicolson, 1971), 206-07.

20. Edith Turner, *Experiencing Ritual: A New Interpretation of African Healing* (Philadelphia: University of Pennsylvania Press, 1992), 163.

21. See Karen Armstrong, *The Lost Art of Scripture: Rescuing the Sacred Texts* (New York: Alfred A. Knopf, 2019), especially pp. 99 and 503. Ludwig Wittgenstein consistently moved "to divorce rituals, ceremonies, and magical practices from the world of belief and doctrine and instead root them in the dispositions, inclinations, and sensibilities common to all humans. And divorce them as well from the world of chance, of haphazard occurrences, and of means-ends activities. ... By so divorcing them, one can see them for what they are—spiritual matters," Michael Puett, "Wittgenstein on Frazer," Chapter 7 in *The Mythology in Our Language: Remarks on Frazer's Golden Bough*, edited by Giovanni da Col, edited and translated by Stephan Palmié (Chicago: Hau Books, 2018), 139.

22. Michael Lambek, "Remarks on Wittgenstein's Remarks on Frazer's Golden Bough: Ritual in the Practice of Life," Chapter 9 in Col and Palmié, 186.

23. "A Curious Old Letter," *Greenfield Gazette & Courier*, 10 September 1877, 1.

24. Adaline M. Tirrell, "Legends of Hampshire Hills—Old Nahum Had His Jokes," *Springfield Republican*, 25 August 1929, 4 E.

25. Francis Gerry Fairfield, "A Century Ago in New England," *Appleton's Journal* 15 (20 May 1876): 654.

26. "Singular Superstition in Western Massachusetts," *Greenville Argus*, 1 October 1869, 1.

27 Douglas Starr, *Blood: An Epic History of Medicine and Commerce* (New York: Alfred A. Knopf, 1998), 5.

28. Karen Gordon-Grube, "Evidence of Medicinal Cannibalism in Puritan New England: 'Mummy' and Related Remedies in Edward Taylor's 'Dispensatory'," *Early American Literature* 28, no. 3 (1993): 185.

29. Gordon-Grube 1993, 205.

30. See Ludwig Wittgenstein, *Philosophical Investigations*, trans. G. E. M. Anscombe (New York: Macmillan, 1958).

31. S. E. Farley, "Popular Prejudices," *Godey's Lady's Book, and Ladies' American Magazine* 24 (May 1842): 282.

32. "Exhumed the Bodies," *Providence Journal*, 19 March 1892, 3.

33. Bruce A. McClelland, *Slayers and Their Vampires: A Cultural History of Killing the Dead* (Ann Arbor: University of Michigan Press, 2006), 5.

34. Hough 1853, 707nB.

35. For a discussion of etic vs. emic genres, see Sabra J. Webber, *Folklore Unbound: A Concise Introduction* (Long Grove, IL: Waveland Press, 2015), 38ff & Simon Bronner, *Folklore: The Basics* (New York: Routledge, 2017), 51, 66–67.

36. Barber 1988, 125.

37. Michael E. Bell, *Food for the Dead: On the Trail of New England's Vampires* (Middletown, CT: Wesleyan University Press, 2011), xi.

38. For my extended discussion of relations between New England and European vampire traditions, see Michael E. Bell, "New England Vampires as Local Variants of a Belief Tradition," *Journal of Vampire Studies* 1, no. 2 (2020): 165-91.

39. "What is Folklore?" The Folklore Wiki, American Folklore Society (website), accessed October 23, 2019, https://www.afsnet.org/page/WhatIsFolklore.

40. "What is Folklore?" American Folklore Society (website).

41. "What is Folklore?" American Folklore Society (website).

42. Barre Toelken, *The Dynamics of Folklore*, revised and expanded ed. (Logan, Utah: Utah State University Press, 1996), 37.

43. Simon Bronner, *Folklore: The Basics* (New York: Routledge, 2017), 46.

44. "What is Folklore?" American Folklore Society (website).

45. Kathryn Coe, Nancy E. Aiken, and Craig T. Palmer, "Once Upon a Time: Ancestors and the Evolutionary Significance of Stories," *Anthropological Forum* 16, no. 1 (2006): 21.

46. See Michael E. Bell, *Food for the Dead: On the Trail of New England's Vampires* (New York: Carroll & Graf, 2001), 202–03.

47. "Died," *Vermont Republican*, 17 February 1817, 3.

48. Quoted in Rockwell Stephens, "They Burned the Vampire's Heart to Ashes," *Vermont Life* 21, no. 1 (1966): 49.

49. "Died," *Vermont Republican*, 17 February 1817, 3.

50. Folklorists term such outsider interpretations the exoteric factor in folklore, as opposed to the insider or esoteric factor. See William Hugh Jansen, "The Esoteric-Exoteric Factor in Folklore," *Fabula: Journal of Folklore Studies* 2 (1959): 205–11 & Bronner 2017: 132-33, 135.

51. "Cumberland Town Council Records," Town Council Meeting of 8 February 1796 (Cumberland, Rhode Island, 1796).

52. John S. Pettibone, "The Early History of Manchester," *Proceedings of the Vermont Historical Society* 1 (1930): 158.

53. D. Michael Quinn, *Early Mormonism and the Magic World View*, revised and enlarged ed. (Salt Lake City: Signature Press, 1998), xv.

Chapter 2: Signs of the Dead Preying on the Living

54. "A Curious Old Letter," *Greenfield Gazette & Courier*, 10 September 1877, 1.

55. Col. Williams established two iron works in West Stockbridge, in 1766 and 1783. A cursory check of the Dwight Collection, which includes some letters of Col. Elijah Williams (1732-1815), housed at the Norman Rockwell Museum in Stockbridge, uncovered no correspondence to or from Justus Forward.

56. I have not located a death record for Martha Dickinson. While her gravestone is in Hill Cemetery in Hatfield, her husband, Joshua Dickinson, is buried in South Cemetery. The inscription on his tombstone discloses why he was interred in Belchertown instead of Hatfield, where he resided: "In Memory of Mr. Joshua Dickinson of Hatfield who died at Belchert March 2d AD 1793 aged 84." The gravestones of the husbands of both Marthas whose bodies were exhumed are in-

cluded in Allan I. Ludwig, *Graven Images: New England Stonecarving and Its Symbols, 1650–1815* (Middletown, CT: Wesleyan University Press, 1966), 416 & 412, Plates 250B & 251A.

57. Louis H. Everts, *History of the Connecticut River Valley* (Philadelphia: J. B. Lippincott, 1879), 525.

58. Daniel White Wells and Reuben Field Wells, *History of Hatfield, Massachusetts* (Springfield, MA: F. C. H. Gibbons, 1910), 207.

59. Wells and Wells, 205.

60. Wolfgang Mieder, Proverbs and Sayings, vol. 3 of *Encyclopedia of American Folklife*, ed. Simon J. Bronner (Armonk, NY: M. E. Sharpe, 2006), 996.

61. Marla R. Miller, *Rebecca Dickinson: Independence for a New England Woman, Lives of American Women* (Boulder, CO: Westview Press, 2014), 102–03.

62. Mark Doolittle, *Historical Sketch of the Congregational Church in Belchertown, Mass.* (Northhampton, MA: Hopkins, Bridgman & Co., 1852), 51.

63. Doolittle, 55.

64. Doolittle, 53-54.

65. Harry S. Stout, *The New England Soul: Preaching and Religious Culture in Colonial New England* (Oxford University Press, 1986), 17.

66. Stanley Jeyaraja Tambiah, *Magic, Science, Religion, and the Scope of Rationality* (Cambridge: Cambridge University Press, 1990), 12.

67. Merton quoted in Tambiah, 12-13.

68. Oscar Handlin and Lilian Handlin, *A Restless People: Americans in Rebellion, 1770–1787* (Garden City, New Jersey: Achor Press/Doubleday, 1982), 109.

69. Dooittle, 54.

70. Richard D. Brown, "The Healing Arts in Colonial and Revolutionary Massachusetts: The Context for Scientific Medicine in 'Medicine in Colonial Massachusetts 1620-1820'," *Publications of The Colonial Society of Massachusetts* 57 (1980): 41, https://www.colonialsociety.org/node/1199, accessed August 12, 2022.

71. James H. Cassedy, "Church Record-Keeping and Public Health in Early New England in 'Medicine in Colonial Massachusetts 1620-1820'," *Publications of The Colonial Society of Massachusetts* 57 (1980): 248, https://www.colonialsociety.org/node/1213 accessed August 12, 2022.

72. Brown, 39.

73. Doolittle, 54.

74. Brown, 42-43.

75. Keith Thomas, *Religion and the Decline of Magic: Studies in Popular Beliefs in Sixteenth and Seventeenth Century England* (London: Weidenfeld and Nicolson, 1971), 229: "Instead of the village sorcerer putting into practice the doctrines of Agrippa or Paracelsus, it was the intellectual magician who was stimulated by the activities of the cunning man into a search for the occult influences which he believed must have underlain them. The period saw a serious attempt to study long- established folk procedures with a view to discovering the principles on which they rested."

76. Thomas, 14.

77. Thomas, 178.

78. David Hackett Fischer, *Albion's Seed: Four British Folkways in America* (New York and Oxford: Oxford University Press, 1989), 128.

79. Tambiah, 31.

80. Richard Godbeer, *Devil's Dominion: Magic and Religion in Early New England* (Cambridge: Cambridge University Press, 1992), 10.

81. Fischer, 113.

82. A series of family deaths began when Forward was a young man and continued after his marriage, with a daughter stillborn and a young son drowned; a string of consumption deaths then

took two teenaged daughters, his married daughter, Martha, and her husband, Pliny Dwight. When Forward wrote his letter, two of his remaining daughters were showing the dreaded symptoms.

83. William Allen, *The American Biographical Dictionary*, 3rd ed. (Boston: John P. Jewett, 1857), 360.

84. Cassedy, 248.

85. Cassedy, 252.

86. Cassedy, 257.

87. Cassedy, 259.

88. Heide Crawford, "The Cultural-Historical Origins of the Literary Vampire in Germany," *Journal of Dracula Studies* 7 (October 2005): 1–2.

89. Deborah A. Surabian, "Soil Characteristics that Impact Clandestine Graves," posted 02/10/2012, https://www.forensicmag.com/article/2012/02/soil-characteristics-impact-clandestine-graves.

90. Henry S. Nourse, *History of the Town of Harvard Massachusetts, 1731–1893* (Harvard, MA: Printed for Warren Hapgood, 1894), 106.

91. Fischer, 75.

92. Karen Gordon-Grube, "Evidence of Medicinal Cannibalism in Puritan New England: 'Mummy' and Related Remedies in Edward Taylor's 'Dispensatory'," *Early American Literature* 28, no. 3 (1993): 201. She elaborated, "While theologians were debating the physical presence (broadly speaking, the Roman Catholic interpretation) vs. the spiritual presence (broadly speaking, the Protestant interpretation—and in any case the Puritan interpretation) of the body and blood of Christ in the Sacrament, a similar debate was being carried on in medicine: the Paracelsians—whose ideas were particularly popular in England among the Puritans—were propagating a 'spiritual' interpretation of healing vs. the prevalent 'physical' interpretation of the Galenists."

93. Richard Sugg, *Mummies, Cannibals and Vampires: The History of Corpse Medicine from the Renaissance to the Victorians* (London and New York: Routledge, 2011), 188.

94. Murray J. Leaf, *Human Organizations and Social Theory: Pragmatism, Pluralism, and Adaptation* (Urbana and Chicago: University of Illinois Press, 2009), vii-viii.

95. Frederick C. Waite, "Grave Robbing in New England," *Bulletin of the American Medical Library Association* 33 (1943): 274: "It is probable that there was grave robbing by physicians or their agents in the colonial period of New England in excess of what the public suspected, because the legal acquisition of dead human bodies was extremely limited." See also Michael Sappol, *A Traffic of Dead Bodies: Anatomy and Embodied Social Identity in Nineteenth-Century America* (Princeton University Press, 2001).

Chapter 3: The Disinterment Was Done under My Personal Supervision

96. B. M. Prince, "The Consumption Vine: Singular Vegetation Found in the Coffins of the Dead," *Chicago Daily Tribune*, 24 October 1885, 6.

97. Alfred Cole and Charles F. Whitman, *A History of Buckfield, Oxford County, Maine from the Earliest Explorations to the Close of the Year 1900* (Buckfield, Maine: C. F. Whitman, 1915), 482.

98. Ezra Morton Prince, *Prince Genealogy* (Bloomington, IL: Prince, 1902), 35–36.

99. E. M. Prince 1902, 31.

100. Cole and Whitman, 165.

101. E. M. Prince 1902, 37.

102. Ronald L. Baker and Simon J. Bronner, "Legends," in *Encyclopedia of American Folklife*, vol. 2, ed. Simon J. Bronner (Armonk, NY: M. E. Sharpe, 2006), 694.

103. Baker and Bronner, 690.

104. Harold D. Levine, "Folk Medicine in New Hampshire," *New England Journal of Medicine* 224, no. 12 (1941): 488.

105. Folkmed Record Number 22_1939, Online Archive of American Folk Medicine, University of California at Los Angeles, Http://www.folkmed.ucla.edu/ accessed 1/7/2009.

106. Ernest Poole, *The Great White Hills of New Hampshire* (Doubleday, 1946), 256.

107. Folkmed Record Number 11_7598, Online Archive of American Folk Medicine, University of California at Los Angeles, Http://www.folkmed.ucla.edu/ accessed/ 1/7/2009.

108. Folkmed Record Number 22_7599, Online Archive of American Folk Medicine, Http://www.-folkmed.ucla.edu/ accessed 1/7/2009.

109. Enoch Hayes Place, *Journals of Enoch Hayes Place*, transcribed by W. E. Wentworth. (Boston: The New England Historic Genealogical Society and the New Hampshire Society of Genealogists, 1908), 1:5.

110. Place, 25-26.

111. Place, 54.

112. Place, 107.

113. Place, 1-2.

114. Genealogical records and vital statistics indicate that Moses and Betsey had ten children (although some records list just seven or eight). *Genealogical and Personal Memoirs Relating to the Families of Boston and Eastern Massachusetts*, 4, edited by William Richard Cutter (New York, Lewis Historical Publishing, 1908), 2027-2028, Google Books. http://www.usgennet.org/usa/nh/town/barnstead/Vital_Records/BarnB/BarnB-D.htm.

115. The transcriber of Place's diary wrote, "Transcription proved difficult in many instances, especially with the nineteenth century style 'J' and 'I,' which are virtually identical," Place, xii.

116. Place, 2.

117. An article about consumption, published in the Rochester [NY] *Democrat and Chronicle* (December 23, 1898), hints that the grave plant motif was part of many consumption rituals, especially in the earlier years: "There was much superstition about the disease in former times, and its infectious character was not understood. Many years ago an aged physician detailed to us an operation, in which he was an unwilling party, to allay a superstitious fear of consumption. Several members of a family had died of consumption, and the idea that a plant was growing in the breast of the latest buried that demanded a fresh victim for its nourishment was broached. The source of this singular superstition was not stated, but it was well known in the early years of this century. Physicians were asked to have remains exhumed so the deadly plant could be destroyed. This was done, nothing was found, and that superstition was relegated to darkness in the section where the examination was made."

118. Hamilton Child, *Gazetteer of Cheshire County, N.H., 1736–1885* (Syracuse, NY: H. Child, 1885), 223.

119. Griffin, 569.

120. [William Sheldon Briggs], "Letter from New York," *New Hampshire Sentinel*, 13 February 1873, 1.

121. [William Sheldon Briggs], "Letter from New York: No. 11," *New Hampshire Sentinel*, 5 June 1873, 1.

122. [William Sheldon Briggs], "Letter from New York. No. 10," *New Hampshire Sentinel*, 22 May 1873, 1.

123. See Peter Benes, ed., "Fortunetellers, Wise-Men, and Magical Healers in New England, 1644–1850," in *Wonders of the Invisible World: 1600–1900*, ed. Peter Benes, Annual Proceedings of the Dublin Seminar for New England Folklife (Boston: Boston University, 1995), 127–48.

124. Clifford C. Wilber, "The Good Old Days: Superstition Worthy of the Dark Ages," *New Hampshire Sentinel*, 28 May 1937.

Chapter 4: The Exhumation Will Take Place on Saturday at 10 AM

125. [Samuel P. Hildreth], "Extracts from the Diary of an Old Physician," *Western Journal of the Medical and Physical Sciences* 10, no. 40 (1837): 500.

126. Samuel P. Hildreth, *Genealogical and Biographical Sketches of the Hildreth Family* (Marietta, OH, 1840), 146–53.

127. John H. Schroeder, "Major Jack Downing and American Expansionism: Seba Smith's Political Satire, 1847–1856," *New England Quarterly* 50, no. 2 (1977): 214.

128. Schroeder, 214.

129. Schroeder, 233.

130. Mary Alice Wyman, *Two American Pioneers: Seba Smith and Elizabeth Oakes Smith* (New York: Columbia University Press, 1927), 29.

131. Schroeder, 216.

132 Schroeder, 229.

133. "Superstition," *Christian Intelligencer and Eastern Chronical* 12 (23 November 1832): 187.

134. Schroeder, 232.

135. Lynn Thorndike, "Mediaeval Magic and Science in the Seventeenth Century," *Speculum* 28, no. 4 (October 1953): 704&702; see also Keith Thomas, *Religion and the Decline of Magic: Studies in Popular Beliefs in Sixteenth and Seventeenth Century England* (London: Weidenfeld and Nicolson, 1971): 230.

136. Mary Andrews Denison, "Old Superstition," *Home Pictures* (New York: Harper & Brothers, 1853): 295.

137. *Second Annual Report of the Board of Health of the City of Boston*, City Document No. 63, City of Boston, Vol. II (1874).

138. Roland Demers, *Modernization in a New England Town: A History of Willington, Connecticut* (Willington, CT: Willington Historical Society, 1983): 132, 272, 291, 335n35, 389-90, 396.

139. The Johnson family's witchcraft persecutions are summarized in Richard Hite, *In the Shadow of Salem: The Andover Witch Hunt of 1692* (Yardley, PA: Westholme, 2018), 207-08.

140. Ronald L. Baker and Simon J. Bronner, "Legends," in *Encyclopedia of American Folklife*, vol. 2, ed. Simon J. Bronner (Armonk, NY: M. E. Sharpe, 2006), 690, 694. David Hackett Fischer, *Albion's Seed: Four British Folkways in America* (New York: Oxford University Press, 1989), "Massachusetts Family Ways" pp. 69-75. Barre Toelken, *The Dynamics of Folklore*, revised and expanded edition (Logan, Utah: Utah State Univer- sity Press, 1996), "Dynamics of Family Folklore" pp 101-09, especially "Vertical Family" pp. 102-03. Steven J. Zeitelin, Amy Kotkin, and Holly Cutting Baker, *A Celebration of American Family Folklore: Tales and Traditions from the Smithsonian Collection* (New York: Pantheon, 1982), 16, noted that families are "selfish in what they choose to remember and pass on. They are willing to remember incidents which come to epitomize the character of a particular family member; and they are willing to remember an occurrence which marks a turning point in their own life or their family history. In this way, each narrative becomes not a rehash of an event, but a distillation of experience. A single episode comes to represent the entirety of a relative's personality; a whole family history is symbolized by a few dramatic turning points. The stories stand for much larger quantities of experience, and families have at their disposal a heightened form of communication which holds the family together."

141. "Bodies to Be Exhumed," *Montrose Democrat*, 26 April 1871, 3.

142. "The Lenox Exhuming," *Montrose Democrat*, 17 May 1871, 1.

143. Spectator, "The Lenox Exhuming," *Montrose Democrat*, 17 May 1871, 1.

144 Ramanthus M. Stocker, *Centennial History of Susquehanna County, Pennsylvania* (Philadelphia: R. T. Peck, 1887), 712.

145. Stocker,174-75, 693; James T. Du Bois and William J. Pike, *Centennial History of Susquehanna County* (Washington, D.C.: Gray & Clarkson, 1888), 154.

Part III: Newsworkers

146. In Paul Starr, *Creation of the Media: Political Origins of Modern Communications* (New York: Basic Books, 2004), 86, Starr noted that, "between 1790 and 1835, while the population grew from 3.9 million to 15 million, the number of newspapers in the United States climbed eleven-fold, from 106 to 1,258." The number of households receiving newspaper subscriptions also increased dramatically, showing a "remarkably broad distribution of newspapers in an era when, according to some historians, newspaper readership was supposedly restricted to a political and economic elite."

147. Starr 2004, 254.

148. Michael Schudson, *Sociology of the News* (New York: W. W. Norton, 2011), xiv.

149. Folklorists have formulated similar notions of core meanings, including Barre Toelken's ideational center, Tristram Coffin's emotional core, and William Wilson's value center. "If the story is to live," Wilson wrote, storytellers "cannot, in the telling of it, depart too far from the value center of the audience whose approval they seek." Wilson quoted in Barre Toelken,*The Dynamics of Folklore*, revised and expanded ed. (Logan, Utah: Utah State University Press, 1996), 137.

150. Daniel C. Hallin, *The "Uncensored War": The Media and Vietnam* (Berkeley: University of California Press, 1986), 116-17.

151. Schudson 2011:181.

152. George S. Carey, "Folklore from the Printed Sources of Essex County, Massachusetts," *Southern Folklore Quarterly* 32 (1968): 18–19.

153. Carey, 20.

154. Ronald L. Baker and Simon J. Bronner, "Legends," in *Encyclopedia of American Folklife*, vol. 2, ed. Simon J. Bronner (Armonk, NY: M. E. Sharpe, 2006), 689.

155. Carey, 20.

156. See Linda Dégh, *The Dialectics of the Legend*, Folklore Preprint Series, vol. 1, no. 6 Bloomington, Indiana, 1973) and Linda Dégh and Andrew Vázsonyi, "Legend and Belief," *Genre* 4 (1971): 281–304.

157. This process often is referred to by folklorists as oicotypification or oikotypification. See Robert A. Georges and Michael Owen Jones, *Folkloristics: An Introduction* (Bloomington, Indiana: Indiana University Press, 1995), 150 and Simon Bronner, *Folklore: The Basics* (New York: Routledge, 2017), 89.

Chapter 5: Wrath of the Devil's Dark Angel

158. *Albany Gazette*, 26 April 1790, 1.

159. Virginia Hewitt Watterson, *Descendants of Captain Thomas Hewitt of Stonington, Connecticut* (Carlsbad, California: V. W. Watterson, 1996), 79.

160. C. M. Day, *History of the Eastern Townships, Province of Quebec, Dominion of Canada* (Montreal: John Lovell, 1869), 161-62, https://www.canadiana.ca/view/oocihm.03779/308? r=0&s=1.

161. Day 1869, 305: "Under the hand and seal of Sir Robert Shore Milnes" and "granted to Hugh Finlay and his associates."

162. Edwin Emery and William Morrell Emery, *History of Sanford, Maine: 1661-1900* (Fall River, MA: Published by the compiler, 1901), 55.

163. *Massachusetts Soldiers and Sailors of the Revolutionary War* (1906) 14:437. A resolve of July 5, 1775, entitled noncommissioned officers and soldiers to a coat (or its monetary value), which was

"considered in the nature of a bounty." *Massachusetts Soldiers and Sailors of the Revolutionary War* 1(1896), xvi.

164. Ancestry.com. U.S., Revolutionary War Rolls, 1775-1783 [database on-line]. Provo, UT, USA: Ancestry.com Operations, Inc., 2007.

165. "Revolutionary War Facts," https://revolutionarywar.us/facts/.

166. W. Woodford Clayton, *History of York County, Maine, with Illustrations and Biographical Sketches of Its Prominent Men and Pioneers* (Philadelphia: Everts & Peck, 1880), 433, Google Books, PDF.

167. "Old-Time Bric-a-Brac," *Springfield Republican*, 4 December 1876, 2.

168. Vital Records of Springfield, Massachusetts to 1850. Boston, Mass.: New England Historic Genealogical Society, 2002. (Online database. AmericanAncestors.org. New England Historic Genealogical Society, 2008.).

169. Mason A. Green, *Springfield, 1636–1886: History of Town and City* (Springfield, MA: C. A. Nichols & Co., 1888), 606.

170. "Yesterday's Local History," *Springfield Republican*, 3 October 1876, 4.

171. Robert N. Holcomb, "The Story of Springfield: Superstitions," *Springfield Republican*, 25 November 1934.

172. Jack Larkin, *The Reshaping of Everyday Life, 1790-1840* (New York: Harper Perennial, 1988): 77.

173. For its requirement specifically in vampire rituals, see T. P. Vukanovic, "The Vampire in the Beliefs and Customs of the Gypsies in the Province of Kosovo-Metohija, Stari Ras and Novopazarski Sandzak, Yugoslavia," pt. 4, *Journal of the Gypsy Lore Society* 40 (1960): 53: "There is a prescription, widely held throughout the Blakans, that absolute silence must be observed during the ceremony." For examples in other kinds of healing rituals, see Wayland D. Hand, "Folk Curing: The Magical Component," in *Magical Medicine: The Folkoric Component of Medicine in the Folk Belief, Custom, and Ritual of the Peoples of Europe and America*, reprint, 1975 (Berkeley and Los Angeles: University of California Press, 1980), 10.

174. "Superstition," *Christian Intelligencer and Eastern Chronical* 12 (23 November 1832): 187.

175. *Rutland Daily Globe*, 14 May 1875, 2.

Chapter 6: The Above Are Facts … Sickening and Horrible

176. William H. Cook, "A Superstitious Cure for Consumption," *Physio-Medical Recorder* 33, no. 9 (September 1869): 268–70.

177. Elliot G. Storke, *History of Cayuga County, New York* (Syracuse, NY: D. Mason & Co., 1879), 447.

178. Storke, 447.

179. John S. Haller, Jr., *Kindly Medicine: Physio-Medicalism in America, 1836–1911* (Kent, OH: Kent State University Press, 1997), 121.

180. Haller, Jr., 103.

181. Haller, Jr., 39.

182. Paul Starr, *Creation of the Media: Political Origins of Modern Communications* (New York: Basic Books, 2004), 169.

183. *New York Reformer*, 4 September 1851, 4.

184. John W. Barber and Henry Howe, *Historical Collections of the State of New York* (New York: S. Tuttle, 1842), 249.

185. Weston Arthur Goodspeed, *History of the Goodspeed Family* (Chicago: W. A, Goodspeed, 1907), 171, Google Books.

186. F. B. Sanborn, *The Life of Henry David Thoreau: Including Many Essays Hitherto Unpublished, and Some Account of His Family and Friends* (Boston: Houghton Mifflin, 1917), archive.org., accessed 8/10/2022, p. 211.

187. "In the spring of 1861, Thoreau's physician advised him for the sake of his health to take a long journey—to the West Indies, to the south of France, or to the Mississippi Valley. Because he feared the muggy heat of the Caribbean and felt that he could not afford the trip abroad, and perhaps because of his deep interest in American fauna and flora, he decided to go to Minnesota." Robert L. Straker, "Thoreau's Journey to Minnesota," *New England Quarterly* 14, no. 3 (September 1941): 549.

188. Straker, 555.

189. For consumption's impact on Concord, Massachusetts—including the families of both the Thoreaus and Emersons—see Constance Manoli-Skocay, "A Gentle Death: Tuberculosis in 19th Century Concord," https://concordlibrary.org/special-collections/essays-on-concord-history/a-gentle-death-tuberculosis-in-19th-century-concord/ accessed 8/14/2022.

190. C. Grant Loomis, "Henry David Thoreau as Folklorist," *Western Folklore* 16 (1957): 97.

191. "Vermont Vital Records, 1760-1954," database with images, FamilySearch (https://familysearch.org/ark:/61903/3:1:S3HT-62R3-DST?cc=1784223&wc=MFVC-4Z9%3A1029368101: 22 May 2014), 004358395 > image 2331 of 6520; Citing Secretary of State. State Capitol Building, Montpelier.

192. "Vermont Vital Records, 1760-1954," database with images, FamilySearch (https://familysearch.org/ark:/61903/3:1:939V-3RKS-9?cc=1784223&wc=MFV4-WZ9%3A1029384501 : 22 May 2014), 004666997 > image 776 of 4129; Citing Secretary of State. State Capitol Building, Montpelier.

193. "Vermont Vital Records, 1760-1954," database with images, FamilySearch (https://familysearch.org/ark:/61903/3:1:939V-3RVX-Z?cc=1784223&wc=MFV4-WZ9%3A1029384501 : 22 May 2014), 004666997 > image 779 of 4129; Citing Secretary of State. State Capitol Building, Montpelier.

194. "Vermont, Town Clerk, Vital and Town Records, 1732-2005," database with images, FamilySearch (https://familysearch.org/ark:/61903/3:1:3QS7-L999-YBT5? cc=1987653&wc=Q8ZY-ZZ3%3A324709901%2C324848601%2C324848602 : 29 November 2018), Bennington > Winhall > Births with index, marriages with index, deaths with index 1857-1976 vol 1-2 > image 117 of 182; citing various town clerks and records divisions, Vermont.

195. "Vermont, Town Clerk, Vital and Town Records, 1732-2005," database with images, FamilySearch (https://familysearch.org/ark:/61903/3:1:3QS7-L999-YB17? cc=1987653&wc=Q8ZY-ZZ3%3A324709901%2C324848601%2C324848602 : 29 November 2018), Bennington > Winhall > Births with index, marriages with index, deaths with index 1857-1976 vol 1-2 > image 118 of 182; citing various town clerks and records divisions, Vermont.

Chapter 7: Eliza Heard and Believed These Stories

196. Hale Collection of Connecticut Cemetery Records, Connecticut Cemetery Inscriptions and Records, http://www.hale-collection.com/.

197. Josiah Gilbert Holland, *History of Western Massachusetts* (Springfield, MA: Samuel Bowles, 1855), 542.

198. Percy W. Bidwell, "The Agricultural Revolution in New England," *American Historical Review* 26, no. 4 (1921): 689.

199. Karen Gordon-Grube, "Evidence of Medicinal Cannibalism in Puritan New England: 'Mummy' and Related Remedies in Edward Taylor's 'Dispensatory'," *Early American Literature* 28, no. 3 (1993): 186.

200. Gordon-Grube 1993, 188.

201. Gordon-Grube 1993, 185.

202. Gordon-Grube 1993, 185.

203. J. G. Frazer, *The Golden Bough: A Study in Comparative Religion*, vol. 1 (1890, reprint; New York: Macmillan, 1894), 9, https://archive.org/details/goldenboughstudy01fraz.

204. Bruce A. McClelland, *Slayers and Their Vampires: A Cultural History of Killing the Dead* (Ann Arbor: University of Michigan Press, 2006), 24.

205. Merriam-Webster Unabridged Dictionary, http://unabridged.merriam-webster.com/unabridged/clairvoyant.

206. See Fitzgerald's ads in *Portland Daily Press*, 12 May 1874, 2, and *Lowell Daily Citizen and News*, 16 November 1874, 2.

207. For a European version of this practice, see Leszek Gardeła, "Face Down: The Phenomenon of Prone Burial in Early Medieval Poland," *Analecta Archaeologica Ressoviensia* 10 (2015): 99-136. http://www.academia.edu/26603921/Garde%C5%82a_L._2015_Face_Down_The_Phenomenon_of_Prone_Burial_in_Early_Medieval_Poland_Analecta_Archaeologica_Ressoviensia_10_99-136.

208. James George Frazer, *Folk-Lore in the Old Testament: Studies in Comparative Religion, Legend and Law* (London: Macmillan and Co., Limited, 1918), v. 3, part 4, p. 253–54.

209. For more on prone burial in a New England context, see Michael E. Bell, *Food for the Dead: On the Trail of New England's Vampires* (New York: Carroll & Graf, 2001), 275–76.

210. D. M. Bennett, *The World's Sages, Thinkers and Reformers*, second edition, revised and enlarged (New York: The Truth Seeker Company, 1876), 1019.

211. Michael Philips, "What is Materialism?" in *Philosophy Now*. https://philosophynow.org/issues/42/What is Materialism.

212. *The Index*, 12 July 1883, p. 13.

213. See Paul Starr, *Creation of the Media: Political Origins of Modern Communications* (New York: Basic Books, 2004), 184.

214. The following relevant motifs are in Stith Thompson, *Motif-Index of Folk Literature: A Classification of Narrative Elements in Folktales, Ballads, Myths, Fables, Mediaeval Romances, Exempla, Fabliaux, Jest-Books and Local Legends*, revised and enlarged ed., reprint, 1932–36 (Bloomington, Indiana: Indiana University Press, 1955–58): A1142.1 - creator's (deity's) voice makes thunder; M447- curse: to be stricken by thunder; F968 - extraordinary thunder; D1812.5.2.3 - hearing thunder on setting forth a good omen.

Chapter 8: My Sisters Were Drawing My Life Away

215. Moncure Daniel Conway, *Demonology and Devil-Lore* (New York: Henry Holt and Company, 1879), 52.

216. Dudley Wright, *Vampires and Vampirism* (London: Will Rider & Son, 1914), 138.

217. Montague Summers, *The Vampire in Europe* (New York: E. P. Dutton and Company, Inc., 1929), 116–17.

218. William S. Simmons, *Spirit of the New England Tribes: Indian History and Folklore, 1620–1984* (Hanover: University Press of New England, 1986), 4.

219. Michael E. Bell, *Food for the Dead: On the Trail of New England's Vampires* (Middletown, CT: Wesleyan University Press, 2011), xix-xxv.

220. George R. Stetson, "The Animistic Vampire in New England," *American Anthropologist* 9, no. 1 (1896): 8–9.

221. Wright, 137-38.

222. Laura C. Rudolph, "Countries and Their Cultures: Sicilian Americans," accessed January 20, 2022, https://www.everyculture.com/multi//Pa-Sp/Sicilian-Americans.html.

223. Guiseppe Pitrè, *Sicilian Folk Medicine*, trans. Phyllis H. Williams (Lawrence, Kansas: Coronado Press, 1971), 9.

224. Joe Garzone, "Hartford Paesani: Social Mobility of Italian Americans in the Twentieth Century," *Germina Veris*, 2 (2015): 25-47, esp. 38.

Chapter 9: She Had Turned Over in the Grave

225. Sound Recording RI81-MB-C7, 18 November 1981, Rhode Island Folklife Archive.

226. Gary R. Butler, "Indexicality, Authority, and Communication in Traditional Narrative Discourse," *Journal of American Folklore* 105, no. 415 (1992):34-56, 41.

227. Butler, 41.

228. Karen Lee Ziner, "Was She a Victim... or a Vampire?" *Providence Journal-Bulletin*, 25 October 1984, 10.

229. Paul F. Eno, "They Burned Her Heart... Was Mercy Brown a Vampire?" *Narragansett Times*, 25 October 1979, 1-SC.

230. Ziner 1984, 10.

231. Narrative motifs include: "ghost-like lights" (E530.1.), "luminous ghosts" (E421.3.), "luminous spirits" (F401.2.), "soul as light" (E742.), and "illusory light" (K1888.), see Stith Thompson, *Motif-Index of Folk Literature: A Classification of Narrative Elements in Folktales, Ballads, Myths, Fables, Mediaeval Romances, Exempla, Fabliaux, Jest-Books and Local Legends*, revised and enlarged ed., reprint, 1932–36 (Bloomington, Indiana: Indiana University Press, 1955–58).

232. Lowry Charles Wimberly, *Folklore in the English and Scottish Ballads*, reprint, 1928 (New York: Dover Publications, Inc., 1965), 82. In one of the Charlemagne cycle of tales, when the head of an innocent beggar is cut off by the deceived and jealous husband, Hugo, "every drop of blood turns to a burning candle." Francis James Child, ed., *The English and Scottish Popular Ballads*, reprint, 1882–98 (New York: Dover Publications, Inc., 1965), v. 2, 93.

233. Paul Barber, *Vampires, Burial, and Death: Folklore and Reality* (New Haven: Yale University Press, 1988), 70.

234. Agnes Murgoci, "The Vampire in Roumania," in *The Vampire: A Casebook*, ed. Alan Dundes, reprint, 1926 (Madison, Wisconsin: University of Wisconsin Press, 1998), 13.

235. "Newport, Rhode-Island, March 30," *American Weekly Mercury* [Philadelphia], No. 122 (April 5-12, 1722), p. 42, Hathi Trust Digital Library, accessed March 30, 2020, https://babel.hathitrust.org/cgi/pt?id=inu.3200000480287&view=1up&seq=48. This event was recorded by Peter Kalm, probably from this tabloid, in *The America of 1750: Peter Kalm's Travels in North America, The English version of 1770*, ed. Adolph B. Benson (New York: Dover Publications, 1966), 669.

236. Jacob Grimm, *Teutonic Mythology*, 4th ed., trans. James Steven Stallybrass (London: George Bell & Sons, 1883), v.3:1183. Grimm wrote that these blue vapors and lights suggest "the fire of the Thunder-god." A Prussian folktale cited in Grimm 1883, v.1:178 includes the expressive phrase for lightning, "he with the blue whip chases the devil" (or giants). Grimm pointed out that "a blue flame was held specially sacred, and people swear by it," as in the curse blau feuer (i.e., blue fire).

237. Grimm 1883, v.3:916.

238. C. Eugene Emery, Jr., "Did They Hear the Vampire Whisper?" *Providence Sunday Journal Magazine*, 28 October 1979, 6–7, 10, 12–13.

239. See Robert A. Georges, "Toward an Understanding of Storytelling Events," *Journal of American Folklore* 84 (1969): 316.

240. Judson Hale, *Inside New England*, reprint, 1982 (New York: Harper & Row, Publishers, 1986), 161.

241. Lisa Gabbert and Paul Jordan-Smith, "Introduction: Space, Place, Emergence," *Western Folklore* 66, no. 3/4 (2007): 227.

242. Ruth Schell, "Swamp Yankee," *American Speech* 38, no. 2 (May 1963): 121.

243. See Frank de Caro, "Review: Cultural Conservation: The Conference," *Journal of American Folklore* 104, no. 411 (1991): 85–91 and Robert Baron and Nicholas R. Spitzer, eds, *Public Folklore*, Publications of the American Folklore Society, New Series (Washington: Smithsonian Institution Press, 1992).

244. Diane Goldstein, "Vernacular Turns: Narrative, Local Knowledge, and the Changed Context of Folklore," *Journal of American Folklore* 128, no. 508 (2015): 139n8.

245. Peck was referring to Florence Parker Simister (1913-1981), who had written extensively on local Rhode Island history, including a history of Exeter that was published three years prior to my 1981 interview with Peck.

246. Stephanie Ressler, email message to author, February 6, 2016.

247. Stephanie Ressler, "Fear of the Unknown? (Undead)," http://www.scribd.com/doc/292665530/Fear-of-the-Unknown-Undead (accessed March 1, 2016).

248. Sabra J. Webber, *Folklore Unbound: A Concise Introduction* (Long Grove, IL: Waveland Press, 2015), 40.

Chapter 10: What Really Happened to Mercy Brown?

249. "Exhumed the Bodies," *Providence Journal*, 19 March 1892, 3.

250. "The Vampire Theory," *Providence Journal*, 21 March 1892, 8.

251. "The Vampire Theory."

252. J. A. MacCulloch, "Vampire," in *Encyclopaedia of Religion and Ethics*, vol. 12, ed. James Hastings, reprint, 1908–26 (New York: Scribner's Sons, 1928), 58.

253. Roy Bongartz, "When the Wind Howls and the Trees Moan," *Providence Sunday Journal Magazine*, 25 October 1981, 5.

254. See Michael E. Bell, *Food for the Dead: On the Trail of New England's Vampires* (New York: Carroll & Graf, 2001), 260–78.

255. letter to the editor, *Pawtuxet Valley Gleaner*, 25 March 1892, 5.

256. *Wickford Standard*, 25 March 1892, 3.

257. "A Rhode Island Country Town," editorial, *Wickford Standard*, 6 May 1892, 2.

258. For a specific example of how folklore (in this case, a blues ballad) frequently assumes a high degree of esoteric knowledge, see D. K. Wilgus and Lynwood Montell, "Clure and Joe Williams: Legend and Blues Ballad," *Journal of American Folklore* 81, no. 322 (1968): 295–315.

259. editorial, *Pawtuxet Valley Gleaner*, 22 April 1892.

260. See Jack Santino, *Halloween and Other Festivals of Death and Life* (Knoxville: University of Tennessee Press, 1994).

261. Karen Lee Ziner, "Was She a Victim... or a Vampire?" *Providence Journal-Bulletin*, 25 October 1984, 10.

262. Ronald L. Baker, "The Influence of Mass Culture on Modern Legends," *Southern Folklore Quarterly* 40 (1976): 367.

263. Baker, 368.

264. Donald Allport Bird, "A Theory for Folklore in the Mass Media: Traditional Patterns in the Mass Media," *Southern Folklore Quarterly* 40 (1976): 286.

265. Norine Dresser, *American Vampires: Fans, Victims, Practitioners* (New York: W. W. Norton, 1989), 13.

266. Nancy Kinder, "The 'Vampires' of Rhode Island," *Yankee*, October 1970, 167.

267. France Carrado Bolderson, "Horrible History at the Boston Boo-Centennial," *Providence Journal-Bulletin*, 11 October 1975, 10.

268. Bruce Fellman, "Things Still Go Bump in South County Night," *Providence Journal*, 31 October 1980, W4.

269. Roy Bongartz, "When the Wind Howls and the Trees Moan," *Providence Sunday Journal Magazine*, 25 October 1981, 5, 7, 10, 14–15.

270. For a for recent example of using EVP (Electronic Voice Phenomena) to record a vampire's grave in Rhode Island, see Thomas D'Agostino, *A History of Vampires in New England* (Charleston, SC: Haunted America, The History Press, 2010), 80.

271. Sound Recording RI81-MB-C7:1, Rhode Island Folklife Archive.

272. Bill Ellis, "Fabulate," in *Folklore: An Encyclopedia of Beliefs, Customs, Tales, Music, and Art*, vol. 1, ed. Thomas A. Green (Santa Barbara, CA: ABC-CLIO, Inc., 1997), 274–75.

273. Harry A. Senn, *Were-Wolf and Vampire in Romania*, East European Monographs, no. 99 (New York: Columbia University Press, 1982), 41.

274. Gaye Tuchman, *Making News: A Study in the Construction of Reality* (New York: The Free Press, 1978), 47–47.

275. Michael Schudson, *Sociology of the News* (New York: W. W. Norton & Company, 2011), 186–87.

276. Donald H. Holly, Jr. and Casey E. Cordy, "What's in a Coin? Reading the Material Culture of Legend Tripping and Other Activities," *Journal of American Folklore* 120, no. 477 (2007): 345.

277. See Bell 2001, chapter 5.

278. Charles Turek Robinson, *The New England Ghost Files: An Authentic Compendium of Frightening Phantoms* (North Attleboro, MA: Covered Bridge Press, 1994), 189–94.

Part V: Historians

279. Franklin B. Hough, *History of St. Lawrence and Franklin Counties, New York, from the Earliest Period to the Present Time* (Albany, NY: Little, 1853), 707nB, referred from p. 109.

280. Hough, Franklin B. 1854. *History of Jefferson County in the State of New York, from the Earliest Period to the Present Time* (Watertown, NY: Sterling & Riddell), 101.

281. Kent C. Ryden and Simon J. Bronner, "New England," in *Encyclopedia of American Folklife*, ed. Simon J. Bronner (Armonk, NY: M. E. Sharpe, 2006), 868.

282. Arnold Collection of the Knight Memorial Library, Providence, Scrapbook 7, p. 8.

283. Edwin Emery, *History of Sanford, Maine: 1661–1900* (Salem, MA: Salem Press, 1901), 485.

Chapter 11: Heathenish Mutilations of the Dead

284. John S. Pettibone, "The Early History of Manchester," *Proceedings of the Vermont Historical Society* 1 (1930): 149.

285. Pettibone, 158.

286. Edwin L. Bigelow and Nancy H. Otis, *Manchester, Vermont, 1761–1961: A Pleasant Land Among the Mountains* (Manchester, VT: Town of Manchester, 1961), 10.

287. Thomas Scott Pearson, *Catalogue of the Graduates of Middlebury College: Embracing a Biographical Register and Directory* (Windson, VT: Vermont Chronicle Press, 1853), 24.

288. Pettibone, 148.

289. Hiram Carleton, *Genealogical and Family History of the State of Vermont* (New York: Lewis Publishing Company, 1903), 325.

290. Pettibone, 165.

291. Pettibone, 156.

292. John S. Pettibone (1786 - 1872), http://www.wikitree.com/wiki/Pettibone-217, accessed 4/13/13.

293. Pettibone, 157.

294. David Lufkin Mansfield, *The History of the Town of Dummerston*, Vermont Historical Gazetteer: A Local History of All the Towns in the State, Civil, Educational, Biographical, Religious and Military (Ludlow, VT: A. M. Hemenway, 1884): 27–28.

295. Mansfield, 27.

296. Mansfield, 212.

297. "Deaths: David Lufkin Mansfield," *New England Historical and Genealogical Register* 59 (1905): 340.

298. Mansfield, 28–32.

299. Mansfield, 29.

300. Michael E. Bell, *Food for the Dead: On the Trail of New England's Vampires* (New York: Carroll & Graf, 2001), 221.

301. Bell, 222.

302. Bell, 224–25.

303. Robert A. Georges, "Toward an Understanding of Storytelling Events," *Journal of American Folklore* 84 (1969): 316.

304. Gary R. Butler, "Indexicality, Authority, and Communication in Traditional Narrative Discourse," *Journal of American Folklore* 105, no. 415 (1992): 53n2.

305. Mansfield, 151.

306. Mansfield, 177.

307. Mansfield, 205.

308. Mansfield, 60.

309. Jane C. Beck, "Traditional Folk Medicine in Vermont," in *Medicine and Healing*, ed. Peter Benes (Boston, MA: Boston University, 1992), 34.

310. Patricia A. Watson, "The 'Hidden Ones': Women and Healing in Colonial New England," in *Medicine and Healing*, ed. Peter Benes (Boston, MA: Boston University, 1992), 33.

311. Beck, 34.

312. Mansfield, 95.

313. Mansfield, 64ff.

314. John C. L. Clark, "The Famous Dr. Stearns," *Publications of the American Antiquarian Society* 45 (1936): 317–424.

315. John Langdon Sibley, *A History of the Town of Union, in the County of Lincoln Maine* (Boston, MA: Benjamin B. Mussey and Co., 1851), 45.

316. Sibley, 17.

317. Sibley, 482–83.

318. Sibley, 482.

Chapter 12: Her Heart Should Be Consumed for the Benefit of Her Sisters

319. Duane Hamilton Hurd, *History of Clinton and Franklin Counties, New York* (Philadelphia: J. W. Lewis, 1880), 292.

320. Duane Hamilton Hurd, *History of Essex County, Massachusetts, with Biographical Sketches of Many of Its Pioneers and Prominent Men* (Philadelphia: J. W. Lewis, 1888), iii.

321. Hurd, *History of Clinton and Franklin Counties*, 3.

322. Hurd, *History of Clinton and Franklin Counties*, 292.

323. Hurd, *History of Clinton and Franklin Counties*, 278.

324. J. F. Gilbert, "Reminiscences of the Graves Family," *Clinton County Farmer*, 23 August 1898, Reminiscences of Old Chazy Given by Descendants of the Early Settlers at a Literary and Musical Entertainment Held at Academy Hall, Plattsburg, NY, http://sites.rootsweb.com/~nyclinto/chazy/chazy.html.

325. Alfred Sereno Hudson, *Colonial Concord*, vol. 1 of *History of Concord, Massachusetts* (Concord, MA: Erudite Press, 1904), 177.

326. Hudson, 176.

327. Hudson, 176.

328. Hudson, 176.

329. Richard Sugg, *Mummies, Cannibals and Vampires: The History of Corpse Medicine from the Renaissance to the Victorians*, 2nd ed. (London and New York: Routledge), 6.

330. Hudson, 177.

331. S. Bulfinch Emmons, *Philosophy of Popular Superstitions and the Effects of Credulity and Imagination Upon the Moral, Social, and Intellectual Condition of the Human Race* (Boston: L. P. Crown, 1853), 172.

332. Emmons, 173.

333. Bernard Whitman, *Lecture on Popular Superstitions* (Boston: Bowles & Dearborn, 1829), 50.

334. Alfred Cole and Charles F. Whitman, *A History of Buckfield, Oxford County, Maine from the Earliest Explorations to the Close of the Year 1900* (Buckfield, ME: C. F. Whitman, 1915), 165.

335. Henry S. Nourse, *History of the Town of Harvard Massachusetts, 1731–1893* (Harvard, MA: Printed for Warren Hapgood, 1894), 104–05.

336. Nourse, ii.

337. Nourse, iii.

338. Folklorists recognize this narrative motif as "G250. Recognition of witches," documented in Stith Thompson's *Motif-index of Folk-Literature*.

339. This narrative is a version of the Märchen (fairy tale) "The Singing Bones" (Grimm No. 28) and, in ballad form, "The Twa Sisters" (Child No. 10).

340. Samuel S. Shaw, "Memoir of Henry Stedman Nourse," *Proceedings of the Massachusetts Historical Society* (Boston: Massachusetts Historical Society, 1904), 292.

341. D. P. Corey, "Henry Stedman Nourse," *The New England Historical and Genealogical Register* 58 (1904): 218.

342. Corey, 217.

343. Nourse, 2:481.

344. Frederic Denison, *Westerly and Its Witnesses* (Providence, RI: J. A. & R. A. Reid, 1878), 255. For my earlier discussion of this case, see Michael E. Bell, *Food for the Dead: On the Trail of New England's Vampires* (Middletown, CT: Wesleyan University Press, 2011), xxxiii-xxxvii.

345. Denison, 286–87.

346. See also *Boston Daily Advertiser*, 21 July 1868:2 and the *Providence Evening Press*, 29 October 1867:3.

347. Denison, 37.

348. Denison, 303.

349. Denison, 161.

350. Denison, 173.

351. Frederic Denison, *Sabres and Spurs: The First Regiment Rhode Island Cavalry in the Civil War* (Central Falls, Rhode Island: The First Rhode Island Cavalry Veteran Association, 1876), 50.

352. Denison, *Sabres and Spurs*, 50–51.

353. Denison, *Sabres and Spurs*, 51.

354. Denison, *Sabres and Spurs*, 52.

355. Karen Gordon-Grube, "Anthropophagy in Post-Renaissance Europe: The Tradition of Medicinal Cannibalism," *American Anthropologist* 90, no. 2 (1988): 406.

356. Karen Gordon-Grube, "Evidence of Medicinal Cannibalism in Puritan New England: 'Mummy' and Related Remedies in Edward Taylor's 'Dispensatory'," *Early American Literature* 28, no. 3 (1993): 200.

357. Sugg, *Mummies, Cannibals and Vampires*, 233.

Part VI: Physicians

358. Louis Mazzari, "Science and Medicine: Health-Care Providers," in *Encyclopedia of New England: The Culture and History of an American Region*, eds Burt Feintuch and David H. Watters (New Haven: Yale University Press, 2005), 1368.

359. Richard A. Meckel, "Science and Medicine: Introduction, Medicine," in *Encyclopedia of New England: The Culture and History of an American Region*, eds Burt Feintuch and David H. Watters (New Haven: Yale University Press, 2005), 1355.

360. Alfred Sereno Hudson, *Colonial Concord*, vol. 1 of *History of Concord, Massachusetts* (Concord, MA: Erudite Press, 1904), 177.

361. Richard Sugg, "Corpse Medicine: Mummies, Cannibals, and Vampires," *The Lancet* 371, no. 9730 (21 June 2008): 2079.

362. Sugg, 2079.

363. Karen Gordon-Grube, "Evidence of Medicinal Cannibalism in Puritan New England: 'Mummy' and Related Remedies in Edward Taylor's 'Dispensatory'," *Early American Literature* 28, no. 3 (1993): 185.

364. Gordon-Grube, 186.

365. Gordon-Grube, 194.

366. Gordon-Grube, 193.

367. Gordon-Grube, 199.

368. Adaline M. Tirrell, "Legends of Hampshire Hills—Old Nahum Had His Jokes," *Springfield Republican*, 25 August 1929, 4 E.

369. "Editorial Department: Medical Superstitions," *American Journal of Homeopathic Materia and Record of Medical Science*, 1 September 1871, 41–43.

370. Francis Gerry Fairfield, "A Century Ago in New England," *Appleton's Journal* 15 (20 May 1876): 652–56.

371. Jacob Grimm, *Teutonic Mythology*, 4th ed., trans. James Steven Stallybrass (London: George Bell & Sons, 1883), 3:1074.

372. See, for example, Motif G10-G49. Cannibals and cannibalism; Motif F360.1. Fairies pursue unbaptized children.

373. Margaret Alice Murray, *The Witch-Cult in Western Europe: A Study in Anthropology* (Oxford: Clarendon Press, 1921), 156.

374. Wallace Notestein, *A History of Witchcraft in England from 1558 to 1718* (Washington: The American Historical Association., 1911), 23.

375. Quoted in Eric J. Trimmer, "Medical Folklore and Quackery," *Folklore* 76, no. 3 (1965): 166.

376. James T. Whittaker, "Medicine and Surgery of the Queen City," *Journal of the American Medical Association* 26 (11 April 1896): 707.

377. Henry S. Nourse, *History of the Town of Harvard Massachusetts, 1731–1893* (Harvard, MA: Warren Hapgood, 1894), 104.

Chapter 13: The Charm Worked No Good and the Patient Died

378. Wayland D. Hand, "Folk Medical Magic and Symbolism in the West," in *Magical Medicine: The Folkloric Component of Medicine in the Folk Belief, Custom, and Ritual of the Peoples of Europe and America*, reprint, 1969 (Berkeley and Los Angeles: University of California Press, 1980), 306.

379. Irvine Loudon, "A Brief History of Homeopathy," *Journal of the Royal Society of Medicine* 99, no. 12 (December 2006): 608.

380. William L. Stone, annotator, *Ballads and Poems Relating to the Burgoyne Campaign* (Albany, NY: Joel Munsell's Sons, 1893).

381. A. W. Holden, "Address Before the Semi-Annual Meeting of the Homeopathic Medical Society of the State of New York," *Transactions of the Homeopathic Medical Society of the State of New York* 9 (1871): 74.

382. Holden, 76.

383. Holden, 85.

384. Holden, 84.

385. Holden, 88.

386. Holden, 73.

387. Editorial department: "Medical superstitions," *American Journal of Homeopathic Materia and Record of Medical Science* (1 September 1871): 41.

388. "Medical Superstitions," 42.

389. "Medical Superstitions," 42.

390. John Clough, "The Influence of the Mind on Physical Organization," *Boston Medical and Surgical Journal* 21 (1840): 411.

391. Clough, 411.

392. See the entry in http://unabridged.merriam-webster.com/unabridged/psychosomatic.

393. Clough, 412.

394. "Obituary," *Boston Medical and Surgical Journal* 102 (29 January 1880): 120.

395. Frederic Kidder and Augustus Addison Gould, *The History of New Ipswich* (Boston: Gould and Lincoln, 1852), 261.

396. Kidder and Gould, 339.

397. Charles Henry Chandler and Sarah Fiske Lee, *The History of New Ipswich, New Hampshire 1735–1914* (Fitchburg, MA: Sentinel Printing Company, 1914), 255.

398. Chandler and Lee, 216. See also Kidder and Gould, 334.

399. This article's structure is a precursor to the collections of superstitions and popular beliefs that began appearing in Britain and the United States during the nineteenth century, following the same numbered entry format. In the next part, we see a move toward the systematic collection, publication, and interpretation of these beliefs that, at this early stage, were gathered haphazardly. As the nineteenth century wore on, they were regarded as more than just rude superstitions that needed to be extinguished: these "lower forms" of culture required explication to fully understand the "higher forms" into which they evolved.

400. W., "Popular Whims and Superstitions, Relating More Especially to the Practice of Medicine in the Nineteenth Century," *Scientific Tract and Family Lyceum* 1 (1 June 1834): 344–47.

401. Henry I. Bowditch, *Consumption in New England or, Locality One of Its Chief Causes* (Boston: Tickner & Fields, 1862).

402. Bowditch, *Consumption in New England*, v.

403. Bowditch, *Consumption in New England*, v.

404. Henry I. Bowditch, *Is Consumption Ever Contagious, or Communicated by One Person to Another in any Manner?* (Boston: David Clapp, 1864), 4.

405. Bowditch, *Is Consumption Ever Contagious?* 4.

406. Bowditch, *Consumption in New England*, 10.

407. Henry I. Bowditch, *Consumption in New England and Elsewhere, or Soil-Moisture One of Its Chief Causes*, 2d ed. (Boston: David Clapp & Son, 1868), 13.

408. Bowditch, *Consumption in New England*, 50.

409. Edwin Emery, *History of Sanford, Maine: 1661–1900* (Salem, MA: Salem Press, 1901), 403.

410. Emery, 403.

Chapter 14: The Village Witch Told Them ... Cut Her Heart Out and Bury It

411. Henry I. Bowditch, *Consumption in New England* (Boston: Tickner & Fields, 1862), 76.

412. John T. Cumbler, *From Abolition to Equal Rights for All: The Making of a Reform Community in the Nineteenth Century* (Philadelphia: University of Pennsylvania Press, 2008), 23–24.

413. Cumbler, 25.

414. Some maintain that Louisiana was the first state to establish a Board of Health—in 1855, as a response to controversy over the causes of yellow fever—but since it "proved not to be a functional organization," Massachusetts generally is awarded that distinction. Lloyd Novick, Cynthia Morrow, and Glen Mays, eds, *Public Health Administration: Principles for Population-Based Management*, 2nd ed. (Sudbury, MA: Jones and Bartlett Publishers, 2008), 11.

415. State Board of Health of Massachusetts, Fourth Annual Report (Boston: Wright & Potter, 1873), 308.

416. *Fourth Annual Report*, 329.

417. *Fourth Annual Report*, 371.

418. *Fourth Annual Report*, 359.

419. *Fourth Annual Report*, 359–60.

420. *Fourth Annual Report*, 365.

421. Bowditch, *Consumption in New England*, 61.

422. *Fourth Annual Report*, 379.

423. *Fourth Annual Report*, 338–39.

424. *Fourth Annual Report*, 326–27.

425. Harold D. Levine, "Folk Medicine in New Hampshire," *New England Journal of Medicine* 224, no. 12 (1941): 488.

426. Levine, 488.

427. U.S. Federal Census Mortality Schedules, 1850-1885 - 1870, Marshall Bowen 66, b. MA, d. April; farmer; "inflation of the bowels."

428. Pauline Swain Merrill, John C. Gowan, and others, *A Small Gore of Land: A History of New Hampton, New Hampshire, Founded In 1777* ([New Hampton, NH]: New Hampton Bicentennial Committee, 1977), 54–55.

429. See, for example, Stith Thompson, *Motif-Index* (1955-1958), narrative folk motifs: G211.2. witch in form of wild beast; G263.4.0.1. illness caused by curse of witch; M341.1. prophecy: death at (before, within) certain time.

430. The story of Granny Hicks appears in other published works, including Charles J. Jordan, *Tales Told in the White Mountains* (Lebanon, NH: University Press of New England, 2003), 85–86 and Richard M. Dorson, *Jonathan Draws the Long Bow: New England Popular Tales and Legends* (Cambridge, Massachusetts: Harvard University Press, 1946), 42.

431. Levine, 487.

432. Richard M. Dorson, *American Folklore*, with revised bibliographical notes, reprint, 1959, The Chicago History of American Civilization (Chicago: University of Chicago Press, 1977), 58.

433. Levine is termed "a former resident of Bristol" in Merrill et al, *A Small Gore of Land*, 53, in which the authors reprised the Bowen narrative as published by Levine.

434. Levine, 492.

435. George Lyman Kittredge, *Witchcraft in Old and New England* (Cambridge, Massachusetts: Harvard University Press, 1929), 233.

436. C. Grant Loomis, "Some Lore of Yankee Genius," *Western Folklore* 6, no. 4 (1947): 344.

Part VII: Evolutionists

437. "[Introduction]," *Journal of American Folklore* 2, no. 4 (1889): 1.

438. "[Introduction]," 2.

439. "[Introduction]," 2.

440. See Sabra J. Webber, *Folklore Unbound: A Concise Introduction* (Long Grove, IL: Waveland Press, 2015), 15–16.

Chapter 15: A Peculiar Kind of Vampirism

441. Jeremiah Curtin, "European Folk-Lore in the United States," *Journal of American Folklore* 2, no. 4 (1889): 58.

442. Edward W. Said, *Orientalism*, reprint, 1978 (Penguin Classic, 2003), xviii.

443. Jeremiah Curtin, *Memoirs of Jeremiah Curtin*, Wisconsin Biography Series (Madison: The State Historical Society of Wisconsin, 1940), 306.

444. Curtin, "European Folk-Lore in the United States," 58–59.

445. Theodore Roosevelt quoted in the foreword to Jeremiah Curtin, *The Mongols: A History* (Boston: Little, Brown, and Company, 1908), ix.

446. Cheryl L. Collins. "Curtin, Jeremiah, and Alma Cardell Curtin"; http://www.anb.org/articles/14/14-01161.html; American National Biography Online July 09, 2008.

447. John G. Gregory quoted in Curtin, *Memoirs*, 6.

448. Curtin, *Memoirs*, 28.

449. Leonard Twynham, "Jeremiah Curtin and Vermont," *The Vermonter* 36, no. 12 (1931): 274–78.

450. Curtin, "European Folk-Lore in the United States," 58.

451. Frederick J. Seaver, *Historical Sketches of Franklin County and Its Several Towns with Many Short Biographies* (Albany, NY: J. B. Lyon Company, 1918), 432.

452. Keith Thomas, *Religion and the Decline of Magic: Studies in Popular Beliefs in Sixteenth and Seventeenth Century England* (London: Weidenfeld and Nicolson, 1971), 224. Regarding its folk foundation, Thomas wrote, "Even the weapon-salve traced its descent from folk practice," 229. See also Amelia Soth's observation: "This is how the weapon-salve fit into Paracelsian philosophy: When you bleed on a blade, his followers argued, your body becomes linked to it. The wound and the weapon 'doe sympathise together, even as wee see one thred extended from one end of a chamber unto the other.' Therefore, to treat the weapon is to treat the wound." in "The Occult Remedy the Puritans Embraced: Why did the Puritans embrace a medical treatment that looked suspiciously like black magic?" *Cabinet of Curiosities*, JSTOR Daily, posted September 19, 2019, accessed April 10, 2020, https://daily.jstor.org/the-occult-remedy-the-puritans-embraced/.

453. Curtin, "European Folk-Lore in the United States," 56.

454. Jeremiah Curtin, *Tales of the Fairies and of the Ghost World: Collected from Oral Tradition in South-West Munster* (Boston: Little, Brown & Co., 1895), 180–91.

455. Curtin, *Tales of the Fairies*, 130.

456. William Shedden Ralston, *Russian Folk-Tales* (New York: Lovell, Adam, Wesson, 1873), 318–22.

457. Jeremiah Curtin, *Myths and Folk-Tales of the Russians, Western Slavs, and Magyars* (London: Simpson Low, Marston, Searle & Rivington, 1891), 92.

458. George M. Foster, "Peasant Society and the Images of Limited Good," *American Anthropologist* 67 (1965): 293–315.

459. David J. Hufford, "Beings Without Bodies: An Experienced-Centered Theory of the Belief in Spirits," in *Out of the Ordinary: Folklore and the Supernatural*, ed. Barbara Walker (Logan: Utah State University Press, 1995), 13–14.

460. Hufford, 19.

461. This is an echo of the widely distributed folk motif of buried treasure sinking from grasp when excavated, which includes motif M448, curse: to sink into the earth.

462. Motifs G303.4.8.1 & G303.6.3.4 - Devil's odor of sulphur.

463. Motif E410.1 - ground trembles or rumbles when ghost arises from grave; burial of human remains in cooking kettles also is a Native American motif. See Carla Cevasco, "This is My Body: Communion and Cannibalism in Colonial New England and New France," *New England Quarterly* 89, no. 4 (December 2016): 556–86.

464. Richard M. Dorson, "Folklore in American Literature: A Postscript," *Journal of American Folklore* 71, no. 280 (1958): 159.

465. Glencarella, unpublished article and personal communication, August 25, 31, September 5, 2015 & March 27, 2017.

466. John McNab Currier, "Contributions to New England Folk-Lore," *Journal of American Folklore* 4, no. 14 (1891): 253.

467. John McNab Currier, "Contributions to the Folk-Lore of New England," *Journal of American Folklore* 2, no. 7 (1889): 291.

468. H. P. Smith and W. S. Rann, eds, *History of Rutland County, Vermont* (Syracuse, NY: D. Mason & Co., 1886), 878–79.

469. Frederick C. Waite, *The First Medical College in Vermont: Castleton, 1818–1862* (Montpelier: Vermont Historical Society, 1949), 124–25.

470. Tom Campbell, *Fighting Slavery in Chicago: Abolitionists, the Law of Slavery, and Lincoln* (Chicago: Amperstand, 2009), 42–43.

471. Moncure Daniel Conway, *Demonology and Devil-Lore* (New York: Henry Holt and Company, 1879), 2:52.

472. Waite, 208.

473. Waite, 163.

474. John McNab Currier, *Song of the Hubardton Raid, Delivered on the 50th Anniversary of the Raid of the Citizens of Hubbardton, Vermont, on Castleton Medical College* (Castleton, VT: Castleton Medical and Surgical Clinic, 1880), 3.

Chapter 16: In New England Consumption is a Spiritual Visitation

475. "Proceedings of the Anthropological Society of Washington." *American Anthropologist*, vol. 15, no. 2, 1913, pp. 347–362. JSTOR, JSTOR, www.jstor.org/stable/659676. "Proceedings of the Anthropological Society of Washington." *American Anthropologist*, vol. 16, no. 1, 1914, pp. 110–134. JSTOR, JSTOR, www.jstor.org/stable/659503.

476. "Third Annual Meeting of the American Folk-Lore Society." *The Journal of American Folklore*, vol. 5, no. 16, 1892, pp. 1–8. JSTOR, JSTOR, www.jstor.org/stable/533442.

477. George R. Stetson, "The Animistic Vampire in New England," *American Anthropologist* 9, no. 1 (1896): 7.

478. Stetson, "Animistic Vampire," 10.

479. District of Columbia Deaths and Burials, 1840-1964, MyHeritage.com [online database]. Lehi, UT, USA: MyHeritage (USA) Inc., https://www.myheritage.com/research/collection-30183/district-of-columbia-deaths-burials-1840-1964, Record: https://www.myheritage.com/research/

record-30183-174408/geo-r-stetson-in-district-of-columbia-deaths-burials, Geo. R. Stetson, Country: USA; State: District of Columbia; Folder: 4025279; Film: 2115944; Image: 1607.

480. George R. Stetson, "Letter to Sidney S. Rider," unpublished letter in Rider Collection, Rider Collection, housed at John Hay Library, Brown University, Providence, box 181, no. 32 (1895), no. 1:1–2.

481. Stetson, "Letter to Sidney S. Rider," no. 2:1–2.

482. Stetson, "Animistic Vampire," 8–9.

483. Moncure Daniel Conway, *Demonology and Devil-Lore* (New York: Henry Holt and Company, 1879), 52.

484. Stetson, "Animistic Vampire," 8.

485. Stetson, "Animistic Vampire," 9–10.

486. Stetson, "Animistic Vampire," 10.

487. Stetson, "Animistic Vampire," 10.

488. For discussions of Tylor's evolutionism in relations to folkloristics, see Sabra J. Webber, *Folklore Unbound: A Concise Introduction* (Long Grove, IL: Waveland Press, 2015), 15–17 and Robert A. Georges and Michael Owen Jones, *Folkloristics: An Introduction* (Bloomington, Indiana: Indiana University Press, 1995), 42–57.

489. Stetson, "Animistic Vampire," 10.

490. Stetson, "Animistic Vampire," 1.

491. Stetson, "Animistic Vampire," 1.

492. See Bilinda Straight, *Miracles and Extraordinary Experiences in Northern Kenya* (Philadelphia: University of Pennsylvania Press, 2007), 23 for a recent vivid example of this genre of marginalization, on another continent: "Both Catholic and Anglican missionaries frequently challenged Samburu [a culture of northern Kenya] explanations of Nkai's [divinity's] presence on empirical grounds while offering an equally fantastic alternative."

493. Murray J. Leaf, *Human Organizations and Social Theory: Pragmatism, Pluralism, and Adaptation* (Urbana and Chicago: University of Illinois Press, 2009), 38. Ludwig Wittgenstein, in "Remarks on Frazer's The Golden Bough," Chapter 2, Translated by Stephan Pamié in *The Mythology in Our Language: Remarks on Frazer's Golden Bough*, edited by Giovanni da Col, edited and translated by Stephan Palmié (Chicago: Hau Books, 2018), 36-38, made this point, writing, "The same savage who, apparently in order to kill his enemy, pierces an image of him, really builds his hut out of wood, and carves his arrow skillfully and not in effigy. ... And magic always rests on the idea of symbolism and of language."

494. Burt Feintuch and David H. Watters, *Encyclopedia of New England Culture* (New Haven, CT: Yale University Press, 2005), 422–23.

495. Wayland D. Hand, ed., *Popular Beliefs and Superstitions from North Carolina*, vol. 6–7 of *Frank C. Brown Collection of North Carolina Folklore* (Durham, North Carolina: Duke University Press, 1964), 7: nos. 501–65.

496. Stetson, "Animistic Vampire," 7.

497. Stetson, "Animistic Vampire," 8.

498. Alan Dundes, ed., *The Vampire: A Casebook* (Madison, Wisconsin: University of Wisconsin Press, 1998), 161. For this argument, see also G. David Keyworth, "Was the Vampire of the Eighteenth Century a Unique Type of Undead-Corpse?" *Folklore* 117, no. 3 (2006): 241–60.

499. Quoted in Montague Summers, *The Vampire: His Kith and Kin* (London: Kegan Paul, Trench, Trubner & Co., Ltd., 1928), 28 and echoed, later in the century, by Voltaire in François Marie Arouet Voltaire, *A Philosophical Dictionary*, vol. 7 of *The Works of Voltaire: A Contemporary Version*, trans. William F. Fleming (New York: Dingwall-Rock, Ltd., 1927), 143–49 ("These vampires were corpses, who went out of their graves at night to suck the blood of the living, either at their throats or stomachs, after which they returned to their cemeteries. The persons so sucked waned,

grew pale, and fell into consumption; while the sucking corpses grew fat, got rosy, and enjoyed an excellent appetite.").

500. Jacob Grimm, *Teutonic Mythology*, 4th ed., trans. James Steven Stallybrass (London: George Bell & Sons, 1883), 4:1455.

501. Stetson, "Animistic Vampire," 2.
502. Stetson, "Animistic Vampire," 1.
503. Stetson, "Animistic Vampire," 1.
504. Stetson, "Animistic Vampire," 2.
505. Stetson, "Animistic Vampire," 2.
506. Stetson, "Animistic Vampire," 2.
507. Stetson, "Animistic Vampire," 2–3.
508. Stetson, "Animistic Vampire," 3.
509. Stetson, "Animistic Vampire," 3.
510. Stetson, "Animistic Vampire," 3.
511. Stetson, "Animistic Vampire," 11–12.
512. Stetson, "Animistic Vampire," 3.
513. Stetson, "Animistic Vampire," 3.

514. Edward B. Tylor, *Primitive Culture: Researches into the Development of Mythology, Philosophy, Religion, Language, Art, and Custom*, vol. 2 (London: John Murray, 1871), 91–98. https://books.google.com.au/books?id=eW11O76PNK4C.

515. Sir James G. Frazer, *Fear of the Dead in Primitive Religion: Lectures Delivered on the William Wyse Foundation at Trinity College, Cambridge, 1932–1933*, [vol. 1] (London: Macmillan, 1933), 142, https://hdl.handle.net/2027/inu.39000005836775.

516. Agnes Murgoci, "The Vampire in Roumania," *Folklore* 37, no. 4 (1926): 320–21.

517. For vampires and werewolves, see, for example, Murgoci, "Vampire in Roumania," 337; and T. P. Vukanovic, "The Vampire in the Beliefs and Customs of the Gypsies in the Province of Kosovo-Metohija, Stari Ras and Novopazarski Sandzak, Yugoslavia," pt. 1, *Journal of the Gypsy Lore Society*, 3rd ser, 36, no. 3–4 (July–October 1957): 129. For being "witch ridden" or "hagged," see David J. Hufford, *The Terror That Comes in the Night: An Experience-Centered Study of Supernatural Assault Traditions* (Philadelphia: University of Pennsylvania Press, 1982), 10–11, 231.

518. J. A. MacCulloch, "Vampire," in *Encyclopaedia of Religion and Ethics*, vol. 12, ed. James Hastings, reprint, 1908–26 (New York: Scribner's Sons, 1928), 589–90.

519. Murgoci, 321.

520. Henry Charles Coote, "Some Italian Folk-Lore," *Folk-Lore Record* 1 (1878): 214.

521. Bruce A. McClelland, *Slayers and Their Vampires: A Cultural History of Killing the Dead* (Ann Arbor: University of Michigan Press, 2006), 23–24.

522. MacCulloch, "Vampire," 589.

523. https://www.1902encyclopedia.com/V/VAM/vampire.html, accessed January 15, 2019.

Chapter 17: Explain It? I Can't. It Is an Old, Old Belief

524. *The Times* (Philadelphia, Pennsylvania), 19 January 1896, 13; *Buffalo Enquirer* (Buffalo, New York), 21 January 1896, 4; *San Francisco Examiner*, 26 January 1896, 31; *Butte Miner* (Butte, Montana), 27 January 1896, 2; *New York World*, 2 February 1896; *Oakland Tribune*, 27 March 1896, 6; *Wilkes Barre Times*, 28 April 1896, 6.

525. Elizabeth Miller, *Dracula: Sense & Nonsense* (Westcliff-on-Sea: Desert Island Books, 2000), 58.

526. Miller, 32.
527. Miller, 33.

528. https://www.findagrave.com/memorial/44663790/rene-bache. Accessed 8/15/19.

529. Vernum Judson Briggs (myheritage.com tree, accessed 5/22/18.)

530. *Pawtuxet Valley Gleaner*, 22 April 1892, 5.

531. *Boston Traveler* (28 March 1896); *Tacoma Daily News* (28 March 1896); *Wilkes-Barre* [PA] *Sunday Leader* (29 March 1896); *Oil City* [PA] *Derrick* (29 March 1896); and *Fort Wayne Weekly Journal* (2 April 1896).

532. *Louisville Courier Journal* (14 February 1896) and *Atlanta Constitution* (17 February 1896).

533. *The Sun* (New York City), https://en.wikipedia.org/wiki/The_Sun_(New_York_City), accessed 1/14/19.

534. Kent C. Ryden and Simon J. Bronner, "New England," in *Encyclopedia of American Folklife*, ed. Simon J. Bronner (Armonk, NY: M. E. Sharpe, 2006), 870.

535. Michael Schudson, *Discovering the News: A Social History of American Newspapers* (New York: Basic Books, Inc., 1978), 102.

536. Rhode Island Department of State, *Manual, with Rules and Orders, for the Use of the General Assembly of the State of Rhode Island* (Providence: Rhode Island Department of State, 1895), 328.

537. Mary Kenyon Huling, *Historical Sketch of the Baptist Church in Exeter, Rhode Island* (Chestnut Hill) (Cranston, RI: Pendleton Press, 1939), 15.

538. Michael Schudson, *The Power of News* (Cambridge, MA: Harvard University Press, 1995), 88.

539. See Martin J. Manning, "Images and Ideas: Yankees," in *Encyclopedia of New England*, eds Burt Feintuch and David H. Watters (New Haven: Yale University Press, 2005), 811–13 for a summary of Yankees and associated regional humor.

540. Daniel G Hoffman, et al., "Folklore in Literature: A Symposium," *Journal of American Folklore* 70, no. 275 (1957): 1.

541. MacEdward Leach, "Folklore in American Regional Literature," *Journal of the Folklore Institute* 3, no. 3 (1966): 387.

542 Leach, 395.

543. For example, see Stith Thompson, *Motif-Index of Folk Literature: A Classification of Narrative Elements in Folktales, Ballads, Myths, Fables, Mediaeval Romances, Exempla, Fabliaux, Jest-Books and Local Legends*, revised and enlarged ed., reprint, 1932–36 (Bloomington, Indiana: Indiana University Press, 1955–58): Motif M422. Curse transferred to another person or thing. See also, E595. Cures by transferring disease to dead.

Chapter 18: A Strange Superstition Was Whispered in Robinson Hollow

544. Adaline M. Tirrell, "Legends of Hampshire Hills—Old Nahum Had His Jokes," *Springfield Republican*, 25 August 1929, 4 E.

545. Adaline M. Tirrell, "Legends of Hampshire Hills—Two Chesterfield Tales," *Springfield Republican*, 15 September 1929, 2 E.

546. Tirrell, "Legends of Hampshire Hills—Two Chesterfield Tales."

547. *Tuskegee News*, 12 June 1902, 4.

548. "Graduates in Oratory," *Boston Herald*, 15 May 1895, 12

549. *Tuskegee News*, 4.

550. When she attended a banquet in honor of WPI's 50th anniversary, she was so identified in "Worcester Polytec," *Boston Evening Transcript*, 9 June 1915, 2:5; Professor of English at Lander College, in April 1925, referred to as Miss Adaline M. Tirrell, in *The State* (Columbia, SC) 20 April 1925, 5; *State* (Columbia, SC) 28 August 1925, 10, provides middle name of Maria.

551. Obituary, *Boston Transcript*, 14 August 1937, 6.

552. Tirrell, "Legends of Hampshire Hills—Two Chesterfield Tales."

553. http://freepages.rootsweb.com/~torrey/genealogy/robinson.htm, accessed 8/16/19.

554. New England Historic Genealogical Society; Boston, Massachusetts; Massachusetts Vital Records, 1840–1911. Ancestry.com. Massachusetts, Death Records, 1841-1915 (database on-line). Provo, UT, USA: Ancestry.com Operations, Inc., 2013.

555. Joseph Brunell, Jr., "Original Record of Joseph Burnell, Jr., 1804–1911, Consisting of an Index and Three Parts," _ Collection, Chesterfield (MA) Town Clerk (1804–1911), 2:17, Http://townofchesterfieldma.com/historical-commission.

556. 1855 Massachusetts State Census.

557. Brunell, "Original Record of Joseph Burnell, Jr., 1804–1911, Consisting of an Index and Three Parts," 1:55.

558. Robinson Hollow Cemetery Memorials, https://www.findagrave.com/cemetery/2331024/memorial-search?page=1#sr-44223171, accessed January 17, 2019.

559. Daniel G Hoffman, et al., "Folklore in Literature: A Symposium," *Journal of American Folklore* 70, no. 275 (1957): 5.

560. Hoffman, et al., 6.

561. Adaline M. Tirrell, "Legends of Hampshire Hills--The Devil in the Candle," *Springfield Sunday Union and Republican*, 8 September 1929, 3E.

562. Adaline M. Tirrell, "Hampshire Hills Legends—Buried in the Old Pasture," *Springfield Republican*, 30 March 1930, 2E.

563. Hoffman, et al., 7.

564. Adaline M. Tirrell, "Hampshire Hills Legends—A Child Led Them Home," *Springfield Republican*, 26 April 1931, 3E.

565. Brunell, "Original Record of Joseph Burnell, Jr., 1804–1911, Consisting of an Index and Three Parts."

566. Joel Engram, "Joel Engram's Book, Kept from 1834 to 1902," _ Collection, Chesterfield (MA) Town Clerk (1834–1902), Http://townofchesterfieldma.com/historical-commission.

567. Bela Stetson, et al., _ Collection, Chesterfield (MA) Town Clerk (1801–92), Http://townofchesterfieldma.com/historical-commission.

568. Olive Cleveland Clarke, "Things That I Remember at Ninety-Five: Memoir of Life on the Mount," _ Collection, Chesterfield (MA) Town Clerk (1881), Http://townofchesterfieldma.com/historical-commission.

569. Brunell, "Original Record of Joseph Burnell, Jr., 1804–1911, Consisting of an Index and Three Parts," 2:22.

570. Engram, "Joel Engram's Book, Kept from 1834 to 1902," 9.

571. Stetson, et al., _ Collection, Chesterfield (MA) Town Clerk, 4.

572. Brunell, "Original Record of Joseph Burnell, Jr., 1804–1911, Consisting of an Index and Three Parts," 1:37.

573. Engram, "Joel Engram's Book, Kept from 1834 to 1902," 16.

574. Stetson, et al., _ Collection, Chesterfield (MA) Town Clerk, 15.

575. Michael E. Bell, *Food for the Dead: On the Trail of New England's Vampires* (New York: Carroll & Graf, 2001), 269–76.

576. Joseph R. Chandler, "Superstitions of New England, No. 3," *The Lady's Book* 6 (April 1833): 166.

577. Joseph R. Chandler, "The Undying One: From Superstitions of New England, No. 4," *The Lady's Book* 6 (May 1833): 214.

578. Joseph R. Chandler, "Superstitions of New England, No. 1," *The Lady's Book* 6 (January 1833): 31–32.

579. Day Otis Kellogg, *New American Supplement to the Latest Edition of the Encyclopaedia Britannica* (New York: Werner, 1897), 2:747.

580. John Thomas Scharf and Thompson Westcott, *History of Philadelphia: 1609–1884* (Philadelphia: L. H. Everts & Co., 1884), 3:1969.

581. "Superstitions in New England" also appeared in the following newspapers: *Salem Gazette*, 22 March 1822, 2; *American Repertory* (St. Albans, Vermont), 9 April 1822, 4; *Republican Advocate* (Batavia, New York), 24 April 1822, 2.

582. letter to the editor, *Old Colony Memorial and Plymouth County Advertiser*, 11 May 1822, 7.

Chapter 19: Sarah Came Every Night Causing Great Pain and Misery

583. George Alfred Hazen, "A Vampire of Seneca Lake," *New York Folklore Quarterly* 6 (1950): 164.

584. Hazen, 164-65.

585. John Corbett, *The Lake Country: An Annal of Olden Days in Central New York* (Rochester, New York: Democrat and Chronicle Print, 1898), 98–99.

586. Corbett, *The Lake Country*, 6.

587. *A Biographical Record of Schuyler County, New York* (New York & Chicago: S. J. Clarke, 1908), 229.

588. *A Biographical Record of Schuyler County*, New York, 230.

589. Corbett, *The Lake Country*, 138–39.

590. Corbett, *The Lake Country*, 95–96.

591. "Cumberland Town Council Records," Town Council Meeting of 8 February 1796 (Cumberland, Rhode Island, 1796).

592. Herbert A. Wisbey, Jr., *Pioneer Prophetess: Jemima Wilkinson, the Publick Universal Friend* (Ithaca, NY: Cornell University Press, 1964), 6.

593. Wisbey, 7.

594. John Corbett, "The Town of Reading: Its Organization, Its Dismemberment for the Town of Starkey, Its Supervisors," *Schuyler County Chronicle*, 26 February 1909, 1.

595. Wisbey, 10–11.

596. Ancestry.com. 1850 United States Federal Census [database on-line]. Provo, UT, USA: Ancestry.com Operations, Inc., 2009. Images reproduced by FamilySearch.

597. https://www.pvhistorian.com/caseybtyler.html, accessed February 13, 2014

598. Sidney S. Rider, "The Belief in Vampires in Rhode Island," *Book Notes* 5:7 (31 March 1888): 37-39.

599. Rider, 39.

600. Russell J. DeSimone and Erik J. Chaput, "Sidney Rider and The Business Of Rhode Island History," accessed April 5, 2014, http://www.providenceri.com/archives/sidney-rider-and- the-business-of-rhode-island-0.

601. Michael E. Bell, *Food for the Dead: On the Trail of New England's Vampires* (New York: Carroll & Graf, 2001), 71–72.

602. Rider, 39.

603. See Bell, *Food for the Dead*, 2001:65-75.

604. Rider, 37.

605. Republished in *Book Notes* 25:6 (21 March 1908): 46-47.

606. "Complimentary Press Notices of Book Notes," *Book Notes* 26:1 (9 January 1909): 8.

Chapter 20: The Dead Gnawing and Feeding upon the Living

607. "Tragic Legend of West Stafford Cemetery," *Hartford Courant*, 28 March 1915, Z10.

608. J. R. Cole, *The History of Tolland County, Connecticut* (New York: W. W. Preston & Co., 1888), iii.

609. Francis Gerry Fairfield, "A Century Ago in New England," *Appletons' Journal* 15 (20 May 1876): 654–55.

610. Fairfield, "Century Ago," 653.

611. Fairfield, "Century Ago," 654.

612. "Fairfield, Francis Gerry," *Appletons' Annual Cyclopaedia and Register of Important Events* (New York: D. Appleton), v. 12 (1887):585, https://babel.hathitrust.org/cgi/pt?id=umn.31951d00546921u&view=1up&seq=606 accessed 9/26/2019.

613. Eugene Lemoine Didier, *A Primer of Criticism* (Baltimore: People's Publishing Company, 1883), 37.

614. Didier, 37.

615. "A Humbug and a New System of Physics," *American Monthly Microscopical Journal* 2, no. 1 (1881): 15.

616. Fairfield attributed Poe's composition of "The Raven" to his suffering from "cerebral epilepsy," a disease that Fairfield asserted created visual and aural hallucinations, as well as habitual lying. See Eliza Richards, "Outsourcing 'The Raven': Retroactive Origins," *Victorian Poetry* 43:2 (2005): 205–21. http://www.jstor.org/stable/40002615, pp. 211-12.

617. Howard, "Howard's Letter," *Boston Globe*, 10 April 1887, 16.

618. "An Opium Eater," *Kansas City Times*, 10 April 1887, 7.

619. Francis G. Fairfield, "Timothy Tot: A Prose Story with Poetic Passages," *The Radical* 10 (1872): 133.

620. Fairfield, "Timothy Tot," 298.

621. Fairfield, "Timothy Tot," 303, 387, 388, 458, 466.

Chapter 21: Thoughts of a Horrid Nature Arose in the Parents

622. Quoted in "Owen Wister's Virginian," Feature at the Bar-D-Ranch, accessed 1/10/20, http://www.cowboypoetry.com/wister.htm.

623. See Alan Jalowitz, biography of Owen Wister, accessed January 10, 2020, https://web.archive.org/web/20090802112457/http://pabook.libraries.psu.edu/palitmap/bios/Wister Owen.html.

624. Owen Wister, "Superstition and the Doctor," *Boston Medical and Surgical Journal* 172, no. 10 (March 11, 1915): 357.

625. Wister, "Superstition," 359.

626. Wister, "Superstition," 360.

627. Wister, "Superstition," 360.

628. Wister, "Superstition," 360–61. Martha R. McPartland, "The Bonesetter Sweets of South County, Rhode Island," *Yankee* (January 1968):80-102, http://hunt4sweet.blogspot.com/2013/03/the-bonesetter-sweets.html, wrote: "Instances naming local doctors who failed to relieve suffering that was later relieved by one of the Sweets have become a part of South County folklore." McPartland described this event in more detail: "Job [Sweet] was born in 1724 and married Jemima Sherman in 1750. He lived all his life in the South County section of Rhode Island. During the Revolutionary War, Dr. Job, as he was called, was sent to Newport to set the bones of French officers, an operation their own doctors would not attempt. After the war, Aaron Burr, later Vice-President of the United States, sent for him to minister to his daughter, Theodosia, who had a dislocated hipbone. Dr. Job, rather reluctantly, journeyed to New York and was there greeted by Colonel Burr, their family doctor, and several other learned medical men. Job was not happy about having an audience. They suggested that a specific hour—ten o'clock the next morning—be set for the operation.

After they had left the house, Job talked soothingly to Theodosia, who was in great pain, and explained to her his methods. When he had eased her fears, he asked her father if he could place his hands on her hip to locate the trouble. Colonel Burr consented and, after a few minutes, Job said to her, 'Now walk around the room,' and much to the surprise of Theodosia and her father she did just that—and without pain. When the medical team arrived the next morning Job was well on his way back to Rhode Island and Theodosia's hip was properly set and on the mend."

629. Wister, "Superstition," 361.

630. Wister, "Superstition," 361.

631. Rhode Island Historical Preservation Commission, *Statewide Historical Preservation Report W-N-I: Narragansett Pier, Narragansett, Rhode Island* (Providence, RI: RIHPC, 1978): 8-9.

632. Brander Matthews, "The New Narragansett Pier," *Harper's Weekly* (July 7, 1906): 958.

633. Rhode Island Historical Preservation Commission, *Statewide Historical Preservation Report W-Nk-1: North Kingstown, Rhode Island.* (Providence, RI: RIHPC, 1979): 28–29, 61.

634. RIHPC Report, *North Kingstown*, 29.

635. An article in the Fort Scott (KS) *Weekly Monitor*, 13 November 1890, 2, noted that "Rev. Edward Everett Hale . . . summers in Rhode Island." *New York Tribune*, 7 August 1902, 8, mentioned that Hale was spending the summer in Matunuck, South Kingstown, Rhode Island.

636. Eric M. Augenstein, "Mary Channing Wister: An Unknown Legend," accessed 1/10/20, http://www.lasalle.edu/commun/history/articles/marychanningwister.htm.

637. Augenstein, "Mary Channing Wister."

638. George T Metcalf, "My Most Unforgetable Character," biography of Harold Metcalf 1860–1923), photocopy of unpublished essay (Providence, RI: Rhode Island Historical Society Library, 1971): 3.

639. See her obituary in the *Philadelphia Inquirer*, 25 August 1913, 1.

640. Metcalf, "My Most Unforgetable Character," 3.

641. Jalowitz, biography of Owen Wister.

642. Quoted in Willard R. Handley, "Wister's Omniscience and Omissions," in *Reading "The Virginian" in the New West*, eds. Melody Graulich and Stephen Tatum (Lincoln: University of Nebraska Press, 2003): 65.

643. Fanny Kemble Wister, ed., *Owen Wister Out West: His Journals and Letters* (Chicago: University of Chicago Press, 1958): 38.

644. Quoted in Fanny Kemble Wister, 259.

645. Walter Prichard Eaton, "In Old South County," *Outing* 60 (June 1912): 268–69.

646. Eaton, 259.

647. Eaton, 275.

648. "Consumption," *Pawtucket Chronicle*, 3 October 1829.

649. Wayland D. Hand, ed., *Popular Beliefs and Superstitions from North Carolina*, vol. 6–7 of *Frank C. Brown Collection of North Carolina Folklore* (Durham, North Carolina: Duke University Press, 1964): 7, no. 5422.

Chapter 22: They Drove a Stake through Her Breast

650. S. E. Farley, "Popular Prejudices," *Godey's Lady's Book, and Ladies' American Magazine* 24 (May 1842): 281.

651. Farley, 281.

652. Farley, 281.

653. Karen Gordon-Grube, "Evidence of Medicinal Cannibalism in Puritan New England: 'Mummy' and Related Remedies in Edward Taylor's 'Dispensatory'," *Early American Literature* 28, no. 3 (1993): 190–91.

654. John Langdon Sibley, *A History of the Town of Union, in the County of Lincoln Maine* (Boston: Benjamin B. Mussey, 1851), 17.

655. Cyrus Eaton, *History of Thomaston, Rockland, and South Thomaston, Maine* (Hallowell, ME: Masters, Smith, 1865), 2:221.

656. S. E. Farley, "Domestic and Social Claims on Woman," *Godey's Lady's Book, and Ladies' American Magazine* 24–25 (1842): 148–49.

657. Farley, "Popular Prejudices," 282.

658. Mary Andrews Denison, *Home Pictures* (New York: Harper & Brothers, 1853), 295–96.

659. Madeleine B. Stern, "Mary Ann Andrews Denison," edited by Edward T. James, Janet Wilson James, Paul S. Boyer, *Notable American Women, 1607-1950: A Biographical Dictionary* (Cambridge, MA: Belknap Press, 1971), 1:462-63.

660. MacEdward Leach, "Folklore in American Regional Literature," *Journal of the Folklore Institute* 3, no. 3 (1966): 388.

661. S. Foster Damon, *Amy Lowell: A Chronicle*, reprint, 1935 (Hamden, CT: Archon Books, 1966), 709–10.

662. Damon, 711.

663. Koen Vermeir, "Vampires as Creatures of the Imagination: Theories of Body, Soul and Imagination in Early Modern Vampire Tracts (1659-1755)," *Diseases of the Imagination and Imaginary Disease in the Early Modern Period*, Brepols Publishers, pp.341-373, 2012, p. 33, halshs-00609387.

664. Vermeir, 2.

665. Edith Wharton, "Bewitched," accessed January 14, 2015, https://ebooks.adelaide.edu.au/w/wharton/edith/here_and_beyond/chapter3.html.

666. Rhode Island Historical Preservation Commission, *Statewide Historical Preservation Report W-Nk-1: North Kingstown*, Rhode Island (Providence, Rhode Island: Rhode Island Historical Preservation Commission, 1979), 28–29.

667. Edith Wharton, *A Backward Glance* (New York: D. Appleton-Century Company, 1934).

668. Gahan Wilson, ed., *First World Fantasy Awards* (Garden City, NY: Doubleday, 1977): 253.

669. Tim Callahan, "Advanced Readings in D&D: Manly Wade Wellman," http://www.tor.com/2013/10/21/advanced-readings-in-dad-manley-wade-wellman/.

670. John Bealle, "Another Look at Charles M. Skinner," *Journal of American Folklore* 53, no. 2 (1994): 114.

671. Manly Wade Wellman, "The Vampire in America," in *First World Fantasy Awards*, ed. Gahan Wilson (Garden City, NY: Doubleday & Company, 1977), 264.

672. Nancy Kinder, "The 'Vampires' of Rhode Island," *Yankee*, October 1970, 114–15, 166–67.

673. Fritz Koch, "R. I. Latter-Day Transylvania?" *Providence Evening Bulletin*, 25 February 1975, 1.

Conclusion: Disease and Death, Fear and Hope

674. Daniel C. Hallin, *The "Uncensored War": The Media and Vietnam* (Berkeley: University of California Press, 1986), 116-17.

675. Kent C. Ryden and Simon J. Bronner, "New England," in *Encyclopedia of American Folklife*, ed. Simon J. Bronner (Armonk, NY: M. E. Sharpe, 2006), 868.

676. Franklin B. Hough, *History of St. Lawrence and Franklin Counties, New York, from the Earliest Period to the Present Time* (Albany: Little & Co., 1853), 707, note B, referred from p. 109.

677. George R. Stetson, "The Animistic Vampire in New England," *American Anthropologist* 9, no. 1 (1896): 1.

678. Stetson, 8.

679. Tom Campbell, *Fighting Slavery in Chicago: Abolitionists, the Law of Slavery, and Lincoln* (Chicago: Amperstand, Inc., 2009), 18.

680. For my extended discussion of relations between New England and European vampire traditions, see Michael E. Bell, "New England Vampires as Local Variants of a Belief Tradition," *Journal of Vampire Studies* 1, no. 2 (2020): 165-91.

681. The Vampire of Korbesz – Hungary, last updated November 2011, accessed April 12, 2020, http://www.shroudeater.com/ckorbes.htm.

682. The Vampire of Amarasesti – Romania, last updated November 19, 2009, accessed April 12, 2020, http://www.shroudeater.com/camarast.htm.

683. The Vampire of Cusmir – Romania, last updated November 19, 2009, accessed April 12, 2020, http://www.shroudeater.com/ccusmir.htm.

684. For specific examples from Greece, see Evangelos Avdikos, "Vampire Stories in Greece and the Reinforcement of Socio-Cultural Norms," *Folklore* 124, no. 3 (2013): 307–26.

685. Michael E. Bell, *Food for the Dead: On the Trail of New England's Vampires* (New York: Carroll & Graf, 2001): 156-77.

686. Paul S. Sledzik and Nicholas Bellantoni, "Brief Communication: Bioarcheological and Biocultural Evidence for the New England Vampire Folk Belief," *American Journal of Physical Anthropology* 94 (1994): 271.

687. For European instances of therapeutic post-mortem decapitation, see Paul Barber, *Vampires, Burial, and Death: Folklore and Reality* (New Haven: Yale University Press, 1988), 13, 25, 61, 78-79, 158, 175-76. Documentation shows that therapeutic post-mortem decapitation arrived in America at least as early as 1872, as reported in "A Queer Story of Superstition," *Chicago Tribune* (February 4, 1872: 8), which records the story of a young Polish woman, living in Berlin, Wisconsin, who died after giving birth. Not long after, her brother's wife became gravely ill. Acting on the belief that the entire family would follow his sister to the grave, he exhumed her body, cut off her head, extracted some blood from it, and administered it in an unspecified liquid to his wife. Her illness then assumed the form of smallpox, from which she fully recovered. For other American examples of post-mortem decapitation, see Bell, *Food for the Dead*, 168-70.

688. Owen Jarus, "20 of the Worst Epidemics and Pandemics in History," *Live Science*, posted March 20, 2020, accessed April 1, 2020, https://www.livescience.com/worst-epidemics-and-pandemics-in-history.html.

689. Centers for Disease Control and Prevention, Coronavirus Disease 2019 (COVID-19), Cases & Latest Updates, Situation Summary, updated March 26, 2020, accessed April 2, 2020, https://www.cdc.gov/coronavirus/2019-ncov/cases-updates/summary.html.

690. See Bilinda S. Straight, *Miracles and Extraordinary Experiences in Northern Kenya* (Philadelphia: University of Pennsylvania Press, 2007), chapter 7.

691. For my earlier, extended discussion of concepts of death as related to the New England vampire tradition, see Michael E. Bell, "Vampires and Death in New England, 1784 to 1892," in *Becoming Dead: The Entangled Agencies of the Dearly Departed*, ed. Bilinda S. Straight, special issue, *Anthropology and Humanism* 31, no. 2 (2006): 124-40.

692. Alan Dundes, "Wet and Dry, the Evil Eye: An Essay in Indo-European and Semitic Worldview," in *Interpreting Folklore*, ed. Alan Dundes (Bloomington, Indiana: Indiana University Press, 1980), 93–133.

693. Barber, *Vampires*, 114.

694. Montague Summers, *The Vampire: His Kith and Kin* (London: Kegan Paul, Trench, Trubner & Co., Ltd., 1928), 84.

695. Summers, *The Vampire: His Kith and Kin*, 84.

696. Bilinda S. Straight, "Becoming Dead: The Entangled Agencies of the Dearly Departed," *Anthropology and Humanism* 31, no. 2 [Special Issues: *Fragile Borders and Common Miracles: Death Reconsidered*] (December 2006): 104–05.

697. Quoted in Summers, *The Vampire: His Kith and Kin*, 39.

698. Margaret Lock, *Twice Dead: Organ Transplants and the Reinvention of Death* (Berkeley: University of California Press, 2002), 75.

699. Wayland D. Hand, ed., *Magical Medicine: The Folkloric Component of Medicine in the Folk Belief, Custom, and Ritual of the Peoples of Europe and America* (Berkeley: University of California Press, 1980), xxv.

700. Hand, xxvi.

701. Michael E. Bell and Simon J. Bronner, "Belief," in *Encyclopedia of American Folklife*, vol. 1, ed. Simon J. Bronner (Armonk, NY: M. E. Sharpe, 2006), 83–84.

702. For a similar argument, citing psychological literature, see Jan L. Perkowski, *The Darkling: A Treatise on Slavic Vampirism* (Columbus, Ohio: Slavica Publishers, 1989), 151–52. In Moncure Daniel Conway, *Demonology and Devil-Lore* (New York: Henry Holt and Company, 1879), Chapter 11, Disease (p. 249), opens with the following paragraph: "A familiar fable in the East tells of one who met a fearful phantom, which in reply to his questioning answered - 'I am Plague: I have come from yon city where ten thousand lie dead: one thousand were slain by me, the rest by fear.' Perhaps even this story does not fully report the alliance between plague and fear; for it is hardly doubtful that epidemics retain their power in the East largely because they have gained personification through fear as demons whose fatal power man can neither prevent nor cure, before which he can only cower and pray."

Index

Abell, Erasmus Darwin 191–92
Adams family 43, 63, 68, 87–88, 180
    Salmacious (Salmatius) 68, 88, 180
Adams, Amasa (Aaron) 159-60
Adams, Ebenezer 110
Africa(n) 3, 102, 116, 167, 199, 215-16, 268, 283
African American 102, 167, 268
  see also Black
Aldrich, Simon Whipple 4, 289
Allen, John Larrabee 185–86
America 3, 5, 19–20, 32, 65, 83, 113, 115–16, 169, 171, 175, 189, 197, 200, 213–14, 217, 252, 284, 288, 291
American 1, 3, 6, 15, 31, 33, 41–42, 50, 65, 74, 83, 92, 96, 99–100, 102, 111, 115–16, 127, 130, 140, 154–56, 166–67, 169, 175–76, 181, 184, 189–91, 194, 197, 199–200, 205–07, 209, 211, 216, 219–20, 225, 236, 248, 252, 257–58, 268, 273, 280, 285, 289
Ames, Galen 75–76
amulet(s) 102, 182, 263, 274
Anglo-Saxon 145, 268, 288
Appleton's Journal 255–57, 259, 261–62
Arnold, James 110
ashes 4, 9, 11, 29, 36, 42, 46, 49, 60–62, 69, 71, 75, 80–82, 86, 88, 91–92, 100–101, 103, 119, 123, 125, 136, 138, 141–42, 149, 161, 170, 173–74, 180–82, 184, 192, 199–200, 202, 204–6, 220, 222, 274, 276, 279, 290
Bache, Rene 219–21
Baker, Ronald 41, 140
Baker, Thomas 155–56
Barber, John (JB) 291
Barber, Paul 14, 126
Bartlett family 183
Beck, Jane 155
Bellantoni, Nicholas 290
Bird, Donald Allport 140
Black 167–69, 220
  see also African American
blood 4, 9, 11, 13, 25–26, 29, 34–36, 43, 49–50, 84, 86–88, 93, 97, 100–101, 103, 112, 119, 123, 125–26, 132, 135–36, 138, 141, 149, 151, 155, 159, 161, 171, 174, 176, 190, 199–200, 202–5, 214–17, 219, 221–23, 225–27, 233, 240, 248, 250, 259, 270, 273, 276, 283, 285, 289–90, 292–93
bloodless 138, 200, 215
bloodletting 189
bloodsucker 141, 217

bloody 29
Boas, Franz 197
boil 175, 204, 217, 222, 226
bone 5, 13, 25, 44, 58, 61, 108-09, 141, 161 170-71, 134–75, 201–02, 237, 264, 269, 273, 278, 280, 282, 290–91
Boralajova, Nicholas 166
Bowditch, Henry 185–86, 189–91
Bowen, Marshall 193-95
Bowles family 165–66, 237
Boynton, Joel 160, 273
    Maria 159–61
Breed family 183
Briggs, E. J. 72
Briggs, V. J. 220
Briggs, William Sheldon 23, 45–47
Britain 171, 174, 185, 264, 271, 291
British 32, 83, 126, 137, 154, 263, 293
Bronner, Simon 15, 41, 145, 221
Brown family
  Edwin 110, 119, 135–36, 138–39, 141–42, 212, 285
  Mercy Lena 16, 20, 119, 123–33, 135–38, 140–43, 184, 212, 220, 222, 264, 266–67, 282, 285, 287
Brown, Reuben 119, 124, 128–31, 133, 136, 287, 17, 20, 31, 53, 57, 119, 121, 123–26, 128–33, 135–38, 140–43, 145, 169, 184, 212, 216–17, 220, 222–24, 226–27, 249–50, 264, 266–67, 269, 282, 285, 287
Brunvand, Jan 15
Burial 21, 42, 183, 241, 290
burial 11–12, 21, 25, 28, 36, 38, 41–42, 43, 58–59, 71, 82, 86, 99, 115, 119, 159, 170, 174, 183, 195, 201, 205, 233, 237, 241, 255, 276, 290
buried 18, 25–26, 29, 34, 36–37, 42, 45–47, 51–52, 58, 69, 77, 79, 81, 84, 87–88, 92–93, 95, 99–100, 108–10, 113, 126–27, 136, 138, 143, 146, 149, 152, 156, 159, 167, 170, 184, 194, 197, 199, 201-2, 204–8, 215, 217, 219, 227, 237, 240, 247, 259, 264, 268, 274–76, 279, 281, 290
burn 4, 6, 9, 11, 13, 18, 20, 29, 36, 38, 40, 46–47, 49, 52, 57, 61-62, 69, 71–73, 75, 77, 79–82, 84–87, 89, 91–93, 96–97, 100-01, 103, 107–9, 112–16, 119, 124–26, 129, 133, 136, 149–52, 156, 159, 161, 164, 167, 170, 173–74, 180–82, 192, 199–200, 202, 206, 208, 217, 220–22, 226–28, 233–34, 236–38, 243–44, 246, 248, 255-56, 259, 261, 264, 268–70, 274, 276, 283–84, 289–90, 292
Burns, Tom 130–31

326

burnt 88, 123, 181, 184, 237–38, 279, 290
Burton, Isaac 20, 149–51
bury 20, 27, 37, 43, 45-46, 49, 57-58, 60, 98, 108, 145, 159, 165, 170, 183, 193, 195, 227, 234, 240, 247, 259, 278
Butler, Gary R. 124–25, 154
Butterfield, Benjamin 75–76
    Nabby 75
cadaver(s) 35, 82, 126, 184-85, 293
Calmet, Dom Augustine 214–15, 217
Canada
  Quebec Province 71
  Stanbridge East 71-74
cannibalism 7, 35, 60, 66, 78, 80, 99, 103, 137, 146, 171, 173–74, 288, 294
Carey, George S. 65–66
Carver, Maria 159–61
Case, Campbell 58, 93–95
Caswell, Phebe Rose 111–12
Chace, Lavina 19, 53
Chandler, Joseph Ripley 40–41, 183, 229, 238–42, 262
charm(s) 32, 102, 163, 263, 180, 233–34, 236, 274
Chestnut Hill Cemetery 119, 126, 130, 141–42, 224, 287
Christian 14, 18, 20, 57, 76, 86, 90, 102, 163, 169, 175-76, 214, 226, 239, 275, 294
Churchill, Albert 59–60
Cincinnati 84, 97, 104
civilization 3, 18–19, 76, 85, 101–2, 137, 197, 211, 213, 216, 221, 257, 288
civilized 19, 81, 102, 169, 197, 214, 267, 269, 287
clairvoyant 98–99, 224
Clark, Olive Cleveland 237
clergy 5, 20, 31–33, 161, 257
Clough, John 181–83
Cobb, Sylvanus 41, 163
coffin 25, 37–38, 40, 45, 52, 58-59, 69, 71, 76–77, 81, 86, 89, 99, 108, 115, 152, 154, 170, 186, 193, 195, 201, 207, 217, 239–40, 259, 260-61, 273, 278–80, 289–91
Cole family 96-97, 165
Cole, J. R. 255-56, 261
Colorado Springs 119, 138
Colvin, Russell 205–6
Congdon, Brister (Bristoe, Bristow) 167–71
Congregational 26, 29–30, 34
Congregationalism 30–32
Connecticut 93, 95, 104, 107, 245
  Colchester 168
  Cornwall Hollow 93-96
  Durham 160
  Griswold 71, 85, 114–15, 283, 290–91
  Hartford 54, 113-16, 161
  Haddam 73
  Jewett City 85-86, 114-15, 252, 283-85
  Litchfield County Litchfield 11, 84, 93, 95, 277
  Lyme 17
  New London County 113–15, 291
  New Haven 33, 169
  Simsbury (Granbury) 26
  Stafford 12, 257, 260-61, 288
  Stafford Springs 259, 261
  Sterling 245
  Stonington 69, 71, 167-68
  West Stafford 12, 174, 251, 255, 258-59, 261
  Willington 15, 51-52
  Windham 54
  Woodstock 54
Connecticut River Valley 25–27, 75, 93–95, 192, 234, 238
consumption 1, 3–7, 9, 11–20, 25, 27, 29, 31–35, 37–40, 42–43, 46–47, 49–55, 57–63, 65–66, 69, 72–73, 75–76, 78–101, 105, 108–9, 111–12, 114–16, 119, 125, 127, 132, 135–38, 141, 146, 149, 151–52, 154–57, 159–67, 171, 173–74, 176–77, 180–87, 189–95, 197, 199–200, 204, 206–9, 211–14, 216, 221–23, 225–28, 233, 235–36, 238–39, 241–42, 244–47, 251–52, 255–56, 259–61, 264, 267–71, 273–76, 278–80, 282, 284, 287–92, 294
consumptive 17, 20, 50, 78, 82, 99–100, 108–9, 113–16, 138, 151, 162, 164, 173, 186, 190–92, 216, 241–42, 270–71, 275–76, 289
contagion 6-7, 44, 62, 77–78, 83, 89, 91, 99, 127, 161-62, 185, 190–91, 221, 223
Conway, Moncure Daniel 107–8, 208–9, 212–13
Cook, William H. 81–83
corpse 4, 7, 9, 11, 13–14, 27-28, 32-37, 40, 42, 45, 50, 52-53, 55, 58–60, 62, 69, 71, 77–79, 87–88, 91–92, 96, 98–101, 112, 116, 119, 121, 125–27, 131, 136–37, 146, 152, 154, 156, 161–63, 170–71, 173–75, 180, 184, 187, 193, 195, 201-03, 205, 214, 216–17, 222, 226, 237–38, 242–43, 248-49, 252, 256, 259–61, 266, 269–71, 276, 280–81, 284, 290, 292–94
Corwin, Thomas 204–5
Cotton, John 30
cremation 80, 100, 112, 138, 159, 170, 212, 233, 238, 246, 256–57, 275, 290
crypt 119, 141, 290
Currier, John McNab 197, 206–7, 209
curse 103, 140-41, 217, 226, 235
Curtin, Jeremiah 197, 199–205, 207, 209, 252, 280
  Alma 200-01
Cushing Cemetery 204–5
Darwin, Charles 213

dead 1–7, 9, 11–14, 20, 25–26, 28–29, 32, 34, 36, 38, 42–44, 46–47, 49–50, 52, 55, 58–62, 70, 73, 76, 79–80, 84, 86–87, 91, 93–94, 97, 100–101, 107–10, 112–15, 126, 129, 132, 135–36, 138–40, 142, 145, 149, 154, 156, 161, 166–67, 170, 175–76, 180, 183–84, 192, 199, 201–2, 212, 215–17, 221–22, 226–28, 233, 236, 238, 240, 242–49, 255–56, 258–61, 266, 268, 276, 278–80, 282–83, 288–90, 293

deadly 99, 258, 289

death 1, 3–4, 7, 9, 11, 14, 17–18, 25–27, 29, 32–34, 39–45, 50, 53, 58, 60–62, 70, 72, 75, 77–78, 81, 83, 88–91, 93–95, 97, 99, 109–10, 125, 128–29, 135, 138–39, 141–44, 149, 151–52, 154, 156–57, 160, 162, 165–67, 173–74, 180–81, 185–87, 192–95, 200–203, 209, 212, 215–16, 222, 226, 233, 235, 237–39, 241–42, 244–47, 255, 257, 259, 261–63, 266, 268, 270, 273–74, 276, 278, 281, 287, 289–94

decapitation 170, 207-08, 290

decay 34, 37-38, 51, 58, 60-62, 78–79, 82, 86, 91, 93, 108, 138, 200, 255, 258, 270, 279, 294

deceased 4, 9, 11–12, 14, 25, 28, 32–33, 43, 47, 51, 58, 81, 86–87, 89, 92, 100, 115–16, 138, 159, 161–62, 173–74, 180–81, 186, 193, 199–200, 203–4, 206-7, 226, 249, 260–61, 274, 276, 289

decompose 14, 33–36, 52, 59, 86, 112, 114, 136, 174, 180–81, 248, 290, 292–93

demon 32, 55, 102, 136, 141, 149, 151 192, 195, 213, 215–17, 289, 291, 293

Denison, Frederic 167-71

Denison, Mary Andrews 53, 252, 275, 277

Dennett family 43–44

Deutsch, James 3

devil 42, 69, 72–73, 107, 156, 163, 205, 208, 212–13, 278, 281

devilry 202

diary 28, 43–44, 49, 173, 256

Dickerman Brook 193, 195

Dickinson family 25–28, 35
 Martha Morton 25, 27-28, 35
 Violet 26–27

Dickinson, Rebecca (Aunt Beck) 28–29, 31, 34

disease 3–4, 6–7, 11–12, 18, 20, 26, 33, 37–38, 42–44, 46, 49, 51–52, 55, 57–58, 60–61, 72–73, 75, 77–78, 80–81, 83–86, 88–89, 91–94, 96–97, 100, 102, 109, 111–15, 123, 125, 127, 135, 138, 152, 156, 159, 161–62, 164, 167, 173–77, 181–83, 185, 190–91, 193–95, 200, 206, 216, 221, 223, 226–28, 233, 239–41, 246, 270, 274, 282–83, 287, 291–92, 294

disinter 16, 37–38, 46–47, 51, 58–59, 62, 69, 77, 82, 84, 87, 92, 96–97, 100-01, 119, 146, 149, 181-82, 184, 186, 199, 201, 204–5, 284, 290

divination 32, 46, 264

divine 13, 32, 99, 102

Divine Spirit 244

Dodge, Roxana (Roxany) 235–36

Doolittle, Mark 29, 31

Dorson, Richard M. 194, 205, 225, 236

Downing, Major Jack 41, 50–51, 131, 224, 260

Dracula 1, 126–27, 219, 252, 259, 268, 277, 281–83, 285

dream(s) 146, 162, 201, 273, 247–49

Duffy, John 3, 5–6

Dundes, Alan 214, 292

Dwight family 25–29, 35–36, 246
 Martha 25, 27-29, 35-36

Dyer, Charles Volney 114–15, 208–9

dying 3, 6, 32, 42–43, 51, 60–62, 71–74, 85, 87–89, 91, 95, 97, 100, 112, 115, 119, 135, 142, 152, 154, 157, 159, 161, 164–65, 167, 180, 233–34, 237, 244, 256, 260–62, 270, 281, 292–94

Earle, Eliza 289

Eaton, Walter Prichard 251, 268–69

Emerson, Ralph Waldo 102, 104, 265

Emery, Edwin 146

Emmons, Samuel Bulfinch 162–63

English 15, 33, 53, 72, 96, 127, 190, 194, 199–200, 217, 224, 234, 258, 283

epidemic(s) 1, 3, 5–6, 33, 88, 119, 131, 154, 165–66, 182–83, 187, 219, 227, 290–91, 293

Europe 3, 13–15, 30, 34, 107, 116, 126, 171, 173, 175, 185, 199, 202, 205, 214, 217, 220, 250, 252, 280–81, 283–85, 289, 291

European 55, 83, 92, 126–27, 136, 171, 175, 190, 199, 203–4, 215, 222–23, 226, 236, 243, 250, 277, 280, 285, 289–91

evil 4, 7, 32, 36, 41, 54, 62, 67, 102, 116, 136, 146, 154, 191, 193, 195, 203, 213, 215–16, 267, 274, 284, 288–89, 291–92

evolution 18, 102, 137, 181, 197, 211, 213–14, 224, 288

evolutionist 197, 209, 211, 215

exhumation 12–13, 15–20, 23, 28, 33–35, 37–44, 46–47, 49–54, 57, 59–62, 66, 69, 71, 73–76, 78–79, 81, 83–88, 92, 95–96, 99, 101–3, 107–10, 112, 115, 119, 121, 123, 125, 128–29, 133, 136-39, 142, 145–46, 149–57, 159–62, 164–70, 175, 178, 181–84, 186, 194–95, 197, 200–201, 205–6, 209, 211-14, 216-17, 220–22, 224, 226-27, 229, 234, 237–38, 245–47, 249, 252, 255–56, 259–62, 264, 267, 269, 275–77, 282, 287

exhume 4, 13, 15, 19-20, 27–28, 39–40, 43–44, 46–47, 49, 51–53, 55, 57–59, 62, 69, 71–72, 75, 80, 82, 86–88, 97–100, 107, 109–10, 112, 114-16, 119, 131, 135–36, 138, 142, 144, 152, 156–57,

159-63, 167-68, 170, 180-81, 183, 185, 207, 212, 227, 238, 243-44, 246, 248-49, 255-56, 259, 261, 264–66, 269, 274, 281-82, 287, 290–91
experiment 19, 52–53, 60, 181, 185, 190, 244, 276
face down 9, 42, 99, 100, 170, 263, 270
See also prone
Fairfield, Francis Gerry 12, 251, 255–62, 280–81, 285, 288–89
　Josephine 257–59
Farley, Mrs. S. E. 13, 251–52, 259–60, 273–75, 284–85
fear 3-4, 17, 28, 61, 77, 100, 102, 128–29, 140, 145, 163, 166-67, 206, 221, 267, 278, 287, 291, 293–94
fire 13, 51–52, 70, 72, 78–79, 86, 93, 108–9, 119, 123, 126, 136, 141, 149, 159, 170, 199, 202, 204, 207, 220, 223, 234, 248, 268, 283
Fischer, David 32, 34
flesh 13, 35, 49, 58, 78, 97, 109, 119, 130, 135, 171, 175, 204, 216, 237, 276, 280, 290
Fobes, Nabby 75
folk 3–4, 7, 9, 11–13, 15, 18, 20, 28–29, 31, 34–36, 41, 50, 54, 66, 78, 91, 99, 102, 116, 123, 126, 130, 140–41, 155, 162, 173, 175, 179, 181, 190, 193–95, 200–203, 205–6, 216, 225, 228, 236–37, 251, 256, 268, 273–74, 281, 283, 288, 292–94
folklife 75, 123, 131–32, 262, 277
folklore 15, 20, 32, 35, 65–66, 75, 78, 111, 116, 140, 153, 163, 193–94, 197, 199–200, 202–3, 206–7, 209, 216–17, 223, 225, 236, 238, 244, 251, 262, 273, 277, 283–84, 287
folkloric 3, 89, 127, 154
folklorist 3, 7, 9, 15, 20, 25, 41, 65-66, 102, 124, 130-31, 140, 154, 179, 194, 196–97, 200, 203, 205–6, 214, 221, 224-25, 236, 283, 288, 292, 294
folktales 78, 126, 203, 205, 236, 249
Ford family 93-95
fortuneteller(s) 78, 201
Forward family 12, 14, 23, 26–37, 89, 287
　Eunice 27
　Mercy 27, 29, 36
France 140, 176
Frank, Glenn 279, 281
Frazer, James George 97, 99, 213
Frost family 61–63
　Benjamin 62
　Palmyra 62
Galenist 13, 34, 173–74, 179–81, 189
Gallup, Joseph A. 204, 207, 209
germ(s) 83, 97, 99, 136
German 13, 34, 37, 41–42, 152, 173, 179, 215

Germany 39, 152, 176, 215
ghost 32, 41, 99, 126, 136, 142–43, 164, 202, 203, 205, 207, 215-17, 222, 257, 260, 276
ghoul 139, 141, 202–3, 233
Glassie, Henry 15
Glencarella, Steve 205–6
Goldstein, Diane 132
Goodspeed, Stephen
Gordon -Grube, Karen 13, 34, 97, 171, 174, 274
grave 5, 11–12, 14, 18, 25–28, 32-34, 36–39, 41–47, 52–53, 58–62, 70, 76–79, 81–82, 86–89, 92–93, 100–101, 107–9, 114-15, 123, 126–29, 135–36, 140–43, 152, 154, 156, 161–62, 165-70, 175, 186, 195–96, 200–203, 206–9, 214–17, 222, 226–27, 229, 234-36, 239–10, 243, 246-50, 256, 259–61, 270, 273–74, 276–77, 279–81, 283, 285, 289–91, 293
gravesite 114, 132, 143, 220
gravestone 4-5, 60, 74, 87–88, 94–96, 107, 124, 142–44, 166, 235, 241, 246, 289
graveyard 201–2, 227, 256
Greece 173, 264
Greek 192, 215, 218
Green, W. N. 58–59
Grimm, Jacob 127, 175, 215
Hahnemann, Samuel 179
hair 4, 37–38, 74, 119, 135, 236, 240, 248–49, 274, 284
Halloween 25, 132, 136, 139, 141, 287
hand 28, 94, 101, 112, 131–32, 154, 160–61, 163, 189, 213, 252, 255, 263, 265, 270, 275–76, 280, 283–84
Hand, Wayland 179, 294
harm 11, 14, 32, 36, 87, 91, 96, 139, 154, 216, 274, 291, 293
Harrington, Charles 108, 110–11
Harris, Thaddeus Mason 69–71
Harris, Rachel 91, 149–50, 206
Harte, Bret 259
Harwood, Francis 26–27
　Myron S. 27
haunt 41, 84, 141-43, 199, 203, 268, 282, 285, 289, 294
Hazard, Thomas Robinson 269
Hazen, George Alfred 243
head 9, 20, 26, 34, 37, 46, 58, 73, 78, 94, 112, 145, 162, 168, 170, 203, 207–8, 217, 222, 226, 243–44, 273–74, 281–82, 284, 288, 290
headless 208
headstone 93, 234
heal 9, 11, 13, 20, 26, 29, 31–32, 34, 41, 43, 62, 83, 97, 155, 161, 173–75, 180, 195, 201-03, 245, 274, 287, 295
healer 15, 26, 46, 175, 195, 287

health 3–4, 6, 11, 31, 33, 41, 53, 71–72, 79–80, 86, 93, 95, 138, 141, 146, 149, 152–53, 155, 173, 177, 179, 189–92, 201, 224, 233, 239–40, 243, 245–46, 248, 257, 263, 274, 294
heart 4, 9, 12–13, 18, 20, 28–29, 43, 46–47, 49–50, 57, 61–62, 77, 81–82, 84–89, 91–93, 96–97, 100–101, 103, 107–8, 112, 115, 119, 123–25, 129, 133, 135–38, 140–42, 149–51, 159, 161–64, 166–67, 171, 173–74, 180, 182, 189, 192–95, 199–202, 205, 216–17, 220–23, 225–28, 233, 243, 245, 248, 252, 256, 259–61, 264, 266–70, 274, 276, 279–80, 284–85, 289–90, 292–93
herbs 4, 31, 49, 83, 156, 161, 173, 195, 274, 278
heredity 6–7, 18, 60, 62, 78, 82–83, 89, 91, 104, 161, 190–91, 264, 270
Hewitt (Huit), Josiah Grant 69–71
hidden 112, 166, 247, 273
Hildreth, Samuel Preston 49–50
Himes, Hortense 119, 138
Hohenheim, Theophrastus von (Paracelsus) 13, 173
Holbrook, Josiah 184
Holden, Austin Wells 179–81
Holmes, Moses 23, 51–54
Holmes, Oliver Wendell 101, 104, 180, 263, 265
homeopathic medicine 179–81
Honko, Lauri 9
hoodoo(ism) 101-2, 104, 165
hope 4–5, 7, 16, 20, 28, 47, 53, 71, 77–78, 81, 89, 93–94, 99, 119, 138, 146, 152, 157, 163, 173, 183-84, 192, 195, 242, 244, 251, 255, 267, 274, 276, 284, 287, 291-92, 294
hopeless 176, 276
horrible 14, 29, 44, 51, 58, 81-82, 84, 86, 92, 96, 107, 109, 135, 140-2, 146, 192, 225, 233, 263, 270, 280
horrified 60, 82, 170–71
horror 4, 57, 70, 127, 243, 278
Hough, Franklin Benjamin 14, 18, 145, 285, 288
Hudson, Alfred Sereno 161–63, 173
Hufford, David 203
Hungary 217, 219, 289
Hunting (Huntting) family 90-91
    Clarissa 90–91
    Newell 90–91
Hurd, Duane Hamilton 159–61
Hyde (Hide), John 193–94
ill 3-4, 9, 12, 28, 37, 43, 53, 55, 71, 73, 94, 116, 135-36, 138, 141-42, 149, 153, 176, 165, 179, 192–93, 199, 204, 233, 235, 243–44, 247, 290, 292
Illinois 84, 235
  Bloomington 37, 39

Chicago 37, 40–41, 114–15, 152, 208–09, 284
immigrant 39–40, 116, 145, 151-52, 189, 200, 270, 285, 289, 291
incinerate 9, 69, 91, 173, 220
Indian (Indigenous American) 49–50, 109–11, 136, 153, 167, 169, 216, 268
see also Native American
infect 1, 3–7, 54–55, 91, 185, 190, 252, 283, 285, 291–92
infotainment 142, 222, 285
ingest 9, 11, 29, 82, 97, 103, 142, 173–74, 184, 206
interred 25, 27, 36, 46, 59–60, 80, 85, 88, 96, 107–8, 110, 130, 135, 143–44, 167, 187, 246, 256, 259, 290
invisible 12–13, 144, 174, 202, 233
JB 290–91
Johnson family
    Isaac 51–54
Johnson, Rebecca 54
Jotham Weed 146
Kimball, Joseph Horace 289
kin 9, 11, 16, 20, 32, 34–35, 41, 91, 99, 102, 119, 129, 137, 168, 174, 222, 252, 275, 281, 287, 289, 294
Kittredge, George Lyman 195
Kittridge, Thomas 49–50
Koch, Robert 1, 99, 221
Koen, Vermeir 280
LaFarge, Grant 265
Larkin, Jack 78
Laughton family 152–55
Lawrence family 89–91
Leach, MacEdward 194, 225, 277
Leaf, Murray 35
Lederer, Inez Stevens 235–36
legend 41, 63, 66, 96, 103, 111, 120, 126, 128, 140–44, 165, 184, 193, 203, 205–6, 216, 223, 225, 229, 234, 237, 245, 249, 255, 257, 261-62, 264, 283–84, 287-289
Levine, Harold D. 42–43, 193–95
liver 12, 20, 25–26, 29, 35, 61–62, 82, 84–85, 89, 91, 93, 108, 119, 136–37, 149, 167, 175, 181, 195, 222, 226, 233, 264, 269, 290
Loomis, C. Grant 196
Lovecraft, H. P. 283–84
Lowell, Amy 252, 259, 277–82, 285
Lowell, James Russell 104, 265
lungs 3, 12, 20, 25–26, 29, 35, 61–62, 80–81, 84, 87, 89, 91–93, 114–16, 136, 138, 149, 156, 159, 161, 167, 174, 181, 184, 190–92, 195, 206, 208, 220, 233, 256, 261, 274, 282, 292–93
MacCulloch, J. A. 136, 217
Macdonald, Deda Belle 111-12

magic 20, 26, 32, 53, 55, 97, 99, 101, 116, 179, 182, 228, 274, 294
magical 11, 31-32, 35, 46, 54-55, 97, 103, 146, 213, 244, 273, 280, 283, 285, 294
magician 53, 116, 176
Maine 78, 104, 131, 162, 185
  Bangor 63
  Bath 39
  Bruswick 186
  Buckfield 40-41, 50-51, 163
  Buxton 186
  Cornish 186
  Farmington 96
  Jay Hill 96
  Lewiston 98–99
  Massabesec 74
  Norway 41, 79, 163
  Oxford 43
  Oxford County 79, 100
  Pittsfield 191
  Portland 41, 50-1, 98-99
  Saco 185-86
  Sanford 74, 146
  South Union 274
  Springvale 186
  Thomaston 274-75
  Turrner 39–40
  Union 156-57, 274
  Vinal Haven (Vinalhaven) 99-100
  Waldboro 274
  Washington Plantation 73
  Waterville 163
  Winthrop 163
malady 38, 82, 162, 186, 242, 247
malaise 294
Mansfield, David Lufkin 151–56
Massachusetts 6, 74, 78, 152, 185, 190-92, 245, 252, 278
  Andover 49-50, 54
  Barnstable 157
  Belchertown (Cold Spring) 12, 25-29, 31, 33-34, 89, 103, 195, 287
  Berkshire County 93, 95, 236, 282
  Boston 30, 46–47, 50, 90, 98, 101, 103–4, 140–41, 155, 157, 182, 189, 211, 234, 241, 265-66, 275, 277, 279
  Braintree 211
  Bridgewater 157
  Bristol County 73
  Brookline 194-95, 279
  Cambridge 277
  Charlestown 70
  Chesterfield 174, 229, 231, 234–35, 237
  Chicopee 192
  Concord 89, 161-62
  Dedham 75
  Dorchester 70
  Dracut 90
  Essex County 49-50, 65
  Hadley 53
  Hampden County 95, 192, 236
  Hampshire County 97
  Hancock 72
  Harvard 34, 163-66, 177, 237
  Hatfield 25, 27–28, 34
  Kingston 40-41, 238, 262
  Lawrence 49
  Lowell 98
  Malden 163, 269
  Methuen 49
  New Boston (South Lancaster) 164
  North Blandford 95
  New Marlbough 12, 66, 93-98
  Northampton 234
  Otis 94–96
  Oxford 155
  Pittsfield 235
  Plymouth County 40, 96, 238–42
  Robinson Hollow 12, 166, 234
  Roxbury 194
  Salem 189
  Sandisfield 94–95
  Springfield 75–76
  Stockbridge 28
  Tolland 95
  Tolland County 255–56, 261
  Waltham 41, 162-63
  Warren 157
  West Chesterfield 234, 237
  Westfield 97, 174
  West Stockbridge 27
  Whately 26–27
  Williamsburg 53–54, 234
  Worcester 70
Mather, Cotton 161
McClelland, Bruce A. 14, 98, 217
McNally, Raymond 285
Mead family 20, 91, 149–51
memorate 41, 103, 277
Metcalf, Harold 119, 135–38, 184, 212, 226–27, 251, 266–67
microbe 1, 91, 99, 162, 221, 291
midnight 70, 141, 257, 276
Miller, Elizabeth 219
Miller, Marla R. 28-29
Millerism 84
miracle 103, 181, 294

miraculous 101, 112, 126, 146, 163, 175, 203, 255–56, 261
monster 41, 202-03, 206, 284
monstrosities 181
monstrous 66, 81, 87
moon 102, 164, 184, 207, 214, 273
mortality 5, 33–34, 53, 55, 61, 84, 235, 241
mortuary 11, 238, 274, 293
Moulton, Jotham 146
mouth 5, 15, 60, 96, 133, 137, 143, 229, 246, 252, 269–70, 278
mummia 35, 97, 174, 184, 288
mummies 264
mummified 119, 135
Munson family 150 149–51
Murgoci, Agnes 216
mysteries 3, 45, 175–76, 256
mysterious 4, 7, 43, 49–50, 84, 128, 131, 141, 164, 167, 174, 181, 193, 201, 205, 268, 282, 291
mystery 6–7, 78, 140, 143, 156, 177, 201, 224, 251, 277
mystical 203
mystifying 7, 146
myth 206, 234
mythology 199–200, 203
Native American 111, 154, 167, 200, 205
  see also Indian
New Hampshire 42, 74, 97, 104, 185, 197
  Barnstead 23, 43–44
  Bath 207
  Bristol 195, 200
  Cheshire County 46, 60
  Epping 194
  Gilmanton 182
  Grafton County 206
  Hampstead 49
  Holderness 193
  Jaffrey 60-63
  Keene 45-46, 60
  Lempster 192
  Loudon 44
  Marlboro 90
  Merrimack 193–94
  Newburyport 274
  New Hampton 193–94
  New Ipswich 60, 63,182-83
  Ringe 192
  Rochester 43
  Salisbury 153
  Strafford 43, 45
  Sutton 192
  Walpole 153
New York
  Albany 104
  Albany County 69
  Amenia 151
  Ballstown (Ballston) 69
  Buffalo 84–85
  Canandaigua 244
  Cayuga County 66, 81-82, 84, 277
  Chazy 47, 79, 103, 159–61
  Clinton County 159–60
  Eagle Mills 92
  Franklin County 14, 145, 149, 159-60, 201, 285
  Glens Falls 179–80
  Grafton 91-92
  Jefferson County 145
  Kelloggsville 81–84
  Lewis County 145
  Malone 201-02
  Palatine 69
  Queensbury 87, 179
  Rensselaer County 92, 229, 250
  Schenectady 283
  Schuyler County 229, 243, 245
  Skaneateles 83
  St. Lawrence County 14, 18, 145, 285
  Stone Arabia 69
  Warrensburg 179
  West Glens Falls (Goodspeedville) 87-88
  White Creek 179
night 70, 72, 88, 126–27, 132, 136, 139, 141–42, 146, 159, 202, 215, 217, 219, 222–23, 247, 249–50, 256–57, 259–61, 279, 281 289–90
nightly 247–48
nightmare 203, 215
Nourse, Henry Stedman 163–66, 177, 237
Nurse, Rebecca 165
Nye family 157
occult 13, 20, 32, 46, 66, 97, 101–2, 174, 216, 247, 263, 283–84
omen 146, 201, 214, 220-21, 257 273
oral 15, 18, 20, 54, 60, 63, 66, 71, 75, 80, 109, 129–31, 140, 142, 150, 154, 160, 164, 173, 182–83, 194–95, 205-7, 212, 214, 229, 234, 236–37, 245, 259, 261–62, 269, 282, 285, 287
orally 62, 66, 100, 104, 149, 154, 164, 236
organ(s) 4, 9, 11–13, 29, 34, 36, 62, 82, 85, 88–89, 91, 93, 97, 136, 138, 173-74, 186, 195, 203, 217, 255–56, 261, 270, 289–90, 292–93
outsider 13–14, 108, 110-11, 120, 131-33, 137, 139, 141, 143, 170, 222, 269, 287
pagan 14, 76, 102, 116, 138, 214
Paracelsian 13, 34-35, 83, 97, 161–62, 171, 173–75, 179-80, 184, 202, 274, 288
Pawtuxet Valley Gleaner 137, 139, 220, 229, 245
Peck, Lewis Everett 20, 119, 123-33, 136, 141–42, 222, 282, 287

Pennsylvania 20, 83, 194
  Harford 57-58
  Lenox 20, 57–60
  Montrose 20, 57, 59–60, 62
  Susquehanna County 59
  West Lenox 60
performance 7, 9, 11, 15, 23, 35, 59-60, 66, 102–3, 129, 139, 141, 146, 183, 233, 251, 256, 260, 275-76, 281, 293-94
performed 5, 11, 14–16, 19-20, 28, 53, 57, 63, 72, 78–79, 83, 88, 91, 97, 108, 110–12, 157, 168, 175, 183, 195, 202, 206, 209, 214, 227, 229, 244, 249, 256–57, 259, 261, 266, 277, 282, 287
performer 19, 120, 290
Pettibone, John S. 149–51, 206
phthisis 82, 185, 191
Pitrè, Giuseppe 116
Place, Enoch Hayes 23, 43-45, 49
Poland 215, 217
Polish 200
Poole, Ernest 42
Powel, Hulda 20, 149–51
Powers, John D. 204, 207, 209
prayer 31-32, 43–44, 94, 173, 270, 281, 284
preemptive ritual 42, 166, 195, 233
prescribe 4, 29, 77–78, 98-99, 112, 141, 155–56, 171, 174, 179–80, 194, 251, 287-88, 290
prescription 42, 53, 83, 155, 162, 173-74, 181, 184, 206, 209
Preston, John 182–83
prey 1, 7, 12, 14, 25, 29, 32, 34, 58, 60, 51, 84, 86, 175, 180, 190, 174, 216, 222, 233, 239–41, 246, 261, 276
Prince, Barbara Miller 39-40, 42
Prince, Job 39–41
Prince, Ezra Morton 38–42, 152, 163
prone 43, 170
  See also face down
prophesies 193
prophetic 201, 249
Protestant 30-31, 116, 294
proverb 28, 51, 92, 116, 156, 274
Providence Journal 53, 107-08, 115, 126, 128, 130, 133, 135-37, 139-42, 212, 216-17, 220, 229, 245, 248
Puritan 30, 35, 97, 145, 174, 221, 233, 236, 288
putrefaction 28, 32, 34, 61, 162, 215, 292–93
quack 4, 15, 50–54, 102, 156, 162, 170, 175–76, 179-80, 195
Quaker 244, 289
Quinn, D. Michael 20
Ranft, Michael 34, 215, 217
Ransom family 17–18, 206
  Frederick 206, 17-18

Ray family 85–87, 114–15, 252, 283–84
reanimated 126, 214, 222, 293
reburied 9, 89, 98, 100, 208, 290
religion 20, 26, 30, 32, 35, 76, 84, 87, 99, 103, 116, 169, 180, 206, 214, 244, 287, 294
religious 30, 32, 35, 43, 102, 116, 151, 163, 180, 213, 238, 244, 248, 257, 275, 294
remedy 3–4, 11, 15, 18, 28, 31, 42, 46, 49–50, 61-62, 71, 77, 92, 94, 96-97, 109, 114, 136, 152, 161-63, 174, 176, 179, 181–82, 189, 192, 195, 221–22, 248, 267, 276, 289
restoration 9, 11, 31, 36, 41, 53, 71–72, 93, 97, 101, 174, 176, 200, 246, 274
reveal 9, 13, 35-36, 43, 52, 66, 93, 104, 119, 123, 133, 150, 170, 189, 197, 199, 203, 223, 235, 238, 249, 255-56, 261, 267, 287, 289–91
revelation 77, 162, 227, 247, 257
revenant 32, 126, 215, 223, 249, 260–62
Revolutionary War 31, 74, 137, 151, 155, 160, 164, 169, 191, 247
Rhode Island 4, 59, 104–5, 115, 121, 145–46, 152, 170, 184–85, 197, 214, 216, 219, 270–71, 285, 288–89
  Coventry 88–89, 143
  Cumberland 19, 53, 244–45
  East Greenwich 220
  Exeter 57, 110, 119, 123–24, 127–28, 130, 132, 135–41, 184, 211–13, 220, 222–24, 247, 249, 266–69, 283, 287
  Foster 229, 245–47
  Jamestown 109
  Kingston 40–41, 108–9, 111, 226–27, 229, 238, 240–42, 245, 262
  Mooresfield 107-08
  Narragansett 100, 107–11, 123, 126–27, 140, 167, 227, 265, 268–69
  Newport 33, 54, 100, 127, 207, 212, 221, 265, 268–69, 282
  North Kingstown 108–10, 119, 138, 212, 223, 226–27, 265–66, 282
  North Smithfield 4, 289
  Peacedale 107–8, 212
  Providence 137, 168–69, 220, 229, 250, 282–84
  Saunderstown 108–12, 212–13, 226, 264–67, 282
  Smithfield 38, 42, 152
  Sodom 123, 222
  South Kingstown 107, 109–10, 213, 223, 226, 245
  Wakefield 109, 248–49
  Washington (South) County 111, 113, 123, 137, 140, 142, 167, 179, 251, 256, 268–69
  Watson's Corner 108–10
  West Greenwich 143
  Westerly 71, 99, 167–69
  Wickford 138, 220, 226, 266

Woonsocket 169
Rider, Sidney 211–12, 223, 229, 247–50
Robinson family 12, 143, 166, 231, 233–37, 269
Romania(n) 126, 216, 290
Rose family 107–13, 115, 212, 245, 249, 265, 282–84
  Benjamin 110, 112-13, 212
  Maria 110, 112
rumor 96, 237, 144, 205
Russia(n) 200, 202-03, 217, 233, 236
Ryden, Kent C. 145, 221
Saugatucket River 108, 110
savage 18, 79, 267-68, 288
savagery 19, 197, 213, 288
scapegoat 14, 98, 163, 205, 216, 292
Schudson, Michael 65, 142, 224
science 1, 4, 6–7, 14, 18–19, 26, 30–31, 33–35, 50, 53, 55, 91, 98–99, 137, 146, 156, 181–82, 184, 189, 197, 199, 207, 211, 213-14, 220, 223, 246-47, 255, 257-58, 263, 265, 288, 291
scientist 18, 26, 31, 137, 173, 185, 206, 214, 268, 291-92
Scotland 38, 42
Scott, Amasa 25–27, 33, 35
scourge 1, 4, 66, 91, 96, 165, 235, 239, 276, 288, 292
Seaver, Frederick 201–2
secret 78, 146, 163, 176
Seneca Lake 229, 243–45
Serbia(n) 166, 215, 217
Shakers 44
shapeshifter 193, 216-17
shroud 71, 77, 81, 182, 217, 240, 260
shroudeaters 215
Sibley, John Langdon 156–57, 274
Sicilian 113–16
silence 37, 41, 70, 77–78, 152, 234, 240, 256, 263, 277, 282
Simmons, William S. 111
skeleton 25, 37–38, 119, 135, 258
Skinner, Charles M. 283–84
skull 13, 38–39, 161, 170–71, 173–74, 184, 290
slavery 39, 43, 169, 252, 268, 277, 289
slaves 167
Slavic 136, 202–3
Slavonia 226
Slavonic 217–18, 293
smallpox 3, 6–7, 70, 127
Smith, Jethro 71–74
Smith, Seba 41, 50–51, 78–79, 131, 163, 224, 260
sorcerer 215, 257, 278
sorcery 101

soul 32, 44-45, 70, 72, 74, 86, 99, 101, 126-27, 139, 154, 176, 215, 217, 221–22, 226, 264, 267–68, 273, 276, 278, 281, 292–94
Spain 96, 291
Spaulding family 39–41, 50, 151–56
  Josiah 152–53, 156
  Leonard 39–40, 151–53, 155
  Reuben 152–53
specter 32, 53, 143, 156, 259, 282
spirit 7, 9, 11, 13, 31–32, 34–36, 54, 71–73, 76, 84, 97–99, 102, 116, 126, 136, 139, 141, 146, 154, 162, 170, 173-74, 179, 197, 202, 213, 214-17, 226, 244-45, 267–68, 274, 288-89, 291–93
spiritualism 66, 72, 84, 259
spiritus 35, 274
See also World Soul
sprout 37, 39, 44, 52, 159, 161
Staples family 19, 53-4, 244
Stearns, Samuel 156
Stetson, George R. 102, 112–13, 197, 209, 211–17, 219–22, 225, 263–64, 269, 285, 288
Stiles, Ezra 33
Stoker, Bram 1, 126, 219, 268, 283
stomach 29, 43-44, 61–62, 193
storytelling 19–20, 41, 119, 129–30, 133, 154, 169, 236, 251, 283, 287
Straight, Bilinda 293
succubus 216
Sugg, Richard 35, 161, 171, 173–74
Summers, Montague 107–8, 283–84, 292–93
sunlight 129
sunrise 88–89, 240
sunset 270
sunshine 74
supernatural 9, 11, 16, 20, 31–32, 41, 65, 78, 99, 116, 127–28, 139, 141, 143, 146, 162–63, 191, 199, 202–3, 205–6, 225, 236–37, 239, 259, 262, 269, 273, 277–78, 282–83, 294
superstition 6-7, 14, 18–19, 40, 44, 46–47, 49, 51, 57-58, 66, 70-71, 75–76, 78, 82, 84–86, 88, 91–93, 96–102, 107, 112–16, 126, 135–36, 138-39, 142, 145–46, 152, 159, 162-64, 167, 169, 171, 174, 179–82, 184, 192, 195, 197, 201, 211–17, 221, 224–26, 233, 236, 238–39, 241, 243, 250-52, 255–56, 259-61, 263-64, 268-71, 273, 275–76, 288, 294
superstitious 14, 49, 57, 71–72, 78, 81-82, 86, 89, 102, 104, 108–9, 111, 131, 139, 159, 166, 170, 176, 181, 186, 215, 241–42, 257, 275
supranormal 32
Swiss 13, 173
Switzerland 205
symbol 11, 13, 34-35, 213-14
symbolism 179

sympathetic 13, 62, 97, 161, 174, 181, 202, 228, 289
sympathies 273
sympathy 135, 246
symptomatic 4
symptom(s) 6, 69, 77, 179, 189, 204, 247, 259, 264, 268, 270, 278, 289
tale 46–47, 72, 74, 85, 93, 115, 128–29, 140, 142–43, 160, 164, 166, 183, 202–5, 223, 233, 236, 238, 241, 243, 247–48, 251, 255, 259, 261, 278, 283–84
Taylor, Edward 97, 174
text(s) 3, 9, 12-13, 16, 20, 38, 42, 51, 57, 59, 69, 71, 79, 86–87, 89, 92, 94, 96-98, 100–101, 107–10, 112, 129, 130-31, 133, 135, 141, 154, 159, 162, 166–67, 169, 173, 180-81, 183, 200, 204–6, 212-13, 217, 221–22, 225, 242, 255–56, 259, 260-62, 288
therapeutic 4–5, 34, 42, 49–50, 71, 166, 170, 179, 183, 186, 236, 282
therapy 3, 11
Thomas, Keith 11, 31
Thompsonian medicine 83
Thoreau, Henry David 89, 192
Thorndike, Lynn 53
Tillinghast family 107, 247, 249
    Sarah 229, 243, 246-49
    Stukeley (Snuffy) 247-49
tincture 101, 162, 173–74, 238
Tirrell, Adaline 12, 229, 234, 236–38
Toelken, J. Barre 15
tomb(s) 51, 77–78, 82, 119, 136, 202–3, 215, 240–41, 252, 276
tombstone 88, 142
Tourje (Tourgee) family 57–60
Transylvania 197, 285
treat 31, 41, 51, 65, 82, 86, 132, 142, 163, 170, 173, 177, 190, 192, 194, 244, 287, 291, 293
treatment 3–4, 6, 11, 17, 31–32, 35, 62, 81–82, 104, 112–13, 125, 131, 151, 155, 171, 173, 175-76, 180, 190, 192, 216–17, 273, 287, 292
troublesome 9, 11, 14, 50, 55, 195, 216, 252, 269
tuberculosis 1, 3, 6–7, 42–43, 54, 92, 99, 113–14, 116, 125, 136, 141, 160, 185, 190, 221, 226, 264, 285, 290–93
Tyler, Casey 229, 245-47
Tylor, Edward B. 213–15
Ukraine 217
undead 119, 214, 249, 252, 294
undecayed 60–62, 88, 199
undecomposed 115
Underwood, Benjamin Franklin 99
unearth(ed) 35, 129, 135, 140
unknown 11, 38, 72–73, 85, 103, 110, 149, 164, 167, 193, 216, 220

unnatural 11, 49, 55, 62, 156–57, 200, 202-03, 238, 289, 293
unquiet 247
Vaughn, Nellie 142–44, 287
Vermont 43, 197, 279
  Addison County 207
  Bellows Falls 89–90
  Bristol 200
  Castleton 208–9
  Chimney Point 74
  Clarendon 208
  Dummerston 40, 151, 153–56
  Hubbardton 207–9
  Londonderry 90
  Manchester 19–20, 91, 149–51, 206
  Middletown 79-80
  Montpelier 45
  Newport 207
  Peru 90
  Rupert 150
  Rutland County 96, 207, 209
  South Woodstock 17
  Stowe 73
  Thetford 73
  Warren 200
  Waterbury 72-73
  Winhall 89-91
  Woodstock 103–4, 192, 199-200, 202-07
victim 6, 37–38, 42, 55, 77, 84, 86, 91, 96, 98, 138, 140, 152, 159, 164-65, 176–77, 183, 192, 203, 206–7, 209, 215, 219, 233, 242, 248, 270–71, 276, 283, 285, 290
vine 37–42, 52, 58, 62, 152, 154, 156, 234
Vinton (Vintin), Joeseph 237–38
viruses 7, 291–92
vital(s) 4, 9, 11–14, 26, 29, 33–34, 40, 43–44, 52, 55, 59-62, 73, 75, 77, 82–83, 85, 90-94, 108–9, 149, 152, 156, 166–67, 169, 174–75, 186, 189–90, 193, 195, 202–03, 224, 233-37, 243, 255–56, 261, 270, 274, 281, 287, 290, 292–93
vitalism 13
vitality 12–13, 36, 61–62, 93, 136, 140, 174–75, 181, 217, 255–56, 260–61, 276, 280, 289
voices 128, 257
Voltaire 287
voodooism 104, 220
Waite, Frederick C. 208–9
Warren, Andrew Oliver 60, 62-63
wasted 12, 25, 174, 227, 233, 243, 261, 280, 282
wasting 49, 93, 107, 115, 156, 203, 262, 282–83, 289
Waterhouse, Henry S. 201–2
Watson, Patricia A. 155
Webster, Daniel 184

Wellman, Manly Wade 252, 282–85
Westfield River 234–35
Weygandt, Cornelius 194–95
Wharton, Edith 252, 265, 277, 281–82, 284–85
Whitman, Bernard 40, 162–63
Whitney, David 20, 57–58, 60
Whitten (Whitton) family 186–87
Wilber, Clifford C. 46–47
Wilkinson, Jemima 244–45
Willard family 166
Williams, Elijah 25–28
Wimberly, Lowry Charles 126
Wisbey, Harold 244–45
wish 34, 58, 75, 82, 132, 137, 161, 164, 216, 246, 257
Wister, Owen 251, 263–67, 269
    Sarah Butler 263, 266
witch(es) 32, 41, 54–55, 96, 126, 163–64, 175-76, 180, 189, 193–95, 203, 207, 216-17, 226, 259, 273, 278, 283
witchcraft 32, 54–55, 72–73, 75, 91–92, 104, 150, 165, 175, 195, 201, 206, 268–69
Wittgenstein, Ludwig 267
wizards 176, 216–17, 226, 257
Woodward, Shepherd (Shepard) 47, 159–61
World Soul 35, 97
Wright, Dudley 107, 115
Wright, Samuel 58–59
Yankee 42, 50–51, 83, 101, 104, 123, 130–31, 141, 145, 195–96, 221, 224, 252, 259–61, 264, 279–80, 285
Young family 123, 128, 165, 229, 244–46
Ziner, Karen Lee 125–26, 128–30, 140

www.ingramcontent.com/pod-product-compliance
Lightning Source LLC
Chambersburg PA
CBHW061747290426
44108CB00028B/2912